Curves, Jacobians, and Abelian Varieties

Recent Titles in This Series

(*Continued in the back of this publication*)

CONTEMPORARY
MATHEMATICS

136

Curves, Jacobians, and Abelian Varieties

Proceedings of an AMS-IMS-SIAM
Joint Summer Research Conference
on the Schottky Problem

Ron Donagi
Editor

American Mathematical Society
Providence, Rhode Island

The AMS-IMS-SIAM 1990 Joint Summer Research Conference on the Schottky Problem was held June 21–27, 1990 at the University of Massachusetts, Amherst, Massachusetts, with support from the National Science Foundation, Grant No. DMS-8918200.

1991 *Mathematics Subject Classification*. Primary 14H42; Secondary 14H40, 14K25, 32E20.

Library of Congress Cataloging-in-Publication Data

Curves, Jacobians, and Abelian varieties: proceedings of a summer research workshop on the Schottky problem/[edited by] Ron Donagi.
    p. cm.—(Contemporary mathematics, ISSN 0271-4132; v. 136)
    Conference held June 21–27, 1990 at the University of Massachusetts, Amherst.
    Includes bibliographical references.
    ISBN 0-8218-5143-8
    1. Curves, Algebraic—Congresses. 2. Jacobians—Congresses. I. Donagi, Ron.
II. Series: Contemporary mathematics (American Mathematical Society); v. 136.
QA565.C87 1992   92-20586
516.3′52—dc20   CIP

Portions of this volume were printed directly from author-prepared copy; portions of this
volume were typeset using $\mathcal{A}_{\mathcal{M}}\mathcal{S}$-TEX,
the American Mathematical Society's TEX macro system.
10 9 8 7 6 5 4 3 2 1   97 96 95 94 93 92

Contents

Preface

This volume represents the proceedings of the 1990 Joint Summer Research Conference on the Schottky Problem, held June 21–June 27, 1990, at the University of Massachusetts, Amherst. The organizing committee consisted of Ron Donagi, Leon Ehrenpreis, Hershel Farkas and Robert Gunning. Most of the papers were submitted by conference participants, and were informally refereed. A couple of additional contributions, which fit our general theme, were invited from non-participants.

The central theme, of finding geometric properties which characterize Jacobians of curves, has been interpreted rather broadly. Thus, one group of papers studies the moduli space of stable vector bundles on a curve, which can be viewed as a non-abelian generalization of the Jacobian. This group includes the works by Fay, Mulase, and Teixidor and Tu. Kempf studies metrics on the Picard bundles, which are stable vector bundles on the Jacobian.

Another generalization of the Jacobian of a curve is given by the Jacobians and Pryms of its covers. These are studied by Accola and Donagi. The connection with stable bundles is intriguing, but not well understood. Both vector bundles and the Abel-Jacobi map for a family of curves are used in Iliev's paper to study the theta divisor in the intermediate Jacobian (which is presumably not a Jacobian) of a certain Fano threefold.

A third group of papers deals with special Jacobians. Berry and Tretkoff, Earle, Martens, and Ries consider Jacobians which are isomorphic or isogeneous to a product. Degenerations of curves and their Jacobians are studied by Gonzalez-Diez, and applied by Jorgenson to the study of Weierstrass points. Farkas discusses Fay's trisecant identity and Gunning's cross-ratios for hyperelliptic Jacobians. A very different and pretty approach to moduli of Riemann surfaces, via Schottky groups and circle packings, is considered by Brooks.

I would like to take this opportunity to thank all the contributors and participants for their efforts in making the conference a success, and especially Leon Ehrenpreis, Hershel Farkas and Robert Gunning, whose planning made the meeting possible. Special thanks go to Donna Harmon, for her patient and tireless work in putting this volume together.

Ron Donagi

Contemporary Mathematics
Volume **136**, 1992

Theta Vanishings for Some Smooth Abelian Coverings of Riemann Surfaces

Robert D. M. Accola

ABSTRACT. Let W be a compact Riemann surface admitting an elementary abelian 2-group of fixed point free automorphisms, $\mathcal{G}$, and assume $\mathcal{G}$ has an additional property, that of being syzygetic. Vanishing properties at half-periods corresponding to the various subgroups of $\mathcal{G}$ are derived for the theta function associated to the Jacobian of W.

1. Introduction

In 1967 H. M. Farkas discovered unusual vanishing properties for the theta function of a smooth 2-sheeted covering of a Riemann surface of genus greater than one [2]. If $W_{2p-1} \to W_p$ is a smooth 2-sheeted covering of compact Riemann surfaces of genera $2p-1$ and p, then the theta function for $J(W_{2p-1})$, the Jacobian of W_{2p-1}, vanishes at u_{p-1} half-periods to order 2, where $u_x = 2^{x-1}(2^x - 1)$. Much earlier, in 1895, Wirtinger discovered a whole variety of points on $J(W_{2p-1})$ where the theta function vanishes to order one, the variety being a translate of the Prym variety by a half-period [6]. In this paper we shall extend the results of Farkas to some smooth Galois coverings where the Galois group is an elementary Abelian 2-group. The coverings will have the additional property of being *syzygetic*, a term which we shall soon explain. It turns out that in these results the quarter-periods on the variety discovered by Wirtinger play a central role.

Let Z_2 stand for a finite group of order 2, and let $(Z_2)^\ell$ stand for an elementary 2-group of order 2^ℓ. We shall consider a smooth (that is, unramified) Galois covering of compact Riemann surfaces $W_{p_0} \to W_{p_\ell}$ of genera p_0 and p_ℓ. The Galois group, $\mathcal{G}$, will be a $(Z_2)^\ell$ and will be syzygetic (still to be explained). Thus $p_0 - 1 = 2^\ell(p_\ell - 1)$.

1991 Mathematics Classification, Primary 14H42, Secondary 30F10.

This paper is in final form and no version of it will be submitted for publication elsewhere.

For example, in the case $\ell = 2$, $p_0 = 4p_2 - 3$, the theta function for $J(W_{4p_2-3})$ will vanish to order 2 at 3 sets of half-periods corresponding to the 3 intermediate coverings $W_{4p_0-3} \to W_{2p_0-1}$. Moreover, this theta function will vanish to order 4 on another set of u_{p_0-2} half-periods.

In the general situation each intermediate covering $W_{p_0} \to W_{p_k}$ ($W_{p_k} = W_{p_0}/\mathcal{H}$, $\mathcal{H} \subset \mathcal{G}$, $\mathcal{H}$ is a $(Z_2)^k$) between W_{p_0} and W_{p_ℓ} will be the source of vanishings for the theta function for W_{p_0} at half-periods. Also the quarter-periods on the varieties of Wirtinger corresponding to intermediate 2-sheeted coverings, $W_{p_k} \to W_{p_{k+1}}$ give rise to theta vanishings for W_{p_0}. The problem of this paper is to classify and then count these half-period theta vanishings for W_{p_0}.

We must assume that the reader is familiar with the theory of theta functions and Riemann surfaces. As references we give [5] and [4] especially Chapter VII.

In the sequel we shall use abbreviations which will be indicated in parentheses after the item abbreviated.

The remainder of this section will serve to introduce notation as well as give a more complete introduction.

Let $\left\{ \begin{matrix} A_1 & A_2 & \dots & A_{p_\ell} \\ B_1 & B_2 & \dots & B_{p_\ell} \end{matrix} \right\}$ be a canonical homology basis (C. H. B.) for W_{p_ℓ}. The covering $W_{p_0} \to W_{p_\ell}$ is determined by a homomorphism from $H_1(W_{p_\ell}, Z)$ onto $(Z_2)^\ell$. Such a homomorphism is determined by the images of the cycles in the C. H. B. Suppose under this homomorphism we have

$$
\begin{matrix}
A_1 \to a_1 & \qquad B_1 \to b_1 \\
A_2 \to a_2 & \qquad B_2 \to b_2 \\
\vdots & \qquad \vdots \\
A_{p_\ell} \to a_{p_\ell} & \qquad B_{p_\ell} \to b_{p_\ell}
\end{matrix}
$$

where the a's and b's generate $(Z_2)^\ell$. We will describe this homomorphism by the symbol

$$
\left\{ \begin{matrix} a_1 & a_2 & \dots & a_{p_\ell} \\ b_1 & b_2 & \dots & b_{p_\ell} \end{matrix} \right\}
$$

Then one can choose a C. H. B. for W_{p_ℓ} so that the homomorphism is represented by the symbol

$$\left\{ \begin{matrix} a_1 & a_2 & \ldots & a_s & 0 & \ldots & 0 & 0 & \ldots & 0 \\ b_1 & b_2 & \ldots & b_s & b_{s+1} & \ldots & b_t & 0 & \ldots & 0 \end{matrix} \right\}$$

where $s + t = \ell$ and (for convenience) $s \leq t$ [4, p. 294]. In this paper we shall consider the case where $s = 0$ and $t = \ell$. This is the meaning of the term *syzygetic*.

The covering $W_{p_0} \to W_{p_\ell}$ determines a homomorphism from $J(W_{p_\ell})$ into $J(W_{p_0})$ obtained by lifting divisors of degree zero from W_{p_ℓ} to W_{p_0} [1, p. 5]. The kernel of this homomorphism, which is again an elementary 2-group of order 2^ℓ and which determines the covering, can be represented by a group, G, of half-integer period characteristics (Per. Ch.). When the C. B. H. on W_p has been chosen as above this group, G, is

$$\left\{ \left(\begin{matrix} 0 & 0 & 0 & \ldots & 0 & 0 & 0 & \ldots & 0 \\ \sigma_1 & \sigma_2 & \sigma_3 & \ldots & \sigma_{\ell-1} & \sigma_\ell & 0 & \ldots & 0 \end{matrix} \right) \middle| \sigma_j = 0 \;\; \text{or} \;\; \tfrac{1}{2} \right\}$$

a syzygetic group of Per. Ch.'s of order 2^ℓ.

If $W_{p_0} \to W_{p_k} \to W_{p_j} \to W_{p_\ell}$, we shall mean by the *array* $W_{p_k} \to W_{p_j}$, the covering indicated and all intermediate coverings. By the *big array* we shall mean the array $W_{p_0} \to W_{p_\ell}$. Thus there will be a Riemann surface in the array $W_{p_k} \to W_{p_j}$ for each subgroup of $\mathcal{G}(W_{p_k}, W_{p_j})$, the Galois group of the covering. (We shall continue to use the symbol $\mathcal{G}$ for $\mathcal{G}(W_{p_0}, W_{p_\ell})$.) By a Riemann surface at the *k-level*, we shall mean a surface which is W_{p_0} modulo a subgroup of $\mathcal{G}$ of order 2^k. If n_k is the number of subgroups of $(Z_2)^\ell$ of order 2^k, then there will be n_k Riemann surfaces at the k-level. If W_{p_k} stands for any surface at the k-level then the genus satisfies: $p_k - 1 = 2^{\ell-k}(p_\ell - 1)$. A surface at the k-level is also determined by a subgroup of G of order $2^{\ell-k}$.

A Per. Ch. with a subscript 2 and integer entries will denote a half-integer Per. Ch. The absence of the subscript 2 will indicate an arbitrary Per. Ch.

Let W_p be a Riemann surface of genus p (≥ 2). Let a C. H. B. be chosen. Then the corresponding theta function with theta characteristic (Th. Ch.) $\begin{bmatrix} g_1 & g_2 & \cdots & g_p \\ h_1 & h_2 & \ldots & h_p \end{bmatrix}$ will be denoted

$$\theta \begin{bmatrix} g_1 & \cdots & g_p \\ h_1 & \ldots & h_p \end{bmatrix}(u) \quad \text{or} \quad \theta \begin{bmatrix} g \\ h \end{bmatrix}(u).$$

If $g_j = \gamma_j/2$ and $h_j = \gamma'_j/2$ where the γ's are 0 or 1, then the Th. Ch. will be denoted

$$\begin{bmatrix} \gamma_1 & \cdots & \gamma_p \\ \gamma_1' & \cdots & \gamma_p' \end{bmatrix}_2 \quad \text{or} \quad [\gamma]_2$$

For a half-integer Th. Ch. $[\gamma]_2$, define $|\gamma| = Exp\{\pi i \sum \gamma_j \gamma_j'\}$ $(= 1$ or $-1)$. $[\gamma]_2$ is called even (resp. odd) if $|\gamma| = 1$ (resp. -1) and $\theta[\gamma]_2(u)$ is even (resp. odd). In dimension p there are u_p $(= 2^{p-1}(2^p - 1))$ odd Th. Ch.'s and g_p $(= 2^{p-1}(2^p + 1))$ even Th. Ch.'s. (Note that $4^j u_k + 2^k u_j = u_{j+k}$.)

We now summarize several consequences of Riemann's vanishing theorem [5, p. 170] which relate vanishings of the theta function to the existence of linear series of W_p. If $\theta \begin{bmatrix} 0 \\ 0 \end{bmatrix}(u)$ vanishes to order $r + 1$ at a point $e \in J(W_p)$ then there is a corresponding complete linear series g^r_{p-1} on W_p. If $2ne \equiv 0$ in $J(W_p)$ then g^r_{p-1} is $((1/2n) - n)$-canonical; that is, $(2n)g^r_{p-1}$ is n-canonical. If n is odd and $ne \equiv 0$ again g^r_{p-1} is $((1/2n) - n)$-canonical. For $n = 1$ or 2 we shall denote the property of being half-canonical by $(\frac{1}{2} - K)$ and the property of being quarter-bicanonical by $(\frac{1}{4} - 2K)$. Since we will be considering linear series corresponding to theta vanishings, all linear series will have degree $p - 1$ unless otherwise stated.

If $[\delta]_2$ is odd then $\theta[\delta]_2(u)$ vanishes at $u = 0$ to order one (in general). Then W_p will admit u_p $(\frac{1}{2} - K)$ g^0_{p-1}'s.

Now we state a lemma which we shall be using throughout this paper. (We denote addition of characteristics by juxtaposition; e.g. $(\alpha)_2 + (\beta)_2 = (\alpha\beta)_2$.)

Lemma 1.1 [1, p. 74]. *Let a C. H. B. be chosen on a Riemann surface W_p. Suppose for $J(W_p)$ there is a Th. Ch. $[\eta]$ and a group G of half-integer Per. Ch.'s of order 2^k so that the theta function $\theta[\eta\sigma](u)$ vanishes at $u = 0$ to order r_σ for each $(\sigma)_2$ in G. Let $W_q \to W_p$ be the smooth Abelian covering corresponding to G. Then the 2^k linear series, $g^{r_\sigma - 1}_{p-1}$, on W_p corresponding to the theta vanishings all lift to W_q to become linearly equivalent and on W_q determine a complete linear series of dimension $(\sum r_\sigma) - 1$. This linear series on W_q will be $(\frac{1}{2} - K)$ if and only if $(2\eta) \, \varepsilon \, G$.*

Now we state the main result of this paper.

Theorem. *Let $W_{p_0} \to W_{p_\ell}$ be a smooth Galois covering of Riemann surfaces with syzygetic Galois group $\mathcal{G}$ isomorphic to $(Z_2)^\ell$. Let n_k be the number of subgroups of $\mathcal{G}$ or order 2^k, $k = 0, 1, 2, \ldots, \ell$. Let $p_k = 2^{\ell-k}(p_\ell - 1) + 1$. Let $u_x = 2^{x-1}(2^x - 1)$. Then W_{p_0} admits the following number of half-canonical $g^{2^k-1}_{p_0-1}$'s:*

$$n_k\left(u_{p_k-k} - (2^{\ell-k} - 1)(4^k - 1)4^{p_{k+1}-(k+1)}\right).$$

We believe that for generic W_{p_ℓ} these linear series are complete and correspond to the only extraordinary vanishings for the theta function of W_{p_0} at half-periods. However, this belief has been verified in only a very few cases where p_ℓ and ℓ are small.

2. $\ell = 2$

We now consider the simplest array where $\ell = 1$. Let $W_{2p-1} \to W_p$ be a 2-sheeted covering where $\mathcal{G}(W_{2p-1}, W_p) = \langle T \rangle$. We choose a C.H.B. on W_p $\left\{ \begin{matrix} A_1 & A_2 & \dots & A_p \\ B_1 & B_2 & \dots & B_p \end{matrix} \right\}$ so that $(\sigma)_2$ is the Per. Ch. determining the covering where

$$(\sigma)_2 = \begin{pmatrix} 0 & 0 & \dots & 0 \\ 1 & 0 & \dots & 0 \end{pmatrix}_2,$$

We then choose the C.H.B. on W_{2p-1}:

$$\left\{ \begin{matrix} A_1' & A_2' & A_3' & \dots & A_p' & A_2'' & A_3'' & \dots & A_p'' \\ B_1' & B_2' & B_3' & \dots & B_p' & B_2'' & B_3'' & \dots & B_p'' \end{matrix} \right\}$$

where each cycle in this C. H. B. covers the corresponding cycle in the C. H. B. on W_p once except for B_1' which covers B_1 twice. The involution T interchanges A_j' and A_j'', B_j' and B_j'' for $j = 2, 3, \dots p$, and leaves the homology classes of A_1' and B_1' fixed. [5, p. 189]. Whenever considering 2-sheeted coverings we shall assume C. H. B.'s chosen as above.

We now derive the results of Farkas [2]. Suppose $[\delta_1]_2$ and $[\delta_2]_2$ are two odd p-dimensional Th. Ch.'s so that

$$[\delta_1]_2 + [\delta_2]_2 = (\sigma)_2$$

If g_{p-1}^0 and h_{p-1}^0 are the $(\frac{1}{2} - K)$ linear series on W_p corresponding to these Th. Ch.'s, then g_{p-1}^0 and h_{p-1}^0 lift to equivalent g_{2p-2}^0's on W_{2p-1}, and determine a $(\frac{1}{2} - K)$ g_{2p-2}^1. The only possible pairs of odd Th. Ch.'s for which this is possible are of the form:

$$\begin{bmatrix} 0 & \delta_2 & \dots & \delta_p \\ 1 & \delta_2' & \dots & \delta_p' \end{bmatrix}_2 \quad \text{and} \quad \begin{bmatrix} 0 & \delta_2 & \dots & \delta_p \\ 0 & \delta_2' & \dots & \delta_p' \end{bmatrix}_2$$

Consequently the number of such pairs is u_{p-1}. The $(2p-1)$-dimensional Th. Ch. for g_{2p-2}^1 is [5, p. 203]

$$\begin{bmatrix} 0 & \delta_2 & \dots & \delta_p & \delta_2 & \dots & \delta_p \\ 1 & \delta_2' & \dots & \delta_p' & \delta_2' & \dots & \delta_p' \end{bmatrix}_2$$

There are $u_p - 2u_{p-1}$ $(= 4^{p-1})$ $(\frac{1}{2} - K)$ g_{p-1}^0's on W_p which lift to $(\frac{1}{2} - K)$ g_{2p-2}^0's on W_{2p-1}.

Now we summarize Wirtinger's results giving a derivation. Let H be the $(2p - 1) \times (2p - 1)$ integer matrix

$$\begin{bmatrix} 1 & 0 & 0 \\ 0 & 0 & I \\ 0 & I & 0 \end{bmatrix}$$

where I is the $(p - 1) \times (p - 1)$ identity matrix. Then the action of T on the C. H. B. on W_{2p-1} is:

$$\begin{bmatrix} H & 0 \\ 0 & H \end{bmatrix}$$

If a point u in $J(W_{2p-1})$ is described by the Per. Ch.

$$\left(\begin{array}{c|ccc|ccc} g_1 & g_2' & \cdots & g_p' & g_2'' & \cdots & g_p'' \\ h_1 & h_2' & \cdots & h_p' & h_2'' & \cdots & h_p'' \end{array} \right)$$

Then Tu is described by

$$\left(\begin{array}{c|ccc|ccc} g_1 & g_2'' & \cdots & g_p'' & g_2' & \cdots & g_p' \\ h_1 & h_2'' & \cdots & h_p'' & h_2' & \cdots & h_p' \end{array} \right)$$

If u has the Per. Ch.

$$\left(\begin{array}{c|ccc|ccc} 0 & g_2 & \cdots & g_p & -g_2 & \cdots & -g_p \\ 0 & h_2 & \cdots & h_p & -h_2 & \cdots & -h_p \end{array} \right)$$

then $Tu = -u$.

If

$$[\delta]_2 = \begin{bmatrix} 1 & 0 & \cdots & 0 & 0 & \cdots & 0 \\ 1 & 0 & \cdots & 0 & 0 & \cdots & 0 \end{bmatrix}_2,$$

then T leaves $[\delta]_2$ unchanged. By the transformation theory for theta functions [5, Ch. III] we have:

$$\theta[\delta]_2(u) = \theta[\delta]_2(Tu) = \theta[\delta]_2(-u) = -\theta[\delta]_2(u).$$

Writing $u = (0, u_2, -u_2)$ where $u_2 \, \varepsilon \, \mathbf{C}^{p-1}$, we see that

$$\theta \begin{bmatrix} 1 & 0 & \cdots & 0 & 0 & \cdots & 0 \\ 1 & 0 & \cdots & 0 & 0 & \cdots & 0 \end{bmatrix}_2 (0, u_2, -u_2) = 0$$

for all u_2 in $\mathbf{C}^{p-1}$. That is,

$$\theta \begin{bmatrix} \frac{1}{2} & g_2 & \cdots & g_p & -g_2 & \cdots & -g_p \\ \frac{1}{2} & h_2 & \cdots & h_p & -h_2 & \cdots & -h_p \end{bmatrix} (0) = 0$$

for all $(p-1)$-dimensional Per. Ch.'s $\left(\begin{smallmatrix} g_2 & \cdots & g_p \\ h_2 & \cdots & h_p \end{smallmatrix} \right)$. This is the *Wirtinger variety* of theta vanishings for the cover $W_{2p-1} \to W_p$. The 4^{p-1} $\left(\frac{1}{2} - K \right)$ g^0_{2p-2}'s lifted from W_p lie on the Wirtinger variety and have Th. Ch.'s:

$$\begin{bmatrix} 1 & \gamma_2 & \cdots & \gamma_p & \gamma_2 & \cdots & \gamma_p \\ 1 & \gamma_2' & \cdots & \gamma_p' & \gamma_2' & \cdots & \gamma_p' \end{bmatrix}_2$$

The $\left(\frac{1}{2} - K \right)$ g^1_{2p-2}'s which arise in the manner described by Farkas are invariant under T. Also T maps the Wirtinger variety onto itself, the only fixed points being the 4^{p-1} half-periods.

Now we investigate $\left(\frac{1}{2} - K \right)$ g^1_{2p-2}'s on W_{2p-1} which are invariant under T. We consider all possibilities since 2-sheeted coverings in the big array give rise to many of the theta vanishings discussed in this paper.

If T acts non-faithfully then each divisor of g^1_{2p-2} is mapped onto itself by T and so g^1_{2p-2} is the lift of a g^1_{p-1} on W_p which may be $\left(\frac{1}{2} - K \right)$ or $\left(\frac{1}{4} - 2K \right)$. Suppose g^1_{p-1} is $\left(\frac{1}{4} - 2K \right)$. Then there is a second $\left(\frac{1}{4} - 2K \right)$ linear series, h^1_{p-1}, where $g^1_{p-1} + h^1_{p-1} \, \varepsilon \, K_p$, the canonical series on W_p. h^1_{p-1} also lifts to a $\left(\frac{1}{2} - K \right)$ h^1_{2p-2} on W_{2p-1} which is linearly equivalent to g^1_{2p-2}. Consequently, if g^1_{p-1} is $\left(\frac{1}{4} - 2k \right)$ then g^1_{2p-2} cannot be complete.

Lemma 2.1. *Let T be the involution for the cover $W_{2p-1} \to W_p$. Let g^1_{2p-2} be a complete $\left(\frac{1}{2} - K \right)$ linear series on W_{2p-1} each divisor of which is invariant under T. Then g^1_{2p-2} is the lift from W_p of a complete $\left(\frac{1}{2} - K \right)$ g^1_{p-1}.*

Now suppose that T permutes the divisors of a $\left(\frac{1}{2} - K \right)$ g^1_{2p-2} non-trivially. Since the divisors of g^1_{2p-2} are a $\mathbf{P}^1$ and T is a non-identity involution acting on this $\mathbf{P}^1$, there are two divisors, D_1 and E_1, invariant under T. D_1 and E_1 are lifts of divisors on W_p, D_0 and E_0. Again there are two possibilities.

If D_0 and E_0 are $\left(\frac{1}{2} - K \right)$ then g^1_{2p-2} arises in the manner described by Farkas, and $2D_0 \equiv 2E_0 \equiv K_p$. However, it may happen that $D_0 + E_0 \equiv K_p$ where D_0 and E_0 are both $\left(\frac{1}{4} - 2K \right)$. We now investigate these two possibilities more closely.

Suppose D_0 and E_0 are $(\frac{1}{2} - K)$ on W_p. Let f be a meromorphic function on W_p with divisor (f) where:

$$(f) = 2D_0 - 2E_0$$

The covering $W_{2p-1} \to W_p$ is the two-sheeted covering of W_p where $\sqrt{f}$ becomes single-valued. On W_{2p-1}

$$(\sqrt{f}) = D_1 - E_1$$

On W_{2p-1} let ω and σ be holomorphic one-forms whose divisors are:

$$(\omega) = 2D_1$$
$$(\sigma) = D_1 + E_1$$

Since (ω) is invariant under T and since $2D_0$ is canonical on W_p, we see that ω is the lift of a holomorphic one-form on W_p; that is, ω is symmetric with respect to T. Also, for some $\lambda \varepsilon \mathbf{C} - \{0\}$, $\omega = \lambda\sigma\sqrt{f}$. We see that σ is anti-symmetric with respect to T. On the other hand, if $D_0 + E_0 = K_p$, we see that σ (as defined above) is symmetric and ω is anti-symmetric.

Lemma 2.2. *Let T be the involution for the covering $W_{2p-1} \to W_p$. Let g^1_{2p-2} be a $(\frac{1}{2} - K)$ linear series invariant under a faithful T action with fixed divisors D_1 and E_1. Suppose ω and σ are holomorphic one-forms on W_{2p-1} with divisors*

$$(\omega) = 2D_1$$
$$(\sigma) = D_1 + E_1$$

Let D_1 and E_1 be the lifts of D_0 and E_0, divisors on W_p. Then there are two possibilities.

(i) ω is symmetric and σ is anti-symmetric with respect to T. In this case D_0 and E_0 are $(\frac{1}{2} - K)$.

(ii) ω is anti-symmetric and σ is symmetric. In this case D_0 and E_0 are $(\frac{1}{4} - 2K)$ and $D_0 + E_0 \equiv K_p$.

Notice that the results of Lemma 2.2 do not depend on g^1_{2p-2} being complete.

For a general Riemann surface of genus p the theta function does not vanish at quarter-periods. However, for two-sheeted coverings in the big array above the ℓth (bottom) level, the situation of Lemma 2.2 (ii) always arises because of the quarter-periods on the Wirtinger variety of theta vanishings.

3. $\ell = 2$

We now consider a smooth Galois covering $W_{4p-3} \rightarrow W_p$ with Galois group, $\mathcal{G}$, isomorphic to $(Z_2)^2$. Let $\mathcal{G} = \{e, T_1, T_2, T_3\}$ and let $W_{4p-3}/\langle T_j \rangle = W^j$, a Riemann surface of genus $2p-1$, $j = 1, 2, 3$. Finally suppose that W_{4p-3} admits a $(\frac{1}{2} - K)$ g^1_{4p-4} invariant under $\mathcal{G}$.

If each divisor in g^1_{4p-4} is carried into itself by all elements of $\mathcal{G}$ ($\mathcal{G}$'s action on g^1_{4p-4}, as a $\mathbf{P}^1$, is *totally unfaithful*) then there is a g^1_{p-1} on W_p which lifts to g^1_{4p-4}. If g^1_{4p-4} is complete then g^1_{p-1} must be $(\frac{1}{2} - K)$, as before.

Suppose T_1 acts unfaithfully on g^1_{4p-4}, but T_2 and T_3 act faithfully. (We shall describe this by saying that $\mathcal{G}$'s action is *partially faithful*.) Then there are divisors D, E in g^1_{4p-4} which are invariant under T_2 and T_3. On W^2 there are divisors D_2, E_2 which lift to D, E, and on W^3 there are divisors D_3, E_3 which lift to D, E. On W^1 there is a g^1_{2p-2} which lifts to g^1_{4p-4}. On W_p there are divisors D_0, E_0 which lift to D, E on W_{4p-3}. D_0 and E_0 lift to D_2, E_2 on W^2 and to D_3, E_3 on W^3. D_0 and E_0 lift to two linearly equivalent divisors in g^1_{2p-2} on W^1. If D_0 and E_0 are $(\frac{1}{2} - K)$ then so are all intermediate divisors. If D_0 and E_0 are $(\frac{1}{4} - 2K)$ then D_2, E_2, D_3 and E_3 are also $(\frac{1}{4} - 2K)$, but g^1_{2p-2} on W^1 is $(\frac{1}{2} - K)$. The possibilities for the linear series will be represented by the following diagrams. First, the diagram of the coverings in the array $W_{4p-3} \rightarrow W_p$:

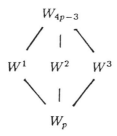

Diagram 1.

Next the linear series on the corresponding Riemann surfaces:

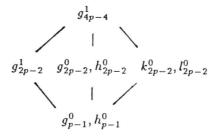

Diagram 2.

In Diagram 2. the D's and E's are shown as linear series of dimension zero.

The two possibilities for the various linear series are $(\frac{1}{2} - K)$ and $(\frac{1}{4} - 2K)$. We diagram these possibilities as follows:

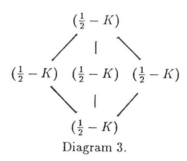

Diagram 3.

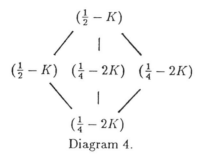

Diagram 4.

The last 3 diagrams illustrate the situation when $\mathcal{G}$'s action on g^1_{4p-4} is partially faithful.

Now we suppose that the action of G on the divisors of g^1_{4p-4} is faithful. g^1_{4p-4} gives a $(4p-4)$-sheeted covering $f : W_{4p-3} \to \mathbf{P}^1$ where f is a meromorphic function. $\mathcal{G}$ descends to be a $(Z_2)^2$ of fractional linear transformations on $\mathbf{P}^1$. We may assume that the 3 involutions on $\mathbf{P}^1$ corresponding to T_1, T_2, and T_3 are, respectively; $z \to -z$; $z \to z^{-1}$; and $z \to -z^{-1}$. The invariant divisors for T_1 thus lie above 0 and ∞ and will be denoted D_1 and E_1. The invariant divisors for T_2 lie above 1 and -1 and will be denoted D_2 and E_2. The invariant divisors for T_3 lie above i and $-i$ and will be denoted D_3 and E_3. $D_j + E_j$ is canonical for $j = 1, 2, 3$. Let σ_j be a holomorphic one-form on W_{4p-3} so that

$$(\sigma_j) = D_j + E_j$$

If T_j^* is the action of T_j on the holomorphic one-forms on W_{4p-3} then

$$T_j^* \sigma_j = \varepsilon_j \sigma_j \qquad (j = 1, 2, 3)$$

where $\varepsilon_j = +1$ or -1.

Lemma 3.1. $\varepsilon_1\varepsilon_2\varepsilon_3 = +1$.

Proof: The function f is z lifted to W_{4p-3} so the actions of T_j on f are as follows:

$$f \circ T_1 = -f$$
$$f \circ T_2 = 1/f$$
$$f \circ T_3 = -1/f$$

By relabeling the divisors, of necessary, it follows that the divisor of f is $D_1 - E_1$; the divisor of $(f - 1)/(f + 1)$ is $D_2 - E_2$; and the divisor of $(f - i)/(f + i)$ is $D_3 - E_3$. We see that

$$((f - 1)(f + 1)) = D_2 + E_2 - 2E_1$$

$$(f/(f^2 - 1)) = (D_1 + E_1) - (D_2 + E_2)$$
$$(f/(f^2 + 1)) = (D_1 + E_1) - (D_3 + E_3)$$

Consequently,

$$\sigma_2 = \lambda_2((f^2 - 1)/f)\sigma_1$$

and

$$\sigma_3 = \lambda_3((f^2 + 1)/f)\sigma_1$$

where $\lambda_2, \lambda_3 \in \mathbf{C} - \{0\}$.

Therefore

$$T_1^*\sigma_2 = \lambda_2((f^2 - 1)/(-f))\varepsilon_1\sigma_1 = -\varepsilon_1\sigma_2$$

and

$$T_1^*\sigma_3 = \lambda_3((f^2 + 1)/(-f))\varepsilon_1\sigma_1 = -\varepsilon_1\sigma_3$$

Similarly,

$$T_2^*\sigma_1 = -\varepsilon_2\sigma_1$$
$$T_2^*\sigma_3 = -\varepsilon_2\sigma_3$$
$$T_3^*\sigma_1 = -\varepsilon_3\sigma_1$$
$$T_3^*\sigma_2 = -\varepsilon_3\sigma_2$$

Thus $T_3^*\sigma_1 = -\varepsilon_3\sigma_1 = T_1^*T_2^*\sigma_1 = T_1^*(-\varepsilon_2\sigma_1) = -\varepsilon_2\varepsilon_1\sigma_1$, and so $\varepsilon_1\varepsilon_2 = \varepsilon_3$. q.e.d.

By Lemma 2.2, if $\varepsilon_j = +1$ then σ_j is the lift of a holomorphic one-form from W^j. D_j and E_j are the lifts of $(\frac{1}{4} - 2K)$ divisors on W_j. Notice that there is no divisor invariant under the full group $\mathcal{G}$. Thus in the case where $\mathcal{G}$ acts faithfully on g^1_{4p-4} the diagram corresponding to the previous Diagram 2 is:

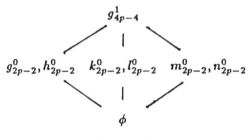

Diagram 5.

Since 2 of the ε_j's can be negative we have two possibilities for the distribution of the $(\frac{1}{2} - K)$'s and $(\frac{1}{4} - 2K)$'s illustrated in the following two diagrams.

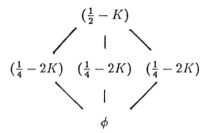

Diagram 6.

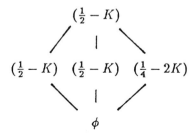

Diagram 7.

The last 3 diagrams illustrate the situation when $\mathcal{G}$'s action on g^1_{4p-4} is faithful. Notice that when $\mathcal{G}$'s action on g^1_{4p-4} is not totally unfaithful, Diagrams 3, 4, 6, and 7 illustrate all possible arrangements of $(\frac{1}{2} - K)$'s and $(\frac{1}{4} - 2K)$'s in the second rows of the diagrams.

In the situation of the preceding diagrams, it is never assumed that g^1_{4p-4} is complete. If we denote by $|g^1_{4p-4}|$ the complete linear series determined by g^1_{4p-4}, then the dimension of $|g^1_{4p-4}|$ must be $1 + 4e$ for $e \geq 0$. For $|g^0_{2p-2}|$ is even dimensional since g^0_{2p-2} corresponds to an odd Th. Ch.

Moreover $|g^0_{2p-2}|$ and $|h^0_{2p-2}|$ have the same dimension since they are interchanged by the involution for the cover $W^1 \to W_p$. If that dimension is $2e$ then the result follows by Lemma 1.1. For an analogous reason $|m^0_{2p-2}|$ and $|n^0_{2p-2}|$ have the same dimension, e', and again by Lemma 1.1 the dimension of $|g^1_{4p-4}| = 1 + 2e'$. Therefore $e' = 2e$.

Lemma 3.2. *In the situation of Diagram 7., the $(\frac{1}{4} - 2K)$ m^0_{2p-2} and n^0_{2p-2} on W^3 determine complete linear series of even dimension (just as do linear series corresponding to odd Th. Ch.'s).*

As a preliminary to the general theorem, we now count the number of $(\frac{1}{2} - K)$ linear series on W_{4p-3} arising from a syzygetic covering $W_{4p-3} \to W_p$, continuing the above notation. By choosing an appropriate C. H. B. on W_p we may assume the covering is determined by the group, G, of Per. Ch.'s generated by $(\sigma_1)_2$ and $(\sigma_2)_2$ where

$$(\sigma_1)_2 = \begin{pmatrix} 0 & 0 & 0 & \ldots & 0 \\ 1 & 0 & 0 & \ldots & 0 \end{pmatrix}_2 \quad \text{and} \quad (\sigma_2)_2 = \begin{pmatrix} 0 & 0 & 0 & \ldots & 0 \\ 0 & 1 & 0 & \ldots & 0 \end{pmatrix}_2 ,$$

We first use Lemma 1.1 to count the number of $(\frac{1}{2} - K)$ g^3_{4p-4}'s on W_{4p-3}. If we let

$$[\delta]_2 = \begin{bmatrix} 0 & 0 & \delta_3 & \ldots & \delta_p \\ 0 & 0 & \delta'_3 & \ldots & \delta'_p \end{bmatrix}_2$$

be an odd Th. Ch., then all Th. Ch.'s

$$\cdot \quad \{ [\delta\sigma]_2 | (\sigma)_2 \, \varepsilon \, G \}$$

are odd. Each odd Th. Ch. of the type $[\delta]_2$ gives rise to a $(\frac{1}{2} - K)$ g^3_{4p-4} on W_{4p-3}. There are u_{p-2} such Th. Ch.'s. We conclude that W_{4p-3} admits u_{p-2} $(\frac{1}{2} - K)$ g^3_{4p-4}'s.

On W_p there are $u_p - 4u_{p-2}$ $(= 6 \cdot 4^{p-2})$ odd Th. Ch.'s not contributing to the g^3_{4p-4}'s on W_{4p-3}. If now we let $[\gamma]_2$ be

$$[\gamma]_2 = \begin{bmatrix} \gamma_1 & \gamma_2 & \gamma_3 & \ldots & \gamma_p \\ 0 & 0 & \gamma'_3 & \ldots & \gamma'_p \end{bmatrix}_2$$

where $(\gamma_1, \gamma_2) \neq (0,0)$ and $[\gamma]_2$ is odd or even, then of the 4 Th. Ch.'s $\{ [\gamma\sigma]_2 | (\sigma)_2 \, \varepsilon \, G \}$ 2 are odd and 2 are even. The two odd Th. Ch. correspond to two $(\frac{1}{2} - K)$ g^0_{p-1}'s on W_p which lift to a g^1_{4p-4} on W_{4p-3}. This g^1_{4p-4} is invariant under the full Galois group $\mathcal{G}$ $(\cong (Z_2)^2)$ and is of the type described by Diagrams 2. and 3. Thus every $(\frac{1}{2} - K)$ g^0_{p-1} on W_p lifts to W_{4p-3} to become part of a g^1_{4p-4} or g^3_{4p-4}. (If the covering is not syzygetic this last statement is false.) The number of $(\frac{1}{2} - K)$ g^1_{4p-4}'s on W_{4p-3} arising from pairs of g^0_{p-1}'s on W_p is $3 \cdot 4^{p-2}$.

Now we count the number of $(\frac{1}{2} - K)$ g^1_{4p-4}'s which are invariant under $\mathcal{G}$ as described by Diagram 5. Referring to Diagram 1. suppose that the covering $W^3 \to W_p$ corresponds to the Per. Ch. $\begin{pmatrix} 0 & 0 & \cdots & 0 \\ 1 & 0 & \cdots & 0 \end{pmatrix}_2$. We choose on W^3 a C. H. B. as described earlier so that the Wirtinger vanishings on W^3 correspond to Th. Ch.'s of the form:

$$\begin{array}{cccccccc} 1 & 2 & \cdots & p & 2 & \cdots & p \\ \left[\begin{array}{c} \frac{1}{2} \\ \frac{1}{2} \end{array}\right. & \begin{array}{c} g_2 \\ h_2 \end{array} & \cdots & \begin{array}{c} g_p \\ h_p \end{array} & \begin{array}{c} -g_2 \\ -h_2 \end{array} & \cdots & \left.\begin{array}{c} -g_p \\ -h_p \end{array}\right] \end{array}$$

where the g's and h's are arbitrary. (Numbers above entries in characteristics will serve as place markers.) For such Th. Ch.'s to give rise to $(\frac{1}{2} - K)$ g^1_{4p-4}'s on W_{4p-3} all the g's and h's must be half-integers except h_2, which must be $1/4$ or $3/4$. For observe that the Per. Ch. for $J(W^3)$ describing the covering $W_{4p-3} \to W^3$ is $(\tilde{\sigma})_2$ where

$$(\tilde{\sigma})_2 = \begin{pmatrix} 0 & 0 & 0 & \cdots & 0 & 0 & 0 & \cdots & 0 \\ 0 & 1 & 0 & \cdots & 0 & 1 & 0 & \cdots & 0 \end{pmatrix}_2$$

Denote the Th. Ch.

$$\begin{array}{ccccccccc} 1 & 2 & 3 & \cdots & p & 2 & 3 & \cdots & p \\ \left[\begin{array}{c} \frac{1}{2} \\ \frac{1}{2} \end{array}\right. & \begin{array}{c} \epsilon_2 \\ h_2 \end{array} & \begin{array}{c} \epsilon_3 \\ \epsilon'_3 \end{array} & \cdots & \begin{array}{c} \epsilon_p \\ \epsilon'_p \end{array} & \begin{array}{c} \epsilon_2 \\ -h_2 \end{array} & \begin{array}{c} \epsilon_3 \\ \epsilon'_3 \end{array} & \cdots & \left.\begin{array}{c} \epsilon_p \\ \epsilon'_p \end{array}\right] \end{array}$$

by $[h_2|\epsilon]$ where the ϵ's are all half-integers. Then for fixed ϵ:

$$[\tfrac{1}{4}|\epsilon] - [-\tfrac{1}{4}|\epsilon] = (\tilde{\sigma})_2,$$

and

$$[\tfrac{1}{4}|\epsilon] + [\tfrac{1}{4}|\epsilon] = (\tilde{\sigma})_2$$

Thus the $(\frac{1}{4} - 2K)$ g^0_{2p-2} and h^0_{2p-2} corresponding to $[\frac{1}{4}|\epsilon]$ and $[-\frac{1}{4}|\epsilon]$ lift to W_{4p-3} to become linearly equivalent. The g^1_{4p-4} so determined is $(\frac{1}{2} - K)$ and is invariant under $\mathcal{G}$ ($\cong (Z_2)^2$). g^0_{2p-2} (and h^0_{2p-2}) is not the lift of a divisor from W_p since the corresponding Th. Ch. is not invariant under the involution for the covering $W^3 \to W_p$. Consequently, we are in the situation described in Diagram 5. The number of g^1_{4p-4}'s on W_{4p-3} determined in this manner equals the number of choices for the half-integers ϵ_j, $\epsilon_{j'}$. This number is 2^{2p-3}.

We now show that for any such g^1_{4p-4}, which is invariant under $\mathcal{G}$, Diagram 7. describes the situation. That is, for any such $(\frac{1}{2} - K)$ g^1_{4p-4} on

W_{4p-3} arising from two $(\frac{1}{4} - 2K)$ g^0_{2p-2}'s on W^3, there is on W^1 (and W^2) a pair of $(\frac{1}{2} - K)$ g^0_{2p-2}'s which lift to W_{4p-3} to become linearly equivalent and determine the given g^1_{4p-4}. To do this we will count on one of the W^j's (any one will do for the purpose of counting) the number of pairs of $(\frac{1}{2} - K)$ g^0_{2p-2}'s which are interchanged by the Galois group $\langle S_j \rangle$ for the covering $W^j \to W_p$ and which lift to W_{4p-3} to become linearly equivalent. We will do this on W^3 with the C. H. B. already chosen. The odd Th. Ch.'s for such pairs of g^0_{2p-2}'s must be interchanged by S_3 and sum to $(\tilde{\sigma})_2$. Here are the pairs:

$$
\begin{array}{cccccccccc}
1 & 2 & 3 & \cdots & p & 2 & 3 & \cdots & p
\end{array}
$$

$$
\begin{bmatrix}
1 & 0 & \gamma_3 & \cdots & \gamma_p & 0 & \gamma_3 & \cdots & \gamma_p \\
1 & 1 & \gamma'_3 & \cdots & \gamma'_p & 0 & \gamma'_3 & \cdots & \gamma'_p
\end{bmatrix}_2 ,
$$

$$
\begin{bmatrix}
1 & 0 & \gamma_3 & \cdots & \gamma_p & 0 & \gamma_3 & \cdots & \gamma_p \\
1 & 0 & \gamma'_3 & \cdots & \gamma'_p & 1 & \gamma'_3 & \cdots & \gamma'_p
\end{bmatrix}_2
$$

and

$$
\begin{bmatrix}
\varepsilon_1 & 1 & \gamma_3 & \cdots & \gamma_p & 1 & \gamma_3 & \cdots & \gamma_p \\
\varepsilon'_1 & 1 & \gamma'_3 & \cdots & \gamma'_p & 0 & \gamma'_3 & \cdots & \gamma'_p
\end{bmatrix}_2 ,
$$

$$
\begin{bmatrix}
\varepsilon_1 & 1 & \gamma_3 & \cdots & \gamma_p & 1 & \gamma_3 & \cdots & \gamma_p \\
\varepsilon'_1 & 0 & \gamma'_3 & \cdots & \gamma'_p & 1 & \gamma'_3 & \cdots & \gamma'_p
\end{bmatrix}_2
$$

where $\binom{\varepsilon_1}{\varepsilon'_1}_2 = \binom{0}{0}_2 \binom{0}{1}_2$ or $\binom{1}{0}_2$. The number of such pairs is 2^{2p-2} $(= 2 \cdot 2^{2p-3})$ and each pair lifts to a $(\frac{1}{2} - K)$ g^1_{4p-4} described in Diagram 7.

Since there are 2^{2p-3} pairs of $(\frac{1}{4} - 2K)$ g^0_{2p-2}'s on each of W^1 and W^2 which must correspond to pairs of $(\frac{1}{2} - K)$ g^0_{2p-2}'s on W^3 as in Diagram 7., *all* such pairs of $(\frac{1}{4} - 2K)$ g^0_{2p-2}'s on W^1 and W^2 must give rise to g^1_{4p-4}'s on W_{4p-3} as described in Diagram 7. The number of g^1_{4p-4}'s of the Diagram 7. type on W_{4p-3} is $3 \cdot 2^{2p-3}$, a set of 2^{2p-3} for each cover $W^j \to W_p$. The situation described in Diagram 6. does not arise.

Thus the total number of $(\frac{1}{2} - K)$ g^1_{4p-4}'s on W_{4p-3} invariant under $\mathcal{G}$ is

$$
3 \cdot 4^{p-2} + 3 \cdot 2^{2p-3} = 9 \cdot 2^{2p-4}
$$

We now count the number of g^1_{4p-4}'s on W_{4p-3} which are invariant only under a subgroup of $\mathcal{G}$ of order 2. The number arising from W^3 by simply pairing odd Th. Ch.'s (the Farkas method) is u_{2p-2}. Of these pairs

$4^{p-1}/2$ are lifted from W_p and 2^{2p-2} give rise to the Diagram 7. situation. Consequently

$$u_{2p-2} - 2^{2p-3} - 2^{2p-2} = u_{2p-2} - 3 \cdot 2^{2p-3}$$

are invariant only under T_3. Since there are 3 coverings $W_{4p-3} \to W^j$, the total number of g^1_{4p-4}'s invariant only under a single automorphism of $\mathcal{G}$ is

$$3(u_{2p-2} - 3 \cdot 2^{2p-3})$$

The total number of g^1_{4p-4}'s obtained by this counting is

$$3(u_{2p-2} - 3 \cdot 2^{2p-3}) + 9 \cdot 2^{2p-4} = 3(u_{2p-2} - 3 \cdot 4^{p-2}).$$

If $p = 2$ this number is $3(6 - 3) = 9$, which it is reassuring to recall as the number previously calculated [1, p. 84]. (If $p = 2$ the hyperelliptic involution on W_2 lifts to W_5 to give a $(Z_2)^3$ generated by 3 elliptic-hyperelliptic involutions.)

For future use we record the following lemma.

Lemma 3.3. *Suppose $W_{4p-3} \to W_p$ is a smooth syzygetic $(Z_2)^2$ covering with Galois group $\mathcal{G} = \{e, T_1, T_2, T_3\}$. Let g^1_{4p-4} be a $(\frac{1}{2} - K)$ linear series on W_{4p-3} arising from $2\left(\frac{1}{4} - 2K\right) g^0_{2p-2}$'s on $W_{4p-3}/\langle T_3 \rangle$ $(= W^3_{2p-1})$ where the two g^0_{2p-2}'s correspond to Wirtinger vanishings on $J(W^3_{2p-1})$ for the covering $W^3_{2p-1} \to W_p$. (g^1_{4p-4} is thus invariant under the faithful $\mathcal{G}$-action.) Then the situation described in Diagram 6. cannot occur.*

4. 2^k-Sets

This section will include a description and definition of the objects we are counting in the context of the big array $W_{p_0} \to W_{p_\ell}$.

First, some additional notation. Recall that W_{p_k} is general notation for a Riemann surface at the k-level in the big array. If $W_{p_i} \to W_{p_j}$ is a covering within the big array recall that $\mathcal{G}(W_{p_i}, W_{p_j})$ denotes the Galois group of this covering. The group of Per. Ch.'s for $J(W_{p_j})$ determining this covering will be denoted $G(W_{p_i}, W_{p_j})$. As before $\mathcal{G}$ and G without additional notation will denote the corresponding groups for the covering $W_{p_0} \to W_{p_\ell}$.

Now we begin the definitions.

I) On a W_{p_k} let a C. H. B. be chosen so that

$$G(W_{p_0} \cdot W_{p_k}) = \left\{ \begin{pmatrix} \overset{1}{0} & \overset{2}{0} & \overset{\cdots}{\cdots} & \overset{k}{0} & \overset{k+1}{0} & \cdots & \overset{p_k}{0} \\ \sigma_1 & \sigma_2 & \cdots & \sigma_k & 0 & \cdots & 0 \end{pmatrix}_2 \middle| \sigma_i = 0 \ \text{ or } \ 1 \right\}$$

Let $[\delta]_2 = \begin{bmatrix} 0 & 0 & \cdots & 0 & \delta_{k+1} & \cdots & \delta_{p_k} \\ 0 & 0 & \cdots & 0 & \delta'_{k+1} & \cdots & \delta'_{p_k} \end{bmatrix}_2$ be odd. (There are $u_{p_k - k}$

such odd Th. Ch.'s.) The set of odd Th. Ch.'s

$$\{[\delta\sigma]_2 \mid (\sigma)_2 \,\varepsilon\, G(W_{p_0}, W_{p_k})\}$$

will be called a 2^k-set of Th. Ch.'s of Type I. The number of 2^k-sets of Th. Ch. of Type I on a fixed W_{p_k} is $u_{p_k - k}$.

II) Again on a W_{p_k} with the same C. H. B. let

$$[\gamma]_2 = \begin{bmatrix} \gamma_1 & \gamma_2 & \cdots & \gamma_k & \gamma_{k+1} & \cdots & \gamma_{p_k} \\ 0 & 0 & \cdots & 0 & \gamma'_{k+1} & \cdots & \gamma'_{p_k} \end{bmatrix}_2$$

where $(\gamma_1, \gamma_2, \ldots, \gamma_k) \neq (0, 0, \ldots, 0)$. Then 2^{k-1} of the Th. Ch.'s

$$\{[\delta\sigma]_2 \mid (\sigma)_2 \,\varepsilon\, G(W_{p_0}, W_{p_k})\}$$

are odd and form a 2^{k-1}-set of Th. Ch.'s of Type II. (2^k-sets of Th. Ch.'s of Type II occur for Riemann surfaces at the $(k+1)$-level.) The number of such 2^{k-1}-sets of Th. Ch.'s of Type II for a fixed W_{p_k} is $(2^k - 1)4^{p_k - k}$.

Definition. $M_k = 4^{p_k - k}$.

Notice that every odd Th. Ch. for a fixed W_{p_k} lies in a 2^k-set of Th. Ch.'s of Type I or in a 2^{k-1}-set of Th. Ch.'s of Type II.

III) Consider a fixed covering $W_{p_k} \to W_{p_{k+1}}$ with C. H. B.'s chosen so that

$$G(W_{p_0}, W_{p_{k+1}}) = \left\{ \begin{pmatrix} 0 & 0 & \cdots & 0 & 0 & \cdots & 0 \\ \sigma_1 & \sigma_2 & \cdots & \sigma_{k+1} & 0 & \cdots & 0 \end{pmatrix}_2 \middle| \sigma_i = 0 \ \text{ or } \ 1 \right\}$$

and chosen according to the perscription in Section 1. For this covering the Wirtinger vanishings for the theta function of W_{p_k} correspond to the Th. Ch.'s:

$$\begin{bmatrix} \frac{1}{2} & g_2 & \cdots & g_{k+1} & \cdots & g_{p_{k+1}} & -g_2 & \cdots & -g_{k+1} & \cdots & -g_{p_{k+1}} \\ \frac{1}{2} & h_2 & \cdots & h_{k+1} & \cdots & h_{p_{k+1}} & -h_2 & \cdots & -h_{k+1} & \cdots & -h_{p_{k+1}} \end{bmatrix}$$

Consider the Th. Ch.'s: $[\eta] =$

$$\begin{bmatrix} \frac{1}{2} & \varepsilon_2 & .. & \varepsilon_{k+1} & \varepsilon_{k+2} & .. & \varepsilon_{p_{k+1}} & \varepsilon_2 & .. & \varepsilon_{k+1} & \varepsilon_{k+2} & .. & \varepsilon_{p_{k+1}} \\ \frac{1}{2} & \eta_2 & .. & \eta_{k+1} & \varepsilon'_{k+2} & .. & \varepsilon'_{p_{k+1}} & -\eta_2 & .. & -\eta_{k+1} & \varepsilon'_{k+2} & .. & \varepsilon'_{p_{k+1}} \end{bmatrix}$$

where the ε's are half-integers and the η's are quarter-integers. If $\langle T \rangle = \mathcal{G}(W_{p_k}, W_{p_{k+1}})$ then $[T\eta] = -[\eta]$, $[T\eta] - [\eta] = (2\eta)$, and $(2\eta) \, \varepsilon \, G(W_{p_0}, W_{p_k})$. For the group $G(W_{p_0}, W_{p_k})$, with respect to the C. H. B. now chosen, is seen to have Per. Ch.'s

$$\begin{pmatrix} 0 & 0 & 0 & \ldots & 0 & 0 & \ldots & 0 & 0 & 0 & \ldots & 0 & 0 & \ldots & 0 \\ 0 & \sigma_2 & \sigma_3 & \ldots & \sigma_{k+1} & 0 & \ldots & 0 & \sigma_2 & \sigma_3 & \ldots & \sigma_{k+1} & 0 & \ldots & 0 \end{pmatrix}_2$$

for $\sigma_i = 0$ or 1. Note that if $[\eta]$ is such a Th. Ch. and $(\sigma)_2 \, \varepsilon \, G(W_{p_0}, W_{p_k})$ then $[\eta\sigma]$ is also such a Th. Ch.

If $2(\eta_2, \eta_3, \ldots, \eta_{k+1}) \neq (0, 0, \ldots, 0)$, then $[\eta]$ will be called a *proper* quarter-integer Th. Ch. A 2^k-*set of Th. Ch.'s of Type III* is a set of Th. Ch.'s

$$\{[\eta\sigma] \mid (\sigma)_2 \, \varepsilon \, G(W_{p_0}, W_{p_k})\}$$

where $[\eta]$ is a proper quarter-integer Th. Ch.

The number of such proper quarter-integer Th. Ch.'s is

$$(4^k - 2^k) 2^{2(p_{k+1} - (k+1)) + k}.$$

Dividing this number by 2^k, we see that the number of 2^k-sets of Th. Ch.'s of Type III for the covering $W_{p_k} \to W_{p_{k+1}}$ is $(4^k - 2^k) M_{k+1}$.

A 2^k-set of Th. Ch.'s of any type corresponds to 2^k linear series which lift to a $(\frac{1}{2} - K)$ linear series on W_{p_0} of dimension at least $2^k - 1$. By a 2^k-*set of linear series* (of Type I, II or III) we shall mean such a set of 2^k linear series. We conjecture that for generic W_{p_ℓ} each of these 2^k linear series has dimension zero. We shall refer to this conjecture as the *dimension zero conjecture*. By definition 2^k-sets of linear series of Types I and III occur on Riemann surfaces at the k-level and those of Type II occur at the $(k + 1)$-level.

Definition. *Two 2^k-sets will be said to be <u>equivalent</u> if they determine the same linear series on W_{p_0}.*

<u>Remark</u>. Two 2^k-sets of linear series on the same Riemann surface are never equivalent.

The object of this paper is to count the number of equivalence classes of 2^k-sets for the big array $W_{p_0} \to W_{p_\ell}$ for $k = 0, 1, 2, \ldots, \ell$. If the dimension zero conjecture holds, then for generic W_{p_ℓ} we are counting half-integer Th. Ch.'s $[\varepsilon]_2$ for W_{p_0} where $\theta[\varepsilon]_2(u)$ vanishes at $u = 0$ to order 2^k.

Lemma 4.1. *Every 2^k-set of linear series is equivalent to a 2^k-set of Type I (at the k-level).*

Proof. A 2^k-set of linear series of Type II occurs on a $W_{p_{k+1}}$. There is a covering $W_{p_k} \to W_{p_{k+1}}$ so that these 2^k linear series lift to become 2^{k-1} linear series $g^1_{p_k-1}$ (not necessarily complete) corresponding to even Th. Ch.'s. There is also such a covering where these 2^k linear series lift to remain 2^k linear series corresponding to odd Th. Ch.'s (which is a 2^k-set of linear series of Type I).

Consider a 2^k-set of linear series of Type III of $(\frac{1}{4} - 2K)$ linear series for the covering $W_{p_k} \to W_{p_{k+1}}$. (We continue the notation used earlier in this section for the definition of such a 2^k-set.) Also consider the covering $W^0_{p_{k-1}} \to W_{p_k}$ determined by the Per. Ch. $(2\eta) \, \varepsilon \, G(W_{p_0}, W_{p_k})$. The 2^k linear series are interchanged in pairs by $\mathcal{G}(W_{p_k}, W_{p_{k+1}})$. We can find 2^{k-1} pairs of divisors, one from each of the 2^k linear series, which are interchanged by $\mathcal{G}(W_{p_k}, W_{p_{k+1}})$. Each such pair of divisors lifts to W^0_{p-1} to determine a $(\frac{1}{2} - K) \, g^1_{p_{k-1}-1}$ (possibly incomplete). Each of these 2^{k-1} $g^1_{p_{k-1}-1}$'s is invariant under $\mathcal{G}(W^0_{p_{k-1}}, W_{p_{k+1}})$. For each such $g^1_{p_{k-1}-1}$ we are in the situation of Diagram 7. (since the situation of Diagram 6. does not arise). Consequently on the other two W_{p_k}'s in the array $W^0_{p_{k-1}} \to W_{p_{k+1}}$ there are 2^k $(\frac{1}{2} - K) \, g^0_{p_k-1}$'s which lift to these 2^{k-1} $g^1_{p_{k-1}-1}$'s. Thus the original 2^k-set of linear series of Type III is equivalent to two other 2^k-sets of linear series of Type I.   q.e.d.

The proof of Lemma 4.1 indicates ways in which 2^k-sets may be equivalent. The purpose of the next section is to prove that these are the only ways in which 2^k-sets can be equivalent.

5. The Basic Lemma

Lemma 5.4 below will allow us in Section 6. to describe completely the equivalence classes of 2^k-sets. Lemma 5.4 will depend on Lemma 5.1 where the syzygetic hypothesis is crucial.

Lemma 5.1. *Assume $\ell = 2$. No $(\frac{1}{2} - K) \, g^0_{p_0-1}$ corresponding to an odd Th. Ch. for W_{p_0} is invariant with respect to $\mathcal{G}$ (a syzygetic $(Z_2)^2$).*

Proof. If such a $g^0_{p_0-1}$ is $\mathcal{G}$-invariant then there is a $g^0_{p_2-1}$ on W_{p_2} which lifts to $g^0_{p_0-1}$. But such a $g^0_{p_2-1}$ is part of a 2-set or 4-set and, therefore, lifts to W_{p_0} to determine a $(\frac{1}{2} - K)$ linear series corresponding to an even Th. Ch.. Contradiction.   q.e.d.

Lemma 5.2. *In the big array consider the coverings:*

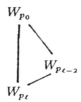

Let $\left\{ \begin{matrix} A_1 & A_2 & \cdots & A_{p_\ell} \\ B_1 & B_2 & \cdots & B_{p_\ell} \end{matrix} \right\}$ be a C.H.B. on W_{p_ℓ} so that

(5.1)

$$G(W_{p_0}, W_{p_\ell}) = \left\{ \begin{pmatrix} 0 & 0 & \cdots & 0 & 0 & \cdots & 0 \\ \sigma_1 & \sigma_2 & \cdots & \sigma_\ell & 0 & \cdots & 0 \end{pmatrix}_2 \middle| \sigma_j = 0 \quad \text{or} \quad 1 \right\}$$

Suppose

$$G(W_{p_{\ell-2}}, W_{p_\ell}) = \left\{ \begin{pmatrix} 0 & 0 & 0 & \cdots & 0 \\ \sigma_1 & \sigma_2 & 0 & \cdots & 0 \end{pmatrix}_2 \middle| \sigma_j = 0 \quad \text{or} \quad 1 \right\}$$

Then there exists a C. H. B. on $W_{p_{\ell-2}}$ (divided into 5 blocks numbered (0), (1), (2), (3), (4)).

(5.2) $\qquad \{block(0) \,|\, block(1) \,|\, block(2) \,|\, block(3) \,|\, block(4)\}$

where:

$$block(0) = \left\{ \begin{matrix} A_0 & A_1' & A_1'' & A_2' & A_2'' \\ B_0 & B_1' & B_1'' & B_2' & B_2'' \end{matrix} \right\}$$

$$block(k) = \left\{ \begin{matrix} A_3^{(k)} & A_4^{(k)} & \cdots & A_{p_\ell}^{(k)} \\ B_3^{(k)} & B_4^{(k)} & \cdots & B_{p_\ell}^{(k)} \end{matrix} \right\} \qquad k = 1,2,3,4$$

so that: i) $G(W_{p_{\ell-2}}, W_{p_\ell})$ leaves invariant the subspace of $H_1(W_{p_{\ell-2}}, Z)$ generated by the 10 cycles in block (0);

ii) $G(W_{p_{\ell-2}}, W_{p_\ell})$ permutes the blocks (1), (2), (3), (4), according to the permutations (12)(34), (13)(24), and (14)(23).

iii) with respect to the C. H. B. (5.2) $G(W_{p_0}, W_{p_{\ell-2}})$ is

(5.3) $\qquad \{(block(0') \,|\, block(1') \,|\, block(2') \,|\, block(3') \,|\, block(4'))_2\}$

where:

$$(5.3) \quad \begin{aligned} block(0') &= \begin{pmatrix} 0 & 0 & 0 & 0 & 0 \\ 0 & 0 & 0 & 0 & 0 \end{pmatrix}_2, \\ block(k') &= \begin{pmatrix} 0 & 0 & \cdots & 0 & 0 & \cdots & 0 \\ \sigma_3 & \sigma_4 & \cdots & \sigma_\ell & 0 & \cdots & 0 \end{pmatrix}_2 \quad k' = 1', 2', 3', 4' \end{aligned}$$

where $\sigma_j = 0$ or 1.

Proof. We shall give here an informal demonstration for this lemma. At the end we shall make a few remarks about a formal proof.

Picture W_{p_ℓ} as follows:

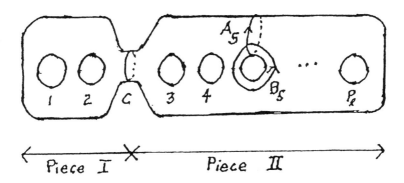

Diagram 8.

where each handle pair A_j, B_j in the C. H. B. corresponds to a hole in the surface. The curve C divides W_{p_ℓ} into Pieces I and II. The lift of C to $W_{p_{\ell-2}}$ is 4 copies of C. Above Piece I is one piece of $W_{p_{\ell-2}}$ of genus 5 bounded by the 4 lifts of C. Above Piece II are 4 copies of Piece II, each of genus $p_\ell - 2$. $\mathcal{G}(W_{p_{\ell-2}}, W_{p_\ell})$ maps the piece of $W_{p_{\ell-2}}$ above Piece I into itself permuting the 4 copies of C. The four other pieces of $W_{p_{\ell-2}}$ are permuted.

A covering $W_{p_{\ell-1}} \to W_{p_\ell}$, determined by the Per. Ch.

$$\begin{pmatrix} 0 & 0 & \cdots & 0 & 0 & 0 & \cdots & 0 \\ 0 & 0 & \cdots & 0 & 1 & 0 & \cdots & 0 \end{pmatrix}_2,$$

with 1 in the jth position, is obtained by cutting W_{p_ℓ} along A_j and cross-identifying the edges of 2 W_{p_ℓ}'s so cut. Performing this operation for

$$\begin{pmatrix} 0 & 0 & \cdots & 0 \\ 1 & 0 & \cdots & 0 \end{pmatrix}_2,$$

and the analogous operation for

$$\begin{pmatrix} 0 & 0 & 0 & \cdots & 0 \\ 0 & 1 & 0 & \cdots & 0 \end{pmatrix}_2$$

on the $W_{p_{\ell-1}}$ so formed (cut $W_{p_{\ell-1}}$ along the two curves above A_2 and cross-identify the edges of two $W_{p_{\ell-1}}$'s so cut) gives our $W_{p_{\ell-2}}$. If $p_\ell = 2$ this process gives a Riemann surface of genus 5 over a Riemann surface of genus 2 with syzygetic Galois group. Since $W_{p_0} = W_{p_{\ell-2}}$ for $\ell = 2$, $(W_{p_0}, W_{p_{\ell-2}}) = \langle e \rangle$, and the Per. Ch. for the identity covering is

$$\begin{pmatrix} 0 & 0 & 0 & 0 & 0 \\ 0 & 0 & 0 & 0 & 0 \end{pmatrix}_2.$$

This corresponds to the 0th block in (5.2).

To obtain a covering $W_{p_{\ell-3}} \to W_{p_\ell}$ corresponding to

$$\left\{ \begin{pmatrix} 0 & 0 & 0 & \cdots & 0 & 0 & 0 & \cdots & 0 \\ \sigma_1 & \sigma_2 & 0 & \cdots & 0 & \sigma_k & 0 & \cdots & 0 \end{pmatrix}_2 \middle| \sigma_j = 0 \text{ or } 1 \right\} \subset G(W_{p_0}, W_{p_\ell})$$

we cut our $W_{p_{\ell-2}}$ along the 4 curves above A_k and cross-identify the edges of two $W_{p_{\ell-2}}$'s so cut. But this is precisely the operation prescribed by the Per. Ch.

$$\bigl(block(0'') \,|\, block(1'') \,|\, block(2'') \,|\, block(3'') \,|\, block(4'') \bigr)_2$$

where:

$$block(0'') = \begin{pmatrix} 0 & 0 & 0 & 0 & 0 & 0 \\ 0 & 0 & 0 & 0 & 0 & 0 \end{pmatrix}_2$$

$$block(j'') = \begin{pmatrix} 0 & 0 & \cdots & 0 & \cdots & 0 \\ 0 & 0 & \cdots & 1 & \cdots & 0 \end{pmatrix}_2 \quad j = 1, 2, 3, 4.$$

to obtain the cover $W_{p_{\ell-3}} \to W_{p_{\ell-2}}$. This completes the demonstration.

A formal proof makes the steps outlined above rigorous. One finds a presentation for the fundamental groups $\pi_1(W_{p_\ell})$ and $\pi_1(W_{p_{\ell-1}})$. The transfer homomorphism corresponds to lifting cycles from W_{p_ℓ} to $W_{p_{\ell-1}}$. The basic problem is to find the presentations. Here we give the presentations where the A_j's and B_j's now represent homotopy classes in $\pi_1(W_{p_\ell})$.

For W_{p_ℓ}: $[A_1, B_1][A_2, B_2]\Delta = e$ where $\Delta = [A_3, B_3][A_4, B_4] \ldots [A_{p_\ell}, B_{p_\ell}]$.

For $(\sigma)_2 = \begin{pmatrix} 0 & 0 & \cdots & 0 \\ 1 & 0 & \cdots & 0 \end{pmatrix}_2$ the one relation for $\pi_1(W_{p_{\ell-1}})$ (as a subgroup of $\pi_1(W_{p_\ell})$) is: (Note: $C^D = DCD^{-1}$).

$$[A_1, B_1^2][A_2, B_2]^{B_1} \Delta^{B_1} [A_2, B_2] \Delta = e$$

This, of course, also exhibits the generators for $\pi_1(W_{p_{\ell-1}})$.

For $(\sigma)_2 = \begin{pmatrix} 0 & 0 & 0 & \cdots & 0 \\ 1 & 1 & 0 & \cdots & 0 \end{pmatrix}_2$ the one relation for $\pi_1(W_{p_{\ell-1}})$ is:

$$[A_1, B_1^2] \left[[B_1, A_1]^{B_1}, [A_2, B_2] B_2 B_1^{-1}\right] [A_2, B_2^2] \Delta^{B_2} \Delta = e.$$

Lemma 5.3. ($\ell \geq 2$) A $2^{\ell-2}$-set of $(\frac{1}{2} - K)$ linear series on a $W_{p_{\ell-2}}$ is not invariant with respect to $\mathcal{G}(W_{p_{\ell-2}}, W_{p_\ell})$.

Proof. For $\ell = 2$ this is a restatement of Lemma 5.1; so assume $\ell \geq 3$.

Let $\mathcal{G}(W_{p_{\ell-2}}, W_{p_\ell}) = \langle T, S \rangle$. Let $W_{p_{\ell-1}} = W_{p_{\ell-2}}/\langle T \rangle$. Let C. H. B.'s be chosen on W_{p_ℓ} and $W_{p_{\ell-2}}$ as described in Lemma 5.2. The action of T on the C. H. B. of (5.2) is to leave invariant the subgroup generated by the cycles in block (0), and to permute the other 4 blocks by an even permutation, say (12)(34).

A Th. Ch. corresponding to the C. H. B. (5.2) can be written

$$[\eta]_2 = [\ \tilde{\eta} \mid \alpha \mid \beta \mid \gamma \mid \delta\ ]_2$$

where $\tilde{\eta}$ is a 5-dimensional characteristic and α, β, γ, and δ are $(p_\ell - 2)$-dimensional.

Suppose $[\eta]_2$ is in a 2^k-set of $(\frac{1}{2} - K)$ Th. Ch.'s invariant with respect to $\langle T, S \rangle$. We shall derive a contradiction. (We shall denote the action of the automorphisms T, S on any appropriate object by the same symbols T, S.)

Then $T[\eta]_2 - [\eta]_2$ and $S[\eta]_2 - [\eta]_2$ must be Per. Ch.'s for $G(W_{p_0}, W_{p_{\ell-2}})$. Now

$$T[\eta]_2 = [T\tilde{\eta} \mid \beta \mid \alpha \mid \delta \mid \gamma]_2$$

and

(5.4) $T[\eta]_2 - [\eta]_2 = (T\tilde{\eta} - \tilde{\eta}|\beta - \alpha|\alpha - \beta|\delta - \gamma|\gamma - \delta)_2$

Comparing this with (5.3) we see

(5.5) $(T\tilde{\eta})_2 - (\tilde{\eta})_2 = (0)_2$

and

(5.6) $(\beta)_2 - (\alpha)_2 = (\delta)_2 - (\gamma)_2$

If S corresponds to the permutation $(13)(24)$ we see that

(5.7) $(S\tilde{\eta})_2 - (\tilde{\eta})_2 = (0)_2$

(5.8) $(\gamma)_2 - (\alpha)_2 = (\delta)_2 - (\beta)_2$

Thus $(\tilde{\eta})_2$ is invariant under $\langle T, S\rangle$ and

(5.9) $(\alpha)_2 + (\beta)_2 + (\gamma)_2 + (\delta)_2 = (0)_2$ or $(\delta)_2 = (\alpha\,\beta\,\gamma)_2$

We arrive at our contradiction by showing that these conditions imply that $[\eta]_2$ is an even Th. Ch.

$[\tilde{\eta}]_2$ is a 5-dimensional Th. Ch. invariant under a syzygetic group $\langle T, S\rangle$. If it were odd this would contradict Lemma 5.1. (One can squeeze the curve C in Diagram 8. to a point to see this.)

Now we consider $[\alpha|\beta|\gamma|\delta]_2$ to show that it also is an even Th. Ch. By (5.6) and (5.8) and the analogous result for ST we have: $(\alpha\,\beta)_2 = (\gamma\,\delta)_2$, $(\alpha\,\gamma)_2 = (\beta\,\delta)_2$, $(\alpha\,\delta)_2 = (\beta\,\gamma)_2$; and all of these are of the form

$$\begin{pmatrix} 0 & 0 & \ldots & 0 & 0 & \ldots & 0 \\ \sigma_3 & \sigma_4 & \ldots & \sigma_\ell & 0 & \ldots & 0 \end{pmatrix}_2$$

Write $(\alpha)_2$ as

$$(\alpha)_2 = \begin{pmatrix} \alpha_3 & \alpha_4 & \ldots & \alpha_\ell & \alpha_{\ell+1} & \ldots & \alpha_{p_\ell} \\ \alpha_3' & \alpha_4' & \ldots & \alpha_\ell' & \alpha_{\ell+1}' & \ldots & \alpha_{p_\ell}' \end{pmatrix}_2$$

and similarly for $(\beta)_2$, $(\gamma)_2$, and $(\delta)_2$. Thus

$$(\alpha\,\beta)_2 = \begin{pmatrix} \alpha_3 + \beta_3 & \ldots & \alpha_\ell + \beta_\ell & \alpha_{\ell+1} + \beta_{\ell+1} & \ldots \\ \alpha_3' + \beta_3' & \ldots & \alpha_\ell' + \beta_\ell' & \alpha_{\ell+1}' + \beta_{\ell+1}' & \ldots \end{pmatrix}_2$$

We see that (all equations are modulo 2).

$$\alpha_j = \beta_j \qquad j = 3, 4, \ldots, p_\ell$$
$$\alpha'_j = \beta'_j \qquad j = \ell+1, \ell+2, \ldots, p_\ell$$

and so

$$\alpha_j = \beta_j = \gamma_j = \delta_j \qquad j = 3, 4, \ldots, p_\ell$$

and

$$\alpha'_j = \beta'_j = \gamma'_j = \delta'_j \qquad j = \ell+1, \ell+2, \ldots, p_\ell$$

Thus $[\alpha|\beta|\gamma|\delta]_2$ can be written

$$[\alpha' + \alpha'' \mid \beta' + \alpha'' \mid \gamma' + \alpha'' \mid \delta' + \alpha'']_2$$

where:

$$\alpha'' = \begin{pmatrix} \alpha_3 & \cdots & \alpha_\ell & \alpha_{\ell+1} & \cdots & \alpha_{p_\ell} \\ 0 & \cdots & 0 & \alpha'_{\ell+1} & \cdots & \alpha'_{p_\ell} \end{pmatrix}_2$$

$$\alpha' = \begin{pmatrix} 0 & \cdots & 0 & 0 & \cdots & 0 \\ \alpha'_3 & \cdots & \alpha'_\ell & 0 & \cdots & 0 \end{pmatrix}_2$$

$$\beta' = \begin{pmatrix} 0 & \cdots & 0 & 0 & \cdots & 0 \\ \beta'_3 & \cdots & \beta'_\ell & 0 & \cdots & 0 \end{pmatrix}_2$$

$$\gamma' = \begin{pmatrix} 0 & \cdots & 0 & 0 & \cdots & 0 \\ \gamma'_3 & \cdots & \gamma'_\ell & 0 & \cdots & 0 \end{pmatrix}_2$$

$$\delta' = \begin{pmatrix} 0 & \cdots & 0 & 0 & \cdots & 0 \\ \delta'_3 & \cdots & \delta'_\ell & 0 & \cdots & 0 \end{pmatrix}_2$$

Since $\delta'_j = \alpha'_j + \beta'_j + \gamma'_j$ for $j = 3, 4, \ldots, \ell$, by (5.9), we see that this is an even Th. Ch.   q.e.d.

Remark. We must apologize for the extremely computational nature of the proofs of Lemmas 5.2 and 5.3. We believe, however, that something like this is necessary. The critical Lemma 5.3 is false for a smooth $(Z_2)^\ell$ covering which is not syzygetic. This makes even more complicated the counting of equivalence classes of $2^{\ell-2}$-sets in the non-syzygetic case.

Lemma 5.4. Let $g_{p_0-1}^{2^k-1+r}$ $(r \geq 0)$ be a complete $(\frac{1}{2} - K)$ linear series on W_{p_0} arising from a 2^k-set in the big array. Then $g_{p_0-1}^{2^k-1+r}$ is not invariant with respect to a subgroup of $\mathcal{G}$ or order 2^{k+2}.

Proof. Assume the contrary. Let $\mathcal{G}(W_{p_0}, W_{p_{k+2}})$ be the subgroup of order 2^{k+2} in question. By Lemma 4.3 we may assume the linear series on W_{p_0}

arises from a 2^k-set of linear series of Type I, on, say, W_{p_k}. Then there is a 2^k-set of $(\frac{1}{2} - K)$ Th. Ch.'s for $J(W_{p_k})$ invariant under $\mathcal{G}(W_{p_k}, W_{p_{k+2}})$. This contradicts Lemma 5.3.   q.e.d.

6. The Count

We first consider 2^k-sets of Type II which occur at the $(k + 1)$-level. No two of them can be equivalent. If two were equivalent, since they could not be on the same Riemann surface, the corresponding $g_{p_0-1}^{2^k-1+r}$ would be invariant with respect to a subgroup of $\mathcal{G}$ of order at least 2^{k+2}.

If such a 2^k-set of linear series occurs on $W_{p_{k+1}}$, there are $2^{k+1} - 1$ Riemann surfaces above $W_{p_{k+1}}$ at the k-level. On $2^k - 1$ of these the 2^k-set lifts to 2^{k-1} $g_{p_k-1}^1$'s and on 2^k of them the 2^k-set lifts to a 2^k-set of Type I. On each $W_{p_{k+1}}$ there are $(2^{k+1} - 1)M_{k+1}$ 2^k-sets of Type II (Section 4), so in the big array there are $n_{k+1}(2^{k+1} - 1)M_{k+1}$ 2^k-sets of Type II. Since $n_{k+1}(2^{k+1} - 1) = n_k(2^{\ell-k} - 1)$ we have:

Lemma 6.1. *The total number of 2^k-sets of Th. Ch.'s of Type II in the big array is*

$$n_k(2^{\ell-k} - 1)M_{k+1}$$

The number of 2^k-sets of Type I equivalent to 2^k-sets of Type II is

$$2^k \cdot n_k(2^{\ell-k} - 1)M_{k+1}$$

A 2^k-set of linear series of Type II lifts to a $g_{p_0-1}^{2^k-1+r}$ invariant with respect to a $(Z_2)^{k+1}$ in $\mathcal{G}$.

Now we consider 2^k-sets of Type III. Again we will see that no two 2^k-sets of Type III are equivalent. To prove this we need a lemma which continues the discussion of Section 3.

Lemma 6.2. *Assume $\ell = 2$. In the array $W_{4p-3} \to W_p$ (Diagram 1.) suppose two proper $(\frac{1}{4} - 2K)$ Th. Ch.'s in the Wirtinger variety for the covering $W^3 \to W_p$ correspond to two linear series which lift to a g_{4p-4}^{1+r} on W_{4p-3}. Then g_{4p-4}^{1+r} is invariant with respect to $\mathcal{G}(W_{4p-3}, W_p)$ if and only if g_{4p-4}^{1+r} is $(\frac{1}{2} - K)$.*

<u>Remark.</u> In general g_{4p-4}^{1+r} will be $(\frac{1}{4} - 2K)$ or $(\frac{1}{2} - K)$.

Proof. Assume the usual C. H. B.'s for the Riemann surfaces in the covering so that

$$G(W_{4p-3}, W_p) = \left\langle \begin{pmatrix} 0 & 0 & 0 & \cdots & 0 \\ 1 & 0 & 0 & \cdots & 0 \end{pmatrix}_2, \begin{pmatrix} 0 & 0 & 0 & \cdots & 0 \\ 0 & 1 & 0 & \cdots & 0 \end{pmatrix}_2 \right\rangle$$

and

$$G(W_{4p-3}, W^3) = \left\langle \begin{pmatrix} 0 & 0 & 0 & \ldots & 0 & 0 & 0 & \ldots & 0 \\ 0 & 1 & 0 & \ldots & 0 & 1 & 0 & \ldots & 0 \end{pmatrix}_2 \right\rangle = \langle (\tilde{\sigma}) \rangle$$

If $[\eta]$ corresponds to a proper $(\frac{1}{4} - 2K)$ Wirtinger vanishing then

$$[\eta] = \begin{bmatrix} \frac{1}{2} & \eta_2 & \cdots & \eta_p & -\eta_2 & \cdots & -\eta_p \\ \frac{1}{2} & \eta_2' & \cdots & \eta_p' & -\eta_2' & \cdots & -\eta_p' \end{bmatrix}$$

where η_j, η_j' are quarter-integers and $(2\eta) \neq (0)$. $[\eta\tilde{\sigma}]$ is also the Th. Ch. for a Wirtinger vanishing, and the pair $[\eta]$, $[\eta\tilde{\sigma}]$ determine linear series which become equivalent when lifted to W_{4p-3}. They determine a $(\frac{1}{2} - K)$ g_{4p-4}^{1+r} if and only if $(2\eta) = (\tilde{\sigma})$. The lemma follows from the equivalence of the following statements. $(\langle T \rangle = \mathcal{G}(W^3, W_p))$ $(2\eta) = (\tilde{\sigma})$. $T[\eta] - [\eta] = (\tilde{\sigma})$. $T[\eta] = [\eta\tilde{\sigma}]$. T interchanges $[\eta]$, $[\eta\tilde{\sigma}]$. g_{4p-4}^{1+r} is $\mathcal{G}(W_{4p-3}, W_p)$-invariant. q.e.d.

Now returning to the claim that no two 2^k-sets of Type III in the big array are equivalent, assume the contrary. If the two 2^k-sets correspond to coverings $W_{p_k}^1 \to W_{p_{k+1}}^1$ and $W_{p_k}^2 \to W_{p_{k+1}}^2$, we first conclude that $W_{p_{k+1}}^1 = W_{p_{k+1}}^2$ in order that the $g_{p_0-1}^{2^k-1+r}$ not be invariant with respect to a $(Z_2)^{k+2}$ in $\mathcal{G}$. Thus we have an array

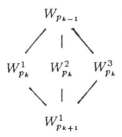

The 2^k-sets on $W_{p_k}^1$ and $W_{p_k}^2$ lift to 2^{k-1} $g_{p_{k-1}}^1$'s on $W_{p_{k-1}}$. Each of these $g_{p_{k-1}}^1$'s is $\mathcal{G}(W_{p_{k-1}}, W_{p_{k+1}}^1)$-invariant. By Lemma 6.2 each of these $g_{p_{k-1}}^1$'s is $(\frac{1}{2} - K)$. Thus we see that for each of these $g_{p_{k-1}}^1$'s we are in the situation of Diagram 6. But by Lemma 3.3 this situation does not occur. Consequently no two 2^k-sets of Type III are equivalent.

Also by considering 2^{k-1} copies of Diagram 7. we see that a 2^k-set of Type III is equivalent to two 2^k-sets of Type I. Since in the big array there are $n_k(2^{\ell-k} - 1)$ coverings of the type $W_{p_k} \to W_{p_{k+1}}$, and by Section 4 each such covering gives $(4^k - 2^k)M_{k+1}$ 2^k-sets of Type III we have the following.

Lemma 6.3. *The number of 2^k-sets of Type III is*

$$n_k(2^{\ell-k} - 1)(4^k - 2^k)M_{k+1}$$

The number of 2^k-sets of Type I equivalent to 2^k-sets of Type III is

$$2n_k(2^{\ell-k} - 1)(4^k - 2^k)M_{k+1}$$

Of the 2^k-sets, those of Types II and III account for all $g_{p_0-1}^{2^k-1+r}$'s on W_{p_0} invariant with respect to a $(Z_2)^{k+1}$ in $\mathcal{G}$. We now add the results of Lemmas 6.1 and 6.3.

Lemma 6.4. *The number of $g_{p_0-1}^{2^k-1+r}$'s (arising from 2^k-sets) invariant under a $(Z_2)^{k+1}$ in $\mathcal{G}$ is*

$$n_k(2^{\ell-k} - 1)(4^k - 2^k + 1)M_{k+1}$$

The number of 2^k-sets of Type I equivalent to 2^k-sets of Types II or III is

$$n_k(2^{\ell-k} - 1)(2 \cdot 4^k - 2^k)M_{k+1}$$

The number of 2^k-sets of Type I is $n_k u_{p_k-k}$. Thus the number of 2^k-sets of Type I invariant only with respect to a $(Z_2)^k$ in $\mathcal{G}$ is

$$n_k u_{p_k-k} - n_k(2^{\ell-k} - 1)(2 \cdot 4^k - 2^k)M_{k+1}$$

We now add this to the result of Lemma 6.4.

Theorem. *The number of $\left(\frac{1}{2} - K\right) g_{p_0-1}^{2^k-1+r}$'s on W_{p_0} arising from 2^k-sets in the big array is*

$$n_k(u_{p_k-k} - (2^{\ell-k} - 1)(4^k - 1)M_{k+1})$$

where $M_{k+1} = 4^{p_{k+1}-(k+1)}$.

7. References

[1] R. D. M. Accola, *Riemann surfaces, theta functions, and abelian auto-morphism groups*, Lecture Notes in Mathematics, No. **483**. Springer-Verlag, Berlin and New York, 1975.

[2] H. M. Farkas, "Automorphisms of compact Riemann surfaces and the vanishing of theta constants", *Bull. Amer. Math. Soc.*, Vol. **73**, 1967, 231-232.

[3] H. M. Farkas, "On the Schottky relation and its generalization to arbitrary genus", *Ann. of Math.*, Vol. **92**, 1970, 57-86.

[4] A. Krazer, *Lehrbuch der Thetafunktionen*, Teubner, Leipzig, 1903 (Chelsea reprint).

[5] H. F. Rauch and H. M. Farkas, *Theta functions with applications to Riemann surfaces*, Williams and Williams, Baltimore, 1974.

[6] W. Wirtinger, *Untersuchungen über Thetafunktionen*, Teubner, Leipzig, 1895.

Department of Mathematics
Brown University
Providence, RI 02912

Contemporary Mathematics
Volume **136**, 1992

The Period Matrix of Macbeath's Curve of Genus Seven

KEVIN BERRY AND MARVIN TRETKOFF

Introduction

The purpose of this paper is to compute a point in the Siegel upper half space of degree seven that is determined by the algebraic curve of genus seven, discovered by A.M. Macbeath, [**3**], that has $PSL(2,8)$ as its group of conformal self-mappings. We do this by applying a refinement of a method due to the second author, [**7**], to information obtained from [**3**]. Some of this work is very tedious and probably could not have been carried out without the aid of a computer. The first author has proved the assertion made in [**7**] that the construction given there is, in fact, an algorithm; and he has implemented it as a Turbo Pascal computer program. This work will appear elsewhere. Finally, we wish to acknowledge the assistance we have received from Professor A.M. Macbeath. Among other things, he showed us how to determine the monodromy group of his curve. In addition, he informed us that in the early 1970's his student, Jennifer Whitworth, calculated the period matrix of his curve. Apparently, her method was different than ours, at least with respect to the topological aspects of the problem. Unfortunately, a tragedy prevented her from submitting her dissertation, and her results remain unpublished.

1991 *Mathematics Subject Classification*. 14H40, 32G20.

This paper is in final form and no version of it will be submitted for publication elsewhere.

Period matrices of Riemann surfaces presented as algebraic curves have been calculated in very few non-trivial cases. As might be expected, these are curves with non-trivial automorphisms. For example, Rauch and Lewittes [4] determined the period matrix of the Klein–Hurwitz curve of genus 3 with 168 automorphisms. Their method was entirely different from the one we use here. Both the Klein–Hurwitz curve and Macbeath's curve admit the maximal number of automorphisms possible for a curve of given genus $g > 1$. Namely, this upper bound is $84(g-1)$, as proved first by A. Hurwitz. In addition, Hurwitz proved that a curve admits the maximal number of possible automorphisms if and only if it can be uniformized by a normal subgroup of finite index in the $(2,3,7)$ triangle group. Curves with this property are called *Hurwitz curves*. Obviously, Hurwitz' characterization of these curves is transcendental, so it is difficult to apply to algebraic curves.

Nevertheless, Hurwitz curves play an interesting role in the theory of Riemann surfaces. For example, see the papers of Earle [2] and Ries [5] in this volume. Here, we only wish to mention the question of whether the entries in the period matrix of such a curve must consist of algebraic numbers. Applying a theorem of C.L. Siegel (see, for example Baker [1], Chapter 6), our calculations show that this is false for the seven by fourteen period matrix we obtain for Macbeath's curve *before* normalization. Upon normalization, we obtain a point in the Siegel upper half space whose entries are rational expressions, with integral coefficients, of the *ratio* of a pair of fundamental periods of a specific elliptic curve. It follows from Schneider's generalization of Siegel's theorem (again, see Baker [1], Chapter 6), that these entries are algebraic if and only if the elliptic curve in question admits complex multiplication. We have not yet determined whether this is, in fact, the case. Thus, it is possible that our period matrix can be simplified.

Section 1: Macbeath's Curve

As noted by Macbeath, the existence of a Hurwitz curve, X, of genus seven is an immediate consequence of Hurwitz' theorem and the fact that $PSL(2,8)$, the simple group of order $504 = (86)(6)$, is a quotient of the $(2,3,7)$ triangle group. In order to find algebraic equations that define X, Macbeath applied Galois theory and algebro-geometric methods. In particular, he found equations for the canonical model of X in six dimensional complex projective space. In this connection, he showed that the seven differentials

(1)
$$d\mu_k = \frac{\zeta^k dz}{\sqrt{P(\zeta^k z)}} \quad , \ k = 0, \ldots, 6,$$

with $\zeta = e^{\frac{2\pi i}{7}}$ and $P(z) = (z-1)(\zeta z - 1)(\zeta^2 z - 1)(\zeta^4 z - 1)$, form a basis for the abelian differentials of the first kind on X. These simple expressions, obviously elliptic differentials, will be very useful in the present work. In addition, we will require Macbeath's presentation of X as a branched covering of the Riemann sphere. Namely, he showed that X can be defined as the Riemann surface of the equation

(2)
$$w = \sqrt{P(z)} + \sqrt{P(\zeta z)} + \sqrt{P(\zeta^2 z)}.$$

Therefore, we may view X as an eight sheeted covering of the sphere that is branched above the seven points $z = \zeta^k$, $k = 0, \ldots, 6$. Finally, it should be noted that the present paper does not utilize the deeper aspects of [3]. In particular, the equations of the canonical model of X and the representation of $PSL(2,8)$ as a group of conformal self-mappings on it, obtained so skillfully by Macbeath, are not used.

As kindly pointed out to us by Professor Macbeath, it is easy to determine the monodromy group of the covering $X/\hat{\mathbb{C}}$ from equation (2). To make matters precise, let $w_1 = w_1(z) = \sqrt{P(z)} + \sqrt{P(\zeta z)} + \sqrt{P(\zeta^2 z)}$ be a solution of (2) that is analytic in a neighborhood of $z = 0$. Thus, we suppose that we have made a choice of branch for each radical occuring in this expression. The radicals will denote these branches for the remainder of the paper. We shall presently see that the following expressions also represent solutions to (2) in a neighborhood of $z = 0$:

$$w_2 = \sqrt{P(z)} + \sqrt{P(\zeta z)} - \sqrt{P(\zeta^2 z)}$$

$$w_3 = \sqrt{P(z)} - \sqrt{P(\zeta z)} + \sqrt{P(\zeta^2 z)}$$

$$w_4 = \sqrt{P(z)} - \sqrt{P(\zeta z)} - \sqrt{P(\zeta^2 z)}$$

$$w_5 = -\sqrt{P(z)} + \sqrt{P(\zeta z)} + \sqrt{P(\zeta^2 z)}$$

$$w_6 = -\sqrt{P(z)} + \sqrt{P(\zeta z)} - \sqrt{P(\zeta^2 z)}$$

$$w_7 = -\sqrt{P(z)} - \sqrt{P(\zeta z)} + \sqrt{P(\zeta^2 z)}$$

$$w_8 = -\sqrt{P(z)} - \sqrt{P(\zeta z)} - \sqrt{P(\zeta^2 z)}.$$

Clearly, the eight expressions $w_1, \ldots, w_8$ define *distinct* analytic functions in a neighborhood of $z = 0$. Moreover, we shall see that each of them can be obtained from w_1 by analytic continuation along an appropriate loop.

Now, let γ_k denote a simple closed curve based at $z = 0$ that winds once counterclockwise about $z = \zeta^k$ and has winding number zero with respect to the remaining branch points, $z = \zeta^j$, $j \neq k$. It is easy to determine the result of continuing $w_1(z)$ along these loops. For example, if we continue $w_1(z)$ along γ_2, the radical $\sqrt{P(\zeta z)}$ changes sign when z returns to $z = 0$, but the other two radicals are unchanged. Therefore, w_1 is replaced by w_3. Similarly, it is easily verified that the sequence $w_5, w_2, w_3, w_6, w_4, w_8, w_7$ is obtained if w_1 traverses $\gamma_0, \gamma_1, \gamma_2, \gamma_3, \gamma_4, \gamma_5, \gamma_6$ in turn. The same reasoning leads to the following permutations of the eight solutions to (2):

$$\pi_0 = (15)(26)(37)(48)$$
$$\pi_1 = (12)(34)(56)(78)$$
$$\pi_2 = (13)(24)(57)(68)$$
$$\pi_3 = (16)(25)(38)(47)$$
$$\pi_4 = (14)(23)(58)(67)$$
$$\pi_5 = (18)(27)(36)(45)$$
$$\pi_6 = (17)(28)(35)(46).$$

Here, π_j denotes the permutation of $w_1, \ldots, w_8$ that results from continuing each of these functions about γ_j. These permutations serve as the input for our algorithm. The group they generate is called the *monodromy group* of the covering.

Section 2: Canonical Homology Basis

Our algorithm starts with a branched covering of the sphere, presented combinatorially by the generators of its monodromy group and determines a canonical homology basis for the covering. Each element in the basis is described as a loop on the sphere that avoids the branch points. These loops have the property that they lift to closed curves on the covering, and the lifts have the required canonical intersection matrix. The geometric and topological aspects of our method have been discussed in [7] and will not be repeated here. However, we wish to repeat an assertion made in [7]: namely, the method is effective and can be implemented on a computer. In fact, the first named author of the present paper has written such a program in Turbo Pascal and applied it to a number of cases, including the Macbeath curve that is the subject of the present paper. Full details, including some refinements and additional applications will appear elsewhere. Here, we only provide the output for the case of Macbeath's curve X.

Now, let $\gamma_{ijk} = \gamma_k^{-1}\gamma_j\gamma_i$, where the product is taken from right to left, so γ_i is traversed first. Here, the three indices i, j, k run from 0 to 6, so we have twenty-one loops on the Riemann sphere that are based at $z = 0$. These loops are the preliminary output from our algorithm. They have the property that they lift to closed paths on X, as can easily be verified by examining the permuations $\pi_0, \ldots, \pi_6$. Next, our algorithm determined a square matrix of degree twenty-one that represents the intersection matrix of these lifts. It seems pointless to reproduce that matrix here, but it may be worthwhile to note that none of its rows or columns is identically zero.

Next, our algorithm picked out fourteen integral linear combinations of the γ_{ijk} whose lifts to X will be denoted by $A_1, \ldots, A_7, B_1, \ldots, B_7$ and have the canonical intersection matrix J. Here, as usual,

$$J = \begin{pmatrix} 0 & I \\ -I & 0 \end{pmatrix},$$

where I is the identity matrix of degree seven. Thus, we would like to use the system $A_1, \ldots, A_7, B_1, \ldots, B_7$ to calculate a period matrix for X. This can, in fact, be done. However, we wish to recall that a "canonical homology basis" must be used in such calculations; it must consist of simple closed curves with a single point in common. Of course, a basis of this type must exist. Moreover, because our system of one-cycles has interection matrix J, it can be obtained from a canonical system by applying a symplectic transformation. According to Siegel [6], page 125, Theorem 5, there is a second canonical homology basis that is related to the given one by this symplectic transformation. Since this second canonical system must be homologous to our system $A_1, \ldots, B_7$, we may use the latter in calculating the period matrix.

We now list the 1-cycles provided by our computer program. In fact, there is an abuse of notation in our list. Namely, although $A_1, \ldots, A_7, B_1, \ldots, B_7$ are 1-cycles on X, we represent each of them as an integral linear combination of the γ_{ijk}, and the latter are loops on the sphere. Nevertheless, this will be useful when we evaluate the 1-forms $d\mu_k$. To make matters precise, we recall that the eight points on X lying above $z = 0$ can be identified with the eight solutions $w_1, \ldots, w_8$ to (2). Now we shall require that the lifts to X of γ_{ijk} begin at w_1 in our definition of $A_1, \ldots, B_7$. Since twelve of our 1-cycles involve more than one γ_{ijk}, we see that our system of 1-cycles on X does not consist of simple closed curves. This makes clear, once again, the need for the discussion in the preceding paragraph. We now present our list:

$$A_1 = \gamma_{013}$$

$$A_2 = \gamma_{031}$$

$$A_3 = \gamma_{054} - \gamma_{045}$$

$$A_4 = \gamma_{062} - \gamma_{054} + \gamma_{026}$$

$$A_5 = \gamma_{142} - \gamma_{062} + \gamma_{054} - \gamma_{026}$$

$$A_6 = \gamma_{103} - \gamma_{156} + \gamma_{142} - \gamma_{062} + \gamma_{054} - \gamma_{013}$$

$$A_7 = \gamma_{253} + \gamma_{235} - \gamma_{103} + \gamma_{142} - \gamma_{124} + \gamma_{054} - \gamma_{026}$$

$$B_1 = \gamma_{026}$$

$$B_2 = \gamma_{045} - \gamma_{026}$$

$$B_3 = \gamma_{124} - \gamma_{045}$$

$$B_4 = \gamma_{156} - \gamma_{026} + \gamma_{013}$$

$$B_5 = \gamma_{165} - \gamma_{124} + \gamma_{062}$$

$$B_6 = \gamma_{235} - \gamma_{165} + \gamma_{142} - \gamma_{062} + \gamma_{054} - \gamma_{026}$$

$$B_7 = \gamma_{206} - \gamma_{235} + \gamma_{103} + \gamma_{124} - \gamma_{054} + \gamma_{031}$$

Section 3: The Period Matrix

Our discussion makes it clear that the entries in our period matrix will be linear combinations of the integrals of $d\mu_k$, $k = 0, \ldots, 6$, along liftings of the loops γ_j, $j = 0, \ldots, 6$, to paths on X. Now, if we lift γ_j to X, beginning at the point w_k, then end point of our lift will be $w_{\pi_j(k)}$. This path is homologous to an easily described path that is more useful in carrying out the necessary integration. For this purpose, we let L_j denote the oriented line segment that starts at $z = 0$ and ends at $z = \zeta^j$. Then, L_j has eight lifts to X, each determined by its initial point. We denote the lifting of L_j to X starting at w_h by $L_j(h)$. Now, it is easy to see that the lift of γ_j to X beginning at w_h is homologous to $L_j(h) - L_j\big(\pi_j(h)\big)$. Therefore, we may integrate $d\mu_k$ along this path instead of along the lift of γ_j This amounts to integration along L_j in the z-plane two times, once in each direction. Naturally, we must choose the appropriate branches for the radicals that appear in $d\mu_k$.

Our work is simplified by the substitution $\tau = \zeta^k z$. Namely, we find that

$$\int_0^{\zeta^j} d\mu_k = \int_0^{\zeta^j} \frac{\zeta^k dz}{\sqrt{P(\zeta^k z)}}$$

$$= \int_0^{\zeta^{j+k}} \frac{d\tau}{\sqrt{P(\tau)}}$$

$$= \int_0^{\zeta^{j+k}} d\mu_0.$$

Therefore, all the entries in our period matrix will be linear combinations of integrals of a single elliptic differential. In fact, they turn out to be integral linear combinations of the fundamental periods ω_1 and ω_2 of the elliptic curve $w^2 = P(z)$. Here,

$$\omega_1 = \int_{\gamma_0} d\mu_0 - \int_{\gamma_3} d\mu_0$$

and

$$\omega_2 = \int_{\gamma_3} d\mu_0 - \int_{\gamma_5} d\mu_0.$$

Again, it seems pointless to print this matrix, which is not normalized, here. Instead, we give the corresponding normalized period matrix, that is, the point in the Siegel upper half space of degree seven. Before doing this, it may be useful to note that we avoided the tedious calculations that were required by using the computer program Mathematica. Finally, we note that the entries in our normalized period matrix are rational quantities involving the ratio $t = \omega_2/\omega_1$ of the periods of the elliptic integral $\int d\mu_0$.

The entries in our period matrix are complicated, so we list each column separately. Since the period matrix is symmetric, a fact that was verified in our computer output, no information is lost. Moreover, to simplify printing, we omit the denominator:

$$-1 + 2t + 19t^2 + 6t^3 - 26t^4 - 8t^5 + 3t^6,$$

which must be inserted *under each entry* of the period matrix.

Column [1]

$$-4(1-t)t(1+t-4t^2-3t^3)$$
$$-(1+t-4t^2-3t^3)(1-3t+2t^2-t^3)$$
$$-2(1-t)t(1+t+2t^2+3t^3)$$
$$-1+11t^2-4t^3-12t^4-2t^5+3t^6$$
$$2t(1-3t-6t^2+3t^3)$$
$$-8(-1+t)t^2(1+t)$$
$$(-1-t+2t^2+t^3)(1+3t-2t^2+3t^3)$$

Column [2]

$$-(1+t-4t^2-3t^3)(1-3t+2t^2-t^3)$$
$$-4t(1-t-6t^2+t^3+3t^4)$$
$$2t^2(-5-t+8t^2+t^3)$$
$$-2t(1+4t-3t^2-5t^3-3t^4)$$
$$-2t(1-5t-10t^2+t^3)$$
$$-2t(1+t)(-1+3t-2t^2+t^3)$$
$$-1-2t-t^2-6t^3-2t^4+4t^5+3t^6$$

Column [3]

$$-2(1-t)t(1+t+2t^2+3t^3)$$
$$2t^2(-5-t+8t^2+t^3)$$
$$t(-3+6t+21t^2-6t^3-22t^4+t^6)$$
$$-2t(1+t-t^2-2t^3-2t^4+t^5)$$
$$-1+4t+11t^2-4t^3-8t^4+6t^5-t^6$$
$$-(1+t)(-1+4t+5t^2-4t^3-2t^4-4t^5+t^6)$$
$$-2(-1+t)t(-2-7t-5t^2+2t^3+t^4)$$

Column [4]

$$-1 + 11t^2 - 4t^3 - 12t^4 - 2t^5 + 3t^6$$
$$-2t(1 + 4t - 3t^2 - 5t^3 - 3t^4)$$
$$-2t(1 + t - t^2 - 2t^3 - 2t^4 + t^5)$$
$$-4t(1 + t - 5t^2 - t^3 + 2t^4)$$
$$-1 + 2t + 19t^2 + 22t^3 - 14t^4 - 12t^5 + 3t^6$$
$$1 - 2t - 11t^2 + 8t^4 + 4t^5 - t^6$$
$$-1 - 4t - 3t^2 + 6t^4 - 3t^6$$

Column [5]

$$2t(1 - 3t - 6t^2 + 3t^3)$$
$$-2t(1 - 5t - 10t^2 + t^3)$$
$$-1 + 4t + 11t^2 - 4t^3 - 8t^4 + 6t^5 - t^6$$
$$-1 + 2t + 19t^2 + 22t^3 - 14t^4 - 12t^5 + 3t^6$$
$$-4t(1 + t)(1 - 4t - 8t^2 + 2t^3)$$
$$-1 + 4t + 9t^2 - 8t^3 - 18t^4 + t^6$$
$$2t(1 + t)(1 - t - 6t^2 + t^3)$$

Column [6]

$$8t^2 - 8t^4$$
$$2t - 4t^2 - 2t^3 + 2t^4 - 2t^5$$
$$1 - 3t - 9t^2 - t^3 + 6t^4 + 6t^5 + 3t^6 - t^7$$
$$1 - 2t - 11t^2 + 8t^4 + 4t^5 - t^6$$
$$-1 + 4t + 9t^2 - 8t^3 - 18t^4 + t^6$$
$$1 - 5t - 9t^2 + 15t^3 + 12t^4 - 10t^5 - 3t^6 + t^7$$

$$\text{Column } [7]$$

$$(-1 - t + 2t^2 + t^3)(1 + 3t - 2t^2 + 3t^3)$$

$$-1 - 2t - t^2 - 6t^3 - 2t^4 + 4t^5 + 3t^6$$

$$-2(-1 + t)t(-2 - 7t - 5t^2 + 2t^3 + t^4)$$

$$-1 - 4t - 3t^2 + 6t^4 - 3t^6$$

$$2t(1 + t)(1 - t - 6t^2 + t^3)$$

$$2(-1 + t)t(1 + t)(-1 - 5t - 2t^2 + t^3)$$

$$-2(1 + 6t + 7t^2 - 10t^3 - 14t^4 + 2t^5 + 3t^6).$$

REFERENCES

1. A. Baker, *Transcendental Number Theory*, Cambridge University Press, 1975.
2. C. J. Earle, *Some Riemann surfaces whose Jacobians have strange product structures*, This Volume.
3. A. M. Macbeath, *On a curve of genus 7*, Proc. London Math. Soc. (3) **15** (1965), 527–542.
4. H. E. Rauch and J. Lewittes, *The Riemann surface of Klein with 168 automorphisms*, *Problems in Analysis* (R. C. Gunning, ed.), Princeton University Press, 1970.
5. J. Ries, *Splittable Jacobi Varieties*, This Volume.
6. C. L. Siegel, *Topics in Complex Function Theory*, Vol. II, Wiley–Interscience, New York, New York, 1971.
7. C. L. Tretkoff and M. D. Tretkoff, *Combinatorial group theory, Riemann surfaces and differential equations*, Contemporary Mathematics **33** (1984), 467–519.

STEVENS INSTITUTE OF TECHNOLOGY, HOBOKEN, NEW JERSEY, 07030

Current address: Marvin Tretkoff, School of Mathematics, Institute for Advanced Study, Princeton, New Jersey, 08540

Contemporary Mathematics
Volume **136**, 1992

The Continued Fraction Parameter in the Deformation Theory of Classical Schottky Groups

Robert Brooks*

In [**B2**], we initiated a study of the space of deformations of classical Schottky groups. Recall that such a group Γ is determined by a configuration $\mathcal{C}$ of circles with disjoint interiors on the sphere S^2, a division of $\mathcal{C}$ into pairs of circles $\{c_{i_1}, c_{i_2}\}$, and for each pair a Möbius transformation A_i which takes the interior of c_{i_1} onto the exterior of C_{i_2}. Then the group Γ generated by the A_i's is a discrete group of Möbius transformations, whose fundamental domain is the common exterior of all the circles in $\mathcal{C}$.

Our approach in [**B2**] was to study the deformation space of the configuration of circles $\mathcal{C}$ itself. To that end, let us say that $\mathcal{C}$ is a packing if the region exterior to all the circles consists only of curvilinear triangles.

We will review this theory briefly in §1 below.

*   Partially supported by the NSF

1991 *Mathematics Subject Classification.* Primary 30F99.
This paper is in final form and no version of it will be submitted for publication elsewhere.

A central building block to our theory was the study of the special case of the configuration shown in Figure 1, consisting of two circles of radius 1/2 centered at $(0,0)$ and $(x,0)$ respectively. The description of how circles pack into the rectangle of this configuration is carried by a function $r(x)$, which is the continued fraction whose summands record the combinatorial data of which circles go where.

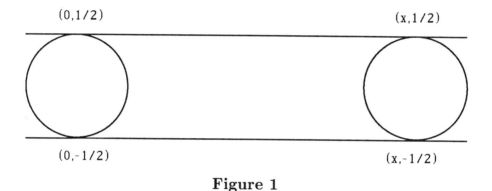

(0,1/2)   (x,1/2)

(0,-1/2)   (x,-1/2)

Figure 1

It was shown in **[B2]** that this function is a continuous, strictly increasing function of x– the strictness statement being the main point, and corresponding to Mostow rigidity in this context.

A graph of the periodic function $r(x) - x$ is shown in Figure 2. On the strength of the evidence of this graph, we raised the question of whether or not $r(x)$ was differentiable, and we conjectured that it was non-differentiable at the countable dense set of points corresponding to the places where $r(x)$ takes on rational values.

Recalling that a continuous, increasing function is differentiable almost everywhere, this is about as non-differentiable behavior as we can expect.

Figure 2: The Function $r(x) - x$

The object of this note is to sketch a proof that this conjecture is correct:

THEOREM A. *$r(x)$ is non-differentiable at the set of values x for which $r(x)$ is rational.*

See [**RS**] for another use of circle packings.

It is a pleasure to thank the organizing committee of the Joint Summer Research Conference on "Schottky Problems" for their invitation to participate, and T. Shiota for conversations which renewed my interest in this problem.

§1: Deformations of Configurations of Circles

Let $\mathcal{C}$ denote a configuration of finitely many circles with disjoint interior on S^2. We will call $\mathcal{C}$ a packing if the region complementary to the interiors of the circles consists of a union of curvilinear triangles.

We record the combinatorics of which circles touch which by the following device: for each circle in $\mathcal{C}$, we place a vertex in S^2, and whenever two circles touch, the corresponding vertices are joined by an edge. The condition that $\mathcal{C}$ is a packing is reflected in the fact that this graph defines a triangulation of S^2.

The interest in packings stems from the following elegant theorem due to Andreev and Thurston:

THEOREM ([A1],[A2],[Th]). *(a) Any two packings which give rise to topologically equivalent triangulations differ by a Möbius transformation.*

(b) Given any triangulation of S^2, there is a packing of circles which realizes it.

To see the uniqueness statement (a), which is the only part we will use, let Γ be the group generated by reflections in all the circles of $\mathcal{C}$, and let Γ_0 be the subgroup of index two consisting of orientation-preserving maps. Denoting by Ω the region of discontinuity of Γ, then Ω/Γ_0 consists of finitely many thrice-punctured spheres, obtained by gluing together two copies of each triangle. Since a thrice-punctured sphere is conformally rigid, the Ahlfors-Bers deformation theory of Kleinian groups tells us that Γ_0, and hence also Γ, is conformally rigid, which establishes (a).

Now suppose we are given an arbitrary configuration of circles $\mathcal{C}$. It is clear by elementary reasoning that one may add finitely many circles to $\mathcal{C}$ so that all complementary regions are either curvilinear triangles or curvilinear rectangles (henceforth, triangles or rectangles).

Each triangle contributes to Ω/Γ_0 a thrice-punctured sphere, as before, and hence contributes nothing to the space of deformations of $\mathcal{C}$. But each rectangle contributes a four-times punctured sphere, such that the cross-ratio of the four points is real (it is invariant under reflections). Hence we have, for any such configuration $\mathcal{C}$, the following description of the deformation space $\mathcal{S}(\mathcal{C})$:

THEOREM. $\mathcal{S}(\mathcal{C})$ *is homeomorphic to* $\mathbb{R}^{\#(rectangles)}$

It was the object of [**B2**] to introduce a parametrization of $\mathcal{S}(\mathcal{C})$ which is well-suited to problems of packing circles. To define it, for each rectangle, let us label the sides of the rectangle in clockwise order by Left, Top, Right, and Bottom. Then we may place inside this rectangle a circle which is one of the following three types:

(i) It touches the top, left, and bottom sides, and is called a horizontal circle.

(ii) It touches the left, top, and right sides, and is called a vertical circle.

(iii) It touches all four sides, and is called an unlikely piece of good luck.

In cases (i) and (ii), the new circle creates a new rectangle, and we proceed inductively to pack more and more circles in, creating smaller and smaller rectangles.

In Figure 3 below, we show this process applied to the rectangle of Figure 1.

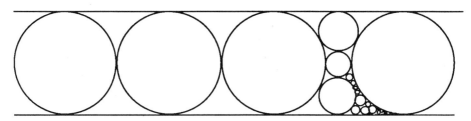

Figure 3: The Filled-In Figure 1

We now wish to record the combinatorics of this infinite packing. To that end, let n_1 be the number of horizontal circles before a vertical circle is required, n_2 the number of vertical circles before a new horizontal circle is required, and so on, and set

$$r(R) = n_1 + \cfrac{1}{n_2 + \cfrac{1}{n_3 + \dots}},$$

the real number whose continued fraction is determined by $n_1, n_2, ..$ Then

THEOREM B. *(a) $r(R)$ varies continuously as $\mathcal{C}$ is deformed continuously.*

(b)

$$\mathcal{S}(\mathcal{C}) \text{ is homeomorphic to } \mathbb{R}^{\#(rectangles)},$$

where a parametrization is given by assigning to each rectangle R its continued fraction parameter $r(R)$.

(c) Therefore, the deformations of $\mathcal{C}$ which can be completed by adding finitely many circles to obtain a packing are dense in the deformations of $\mathcal{C}$ (they contain the configurations whose coordinates are all rational).

We mention here that these results generalize to give moduli for deformations of arbitrary geometrically finite Kleinian groups **[B1]**. One then has:

THEOREM (**[B1]**). *(a) Let Γ be a geometrically finite Kleinian* *group. Then there exist arbitrarily small deformations Γ_ε of Γ, such that Γ_ε is contained in a cofinite-volume group Γ^*.*

(b) Let Γ be a geometrically finite Kleinian group without cusps. Then there are arbitrarily small deformations Γ_ε of Γ, such that Γ_ε is contained in a cocompact group Γ^*_ε.

THEOREM ([**B1**]). *Let S be a compact Riemann surface. Then the set of deformations S_ε of S which admit a packing of circles is dense in the space of all deformations of S.*

In understanding the above theorem, it is worth keeping in mind that the generalization of part (a) of the Andreev-Thurston theorem is that the combinatorics of a packing of circles on a hyperbolic surface determines both the circles themselves and the conformal structure on the surface, see [**Th**] for details.

§2: The Function $r(x)$

An important building block in the proof of Theorem B is the special configuration shown in Figure 1, consisting of two circles of radius $1/2$ centered at $(0,0)$ and $(0,x)$ respectively, and the two lines $y = 1/2$ and $y = -1/2$. Denote by $r(x)$ the continued fraction parameter for this configuration. Then Theorem B in this case says that $r(x)$ is a continuous, strictly increasing function of x.

The object of this section is to study the differentiability properties of $r(x)$. To that end, it is worth while considering the following picture:

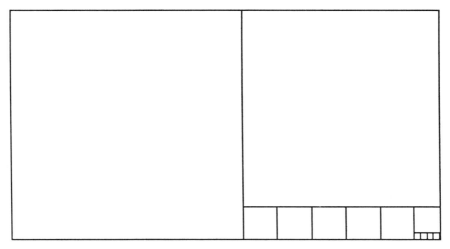

Figure 4: A packing of a rectangle by squares

Consider a rectangle with height 1 and width x. Now imagine "packing" the rectangle with horizontal and vertical squares, in a manner similar to §1. If we denote by $r_{sq}(x)$ the continued fraction parameter for this situation, we see that

$$r_{sq}(x) = x,$$

which is just the definition of continued fractions.

Suppose that we didn't know that $r_{sq}(x) = x$. It would still be possible to compute $r'_{sq}(0)$ in the following way: if $\varepsilon < 1$, then

$r_{sq}(\varepsilon) = 0 + \frac{1}{k+\ldots}$, where k is the number of squares of width ε which can be packed into an $\varepsilon \times 1$ rectangle. Therefore, $k = [\frac{1}{\varepsilon}]$, and we see that

$$\frac{1}{\varepsilon} - 1 < k < \frac{1}{\varepsilon},$$

so that

$$\frac{1}{(1/\varepsilon + 1)} < r_{sq}(\varepsilon) < \frac{1}{1/\varepsilon - 1}.$$

Taking derivatives of both sides at $\varepsilon = 0$, we see by the squeeze law that $r'_{sq}(0) = 1$.

In order to repeat this argument in the present context, we need a good method for calculating how many circles can be packed into a narrow space. To that end, consider the infinite packing of circles shown in Figure 5:

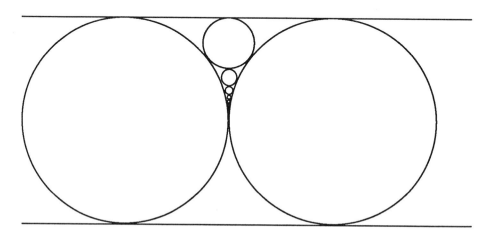

Figure 5

Denoting by s_i the y-coordinate of the bottom of the i-th circle, how could we calculate s_i? If we denote by r_i the radius of the i-th circle, so that $s_i = s_{i-1} - 2r_i$, we could calculate r_i by the condition that the distance between the centers of the i-th circle and the left-hand circle be the sum of the two radii, that is, $1/2 + r_i$, so that

$$(1/2)^2 + (s_{i-1} - r_i)^2 = (1/2 + r_i)^2,$$

or, in other words,

$$r_i = \frac{s_{i-1}^2}{(1 + 2s_{i-1})},$$

so that

$$s_i = \frac{s_{i-1}}{1 + 2s_{i-1}}.$$

A quicker route to this formula is to notice that the i-th circle is the image of the $(i-1)$-st circle under a Möbius transformation which preserves the left-hand and right-hand circles, and so must have 0 as a parabolic fixed point, and which sends the top line to the first circle. From this it is easy to calculate this Möbius transformation as

$$A(z) = \frac{z}{-2iz + 1},$$

so that

$$A(iy) = i\left(\frac{y}{2y + 1}\right),$$

which gives the above formula, noting that $A(s_{i-1}) = s_i$.

Observe that we can now calculate inductively that $s_k = \frac{1}{2k}$.

ROBERT BROOKS

Let's repeat this calculation, now with the lefthand and right-hand circles centered at $(-1/2 - \varepsilon, 0)$ and $(1/2 + \varepsilon, 0)$ respectively. The Möbius transformation A_ε is then given by

$$A_\varepsilon(z) = \frac{z - 2i(\varepsilon + \varepsilon^2)}{-2iz + 1} = \begin{pmatrix} \frac{1}{1+2\varepsilon} & \frac{-2i(\varepsilon+\varepsilon^2)}{1+2\varepsilon} \\ \frac{-2i}{1+2\varepsilon} & \frac{1}{1+2\varepsilon} \end{pmatrix}(z).$$

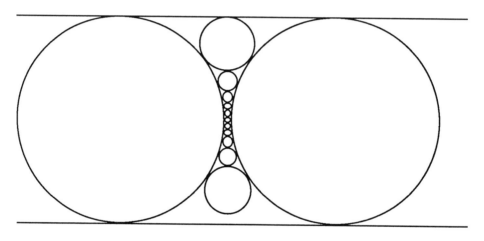

Figure 6

This is an elliptic transformation with fixed points $\pm\sqrt{\varepsilon + \varepsilon^2}$, rotating through angle θ with $\cos(\theta/2) = \frac{1}{1+2\varepsilon}$.

If we denote by k the maximum number of circles between the two lines, we see that

$$k + 2 = [\frac{2\pi}{\theta}] = [\frac{\pi}{\arccos(\frac{1}{1 + 2\varepsilon})}],$$

since each circle, including the two lines, cuts off angle θ as viewed from the fixed points.

Thus,

$$\frac{\pi}{\arccos(\frac{1}{1+2\varepsilon})} - 3 \le k \le \frac{\pi}{\arccos(\frac{1}{1+2\varepsilon})} - 2,$$

so that

$$\frac{1}{\frac{\pi}{\arccos(\frac{1}{1+2\varepsilon})} - 1} \le r(1+2\varepsilon) \le \frac{1}{\frac{\pi}{\arccos(\frac{1}{1+2\varepsilon})} - 3}.$$

Taking derivatives at $\varepsilon = 0$, we see that

MAIN LEMMA. $r'(1) = \infty$.

To handle the general case of calculating $r'(x)$, where $r(x)$ is rational, we first look at the configuration of four circles consisting of the right-hand circle, the last circle placed in the configuration, the bottom line, and the other circle tangent to the first two. After a linear fractional transformation, we can arrange it so the first three of these circles are as in Figure 5. Note that ε in this picture is related smoothly to ε in the old picture, so that whether or not $r'(x) = \infty$ does not depend on which picture we use.

The same argument as in the Main Lemma gives an upper bound for the number k of vertical circles which can be packed in when the left-hand and right-hand circles are separated by ε.

Unfortunately, we must consider not only how these circles move, but how all the circles in the packing move when the right-hand circle is moved by ε.

But that is easily done. According to the description of [**B2**], as the right-hand circle moves to the right, all the horizontal circles get smaller and roll to the left, while all the vertical circles get larger and roll down. In particular, the above upper estimate for k is still valid.

It now follows that $r'(x) = \infty$, as before.

REFERENCES:

[A1]   E.M. Andreev, "Convex Polyhedra of Finite Volume in Lobačevskii Space," Math Sb. (N.S.) 83 (125) (1970), pp. 256-260. English trans., Math USSR Sb. 12 (1970),pp.255-259.

[A2]   E.M. Andreev, "Convex Polyhedra in Lobačevskii Space," Math Sb. (N.S.) 81 (123) (1970), pp. 445-478 . English trans., Math USSR Sb. 12 (1970),pp. 413-440.

[B1]   R. Brooks,"Circle Packings and Co-Compact Extensions of Kleinian Groups," Inv. Math. 86 (1986),pp. 461-469.

[B2]   R. Brooks, "On the Deformation Theory of Classical Schottky Groups," Duke Math J. 52 (1985), pp. 1009-1024.

[RS]   B. Rodin and D. Sullivan, The Convergence of Circle Packings to the Riemann Mapping," J. Diff. Geom. 26 (1987), pp. 349-360.

[Th]   W. Thurston,Geometry and Topology of 3-manifolds, Princeton Univ. lecture notes.

Department of Mathematics
University of Southern California
Los Angeles, California 90089-1113

Email: rbrooks@mtha.usc.edu

Contemporary Mathematics
Volume **136**, 1992

The Fibers of the Prym Map

Ron Donagi

Department of Mathematics
University of Pennsylvania
Philadelphia, PA 19104

Research partially supported by NSF grant No. DMS90-08113. This paper is in final form and no version of it will be submitted for publication elsewhere.
1991 *Mathematics Subject Classification*. Primary 14H42.

Introduction

The Prym map

$$\mathcal{P} : \mathcal{R}_g \to \mathcal{A}_{g-1}$$

sends a pair $(C, \tilde{C}) \in \mathcal{R}_g$, consisting of a curve $C \in \mathcal{M}_g$ and an unramified double cover $\tilde{C}$, to its Prym variety

$$P = \mathcal{P}(C, \tilde{C}) := \ker^0(\mathrm{Nm} : J(\tilde{C}) \to J(C)).$$

Prym varieties and the Prym map are central to several approaches to the Schottky problem, e.g. [B1], [D3-D5], [Deb1], [vG], [vGvdG], [I], [M2], [W]. The purpose of this work is to describe the fibers of the Prym map. When $g = 5$ or 6, these fibers turn out to have some beautiful, and perhaps unexpected, structure. We spend much of our effort in §§4, 5 on analyzing the picture in these two cases, both generically and over some of the natural special loci in $\mathcal{A}_4$ and $\mathcal{A}_5$. In §6 we summarize some of what is known in other genera.

Here are some of the results. When $g = 6$, the map is generically finite of degree 27 [DS]. We show that its monodromy group equals the Weyl group WE_6, and that the general fiber has on it a structure which is equivalent to the incidence correspondence on the 27 lines on a non-singular cubic surface (Theorem (4.2)). The map fails to be finite over some of the interesting loci in $\mathcal{A}_5$, such as $\mathcal{J}_5$ (5-dimensional Jacobians) and $\mathcal{C}$ (intermediate Jacobians of cubic threefolds). Finiteness is restored when $\mathcal{P}$ is compactified (§1.3) and blown up (§4.2); the resulting finite fibers can be described very explicitly ((4.6), (4.7)). A similarly explicit description of the fibers is available over the locus of intermediate Jacobians of Clemens' quartic double solid ((4.8), following [C1], [DS]). The latter is the branch locus of $\mathcal{P} : \mathcal{R}_6 \to \mathcal{A}_5$ [D6].

When $g = 5$, we show (Theorems (5.1)-(5.3)) that the fiber $\mathcal{P}^{-1}(A)$, over generic $A \in \mathcal{A}_4$, is a double cover $\widetilde{F(X)}$ of the Fano surface $F(X)$ of lines on a cubic threefold X. The correspondence between $A \in \mathcal{A}_4$ and the pair $(X, \delta) \in \mathcal{RC}^+$ consisting of the cubic threefold X and the non zero, "even", point δ of order 2 in $J(X)$, is a birational equivalence of the moduli spaces. It also turns out that $\mathcal{R}_5$ has an involution λ which commutes with $\mathcal{P}$, inducing the sheet interchange on the double cover $\widetilde{F(X)}$. This λ is quite exotic; for example, it interchanges double covers of trigonal curves with "Wirtinger" double covers of nodal curves (5.14). Again, we can describe the fiber in more detail over the three distinguished divisors in $\bar{\mathcal{A}}_4$: Jacobians, the boundary (= degenerate abelian varieties), and the locus θ_{null} of abelian varieties with a vanishing thetanull: in all three cases, the cubic threefold becomes nodal, and the covers $(C, \tilde{C})$ in the fiber can be described. A particularly pretty picture arises for $\bar{\mathcal{P}}^{-1}(A)$, where $A \in \mathcal{A}_4$ is the unique 4-dimensional, non-hyperelliptic PPAV with 10 vanishing thetanulls. Varley [V] showed that all Humbert curves (with their natural double covers) are in this fiber. We observe that the corresponding cubic threefold is Segre's 10-nodal cubic (4.8); this leads quickly to a complete description of the whole fiber, (5.17).

For other values of g, the picture does not seem to be quite as rich. For $g \le 4$, one can give a rather elementary description of the fibers using Masiewicki's criterion [Ma] and Recillas' trigonal construction [R]. When $g \ge 7$, the map is generically injective ([FS], [K], [W]), but we show that it is never injective (§6).

The main tool used to analyze the Prym map is the tetragonal construction (§2.5), a triality on the locus of curves with a g_4^1 in $\mathcal{R}_g$, which commutes with $\mathcal{P}$. We exploit it consistently, together with standard facts [ACGH] on the existence of g_4^1's on curves of low genus, to establish the various structures on the fibers of $\mathcal{P}$. In genus 5 this fits into a larger symmetry, indexed by the finite projective plane $\mathbf{P}^2(\mathbf{F}_2)$, which we describe in §5.2 and use to find the cubic threefold.

Almost all the results in this work were announced in [D1]. Since then, several preliminary manuscripts have circulated, but most of these results have not been published before. Several interesting recent developments concerning closely related questions, especially Clemens' notes [C2] and Izadi's thesis [I], convinced me that these ideas may still be useful, and should be published. The present work, then, provides the details for almsot everything in [D1]. The main exception are the results on the Andreotti-Mayer locus, which have since appeared (in a corrected form) in [Deb1] and in [D5]. I include here only the underlying idea, which is the systematic application of the tetragonal construction to double covers of bielliptic curves (§3).

As mentioned above, several beautiful extensions of our results have recently been obtained by Clemens [C2] and Izadi [I]. Their basic idea is that the cubic threefold X associated to an abelian variety $A \in \mathcal{A}_4$ can be realized concretely inside the van Geemen-van der Geer linear system Γ_{00} on A, through use of Clemens' quartic double solids. The period map $\mathcal{J}$ for these is analyzed in [D6], and the fiber $\mathcal{J}^{-1}(A)$ turns out to be a certain cover of the cubic threefold X. Clemens constructs a map

$$c : \mathcal{J}^{-1}(A) \to \Gamma_{00}$$

whose image is X. He conjectures, and Izadi proves, that the projective dual X^* of X can be recovered as an irreducible component of the branch locus of the rational map from A to Γ_{00}^* determined by the linear system Γ_{00}. This concrete model of X leads to several interesting applictions:

• Over $A \in \partial \mathcal{A}_5$, which is a $\mathbf{C}^*$-extension of $A_0 \in \mathcal{A}_4$, Izadi obtains the cubic surface (of Theorem (4.2)) as hyperplane section of the cubic threefold X of A_0. (cf. (4.9) for some more details.)

• The Abel-Prym models of the six genus-5 curves making up a $\mathbf{P}^2(\mathbf{F}_2)$-diagram (§5.2) can be realized as the intersection $\Theta_a \cap \Theta_{-a} \cap H$ of two theta-translates with a divisor in Γ_{00}

• Izadi is able to describe precisely where our birational map $\mathcal{A}_4 \sim \mathcal{RC}^+$ fails to be an isomorphism.

• She is also able to verify some of the [vGvdG] conjectures in genus 4.

A second area of current activity is conjecture (6.5.1), which says that all non-injectivity of the Prym maps is due to the tetragonal construction. For non-hyperelliptic, non-trigonal and non-bielliptic curves of genus ≥ 13, this was proved in [Deb2]. The generic bielliptic case, $g \geq 10$, is in [N]. Radionov [Ra] has recently proved that for $g \geq 7$ the graph of the tetragonal construction provides at least an irreducible component of the non-injectivity locus of $\mathcal{P}$.

Some of the results of the present work were used in [D3] and [D4] to study the Schottky-Jung loci. This leads to a proof, which I hope to publish in the near future, of the Schottky-Jung conjecture in genus 5, i.e. that the Schottky-Jung equations in genus 5 characterize Jacobians. An exciting new idea in [vGP] is the interpretation of Schottky-Jung and tetragonal-type identities via rank-2 vector bundles; we wonder whether the results on the geometry of the Prym map will also admit interpretations in terms of the geometry of the moduli space of vector bundles.

It is a pleasure to acknowledge many beneficial conversations on the

subject of Prym varieties which I have had over the years with Arnaud
Beauville, Roy Smith, Robert Varley, and especially Herb Clemens,
who introduced me to Prym geometry and to his double solids, and
who has had a profound motivating effect on my thinking.

Notation
Moduli Spaces:

$\mathcal{M}_g$:	curves of genus g.
$\bar{\mathcal{M}}_g$:	the Deligne-Mumford compactification.
$\mathcal{A}_g$:	g-dimensional principally polarized abelian varieties (PPAV).
$\mathcal{J}_g$:	the closure in $\mathcal{A}_g$ of the locus of Jacobians.
$\mathcal{Q} \subset \mathcal{M}_6$:	plane quintic curves.
$\mathcal{C} \subset \mathcal{A}_5$:	(Intermediate Jacobians of) cubic threefolds.
$\mathcal{R}\mathcal{A}_g$:	pairs (A, μ), $A \in \mathcal{A}_g$, $\mu \in \mathcal{A}_2$ a non-zero point of order 2.
$\mathcal{R}_g, \mathcal{R}\mathcal{Q}, \mathcal{R}\mathcal{C}$:	the pullback of the cover $\mathcal{R}\mathcal{A}_g \to \mathcal{A}_g$ to $\mathcal{M}_g, \mathcal{Q}, \mathcal{C}$ respectively.
$\overline{\overline{\mathcal{R}}}_g$:	the Deligne-Mumford compactification of $\mathcal{R}_g$.
$\bar{\mathcal{R}}_g$:	the open subset of $\overline{\overline{\mathcal{R}}}_g$ of Beauville-allowable double covers (§1.3).

Maps

$\mathcal{P} : \mathcal{R}_g \to \mathcal{A}_{g-1}$:	the Prym map.
$\bar{\mathcal{P}} : \bar{\mathcal{R}}_g \to \mathcal{A}_{g-1}$:	Beauville's proper version of $\mathcal{P}$.
$\overline{\overline{\mathcal{P}}} : \overline{\overline{\mathcal{R}}}_g \to \bar{\mathcal{A}}_{g-1}$:	a compactification of $\mathcal{P}$, where $\bar{\mathcal{A}}_{g-1}$ denotes (Satake's compactification, or) an appropriate toroidal compactification.
$\Phi, \Psi, \varphi, \psi$:	canonical, Prym canonical, Abel-Jacobi and Abel-Prym maps of a curve.

We work throughout over the complex number field **C**.

§1   Pryms.

§1.1   Pryms and parity.

Let
$$\pi : \widetilde{C} \to C$$

be an unramified, irreducible double cover of a curve $C \in \mathcal{M}_g$. The
genus of $\widetilde{C}$ is then $2g - 1$, and we have the Jacobians

$$J := J(C), \qquad \widetilde{J} := J(\widetilde{C})$$

of dimensions g, $2g - 1$ respectively, and the norm homomorphism

$$\mathrm{Nm} : \widetilde{J} \longrightarrow J.$$

Mumford shows [M2] that

$$\mathrm{Ker}(\mathrm{Nm}) = P \cup P^-$$

where $P = \mathcal{P}(C, \tilde{C})$ is an abelian subvariety of $\tilde{J}$, called the Prym variety, and P^- is its translate by a point of order 2 in $\tilde{J}$. The principal polarization on $\tilde{J}$ induces twice a principal polarization on the Prym. This appears most naturally when we consider instead the norm map on line bundles of degree $2g - 2$,

$$\mathrm{Nm} : \mathrm{Pic}^{2g-2}(\tilde{C}) \to \mathrm{Pic}^{2g-2}(C).$$

Let $\omega_C \in \mathrm{Pic}^{2g-2}(C)$ be the canonical bundle of C.

Theorem 1.1 (Mumford [M1], [M2])

(1) The two components P_0, P_1 of $\mathrm{Nm}^{-1}(\omega_C)$ can be distinguished by their parity:

$$P_i = \{L \in \mathrm{Nm}^{-1}(\omega_C) \mid h^0(L) \equiv i \quad \mathrm{mod.}\ 2\}, \qquad i = 0, 1.$$

(2) Riemann's theta divisor $\tilde{\Theta}' \subset \mathrm{Pic}^{2g-2}(\tilde{C})$ satisfies

$$\tilde{\Theta}' \supset P_1$$

and

$$\tilde{\Theta}' \cap P_0 = 2\Xi'$$

where $\Xi' \subset P_0$ is a divisor in the principal polarization on P_0.

§1.2   Bilinear and quadratic forms.

Let $X \in \mathcal{A}_g$ be a PPAV, and let Y be a torser (=principal homogeneous space) over X. By theta divisor in Y we mean an effective divisor whose translates in X are in the principal polarization. X acts by translation on the variety Y' of theta divisors in Y, making Y' also into an X-torser. In X' there is a distinguished divisor

$$\Theta' := \{\Theta \subset X \mid \Theta \ni 0\} \subset X'$$

which turns out to be a theta divisor, $\Theta' \in X''$. In particular, we have a natural identification $X'' \approx X$ sending Θ' to 0. Let X_2 be the subgroup of points of order 2 in X. Inversion on X induces an involution on X'; the invariant subset X_2', consisting of symmetric theta divisors in X, is an X_2-torser. Let $\langle\, ,\, \rangle$ denote the natural $\mathbf{F}_2$-valued (Weil) pairing on X_2. On X_2' we have an $\mathbf{F}_2$-valued function

$$q = q_X : X_2' \to \mathbf{F}_2$$

sending $\Theta \in X'$ to its multiplicity at $0 \in X$, taken mod. 2.

Theorem 1.2 [M1] The function q_X is quadratic. Its associated bilinear form, on X_2, is $\langle\,,\,\rangle$. When (X, Θ) vary in a family, $q_X(\Theta)$ is locally constant.

When X is a Jacobian $J = J(C)$, these objects have the following interpretations:

$$
\begin{aligned}
J' &\approx \operatorname{Pic}^{g-1}(C) & \text{(use Riemann's theta divisor)} \\
J_2 &\approx \{L \in \operatorname{Pic}^0(C) = J \mid L^2 \approx \mathcal{O}_C\} \approx H^1(C, \mathbf{F}_2) & \text{(semi periods)} \\
J_2' &\approx \{L \in \operatorname{Pic}^{g-1}(C) \mid L^2 \approx \omega_C\} & \text{(theta characteristics)} \\
q(L) &\equiv h^0(C, L) \bmod.\ 2 & \text{(by Riemann-Kempf)}
\end{aligned}
$$

Explicitly, the theorem says in this case that for $\nu, \sigma \in J_2$ and $L \in J_2'$:

$$(1.3) \quad \langle \nu, \sigma \rangle \equiv h^0(L) + h^0(L \otimes \nu) + h^0(L \otimes \sigma) + h^0(L \otimes \nu \otimes \sigma)$$

$$\text{mod. 2.}$$

We note that non-zero elements $\mu \in J_2$ correspond exactly to irreducible double covers $\pi : \tilde{C} \to C$. Let X be the Prym $P = \mathcal{P}(C, \tilde{C})$, which we also denote $P(C, \mu)$, $P(C, \tilde{C})$, $P(\tilde{C}/C)$ etc. Now the divisor $\Xi' \subset P_0$ of Theorem 1.1 gives a natural identification

$$P' \approx P_0 \subset \tilde{J}'.$$

The pullback

$$\pi^* : J \longrightarrow \tilde{J}$$

sends J_2 to $\tilde{J}_2$. Since $\operatorname{Nm} \circ \pi^* = 2$, we see that

$$\pi^*(J_2) \subset P_2 \cup P_2^-.$$

Let $(\mu)^\perp$ denote the subgroup of J_2 perpendicular to μ with respect to $\langle\,,\,\rangle$.

Theorem 1.4 [M2]

(1) For $\tau \in J_2$, $\pi^*\tau \in P_2$ iff $\tau \in (\mu)^\perp$.

(2) This gives an exact sequence

$$0 \to (\mu) \to (\mu)^\perp \xrightarrow{\pi^*} P_2 \to 0.$$

(3) In (2), π^* is symplectic, i.e.

$$\langle \nu, \sigma \rangle_J = \langle \pi^*\nu, \pi^*\sigma \rangle_P, \qquad \nu, \sigma \in (\mu)^\perp \subset J_2.$$

This equality of bilinear forms can be refined to an equality of quadratic functions. The identifications

$$J' \approx \operatorname{Pic}^{g-1}(C), \qquad \tilde{J}' \approx \operatorname{Pic}^{2g-2}(\tilde{C})$$

convert the pullback

$$\pi^* : \operatorname{Pic}^{g-1}(C) \to \operatorname{Pic}^{2g-2}(\tilde{C})$$

into a map of torsors

$$\pi^{*\prime} : J' \to \tilde{J}'$$

over the group homomorphism

$$\pi^* : J \to \tilde{J}.$$

Let

$$(\mu)^{\perp'} := (\pi^{*\prime})^{-1}(P_2').$$

the refinement is:

Theorem 1.5 [D4]

(1) $(\mu)^{\perp'}$ is contained in J_2' and is a $(\mu)^{\perp}$-coset there.

(2) $\pi^{*\prime} : (\mu)^{\perp'} \to P_2'$ is a map of torsors over $\pi^* : (\mu)^{\perp} \to P_2$.

(3) In (2), $\pi^{*\prime}$ is orthogonal, i.e.

$$q_J(\nu) = q_P(\pi^{*\prime}\nu), \qquad\qquad \nu \in (\mu)^{\perp'}.$$

§1.3   The Prym Maps.

Let $\mathcal{R}_g$ be the moduli space of irreducible double covers $\pi : \tilde{C} \to C$ of non-singular curves $C \in \mathcal{M}_g$. Equivalently, $\mathcal{R}_g$ parametrizes pairs (C, μ) with $\mu \in J_2(C) \backslash (0)$, a semiperiod on C. The assignment of the Prym variety to a double cover gives a morphism

$$\mathcal{P} : \mathcal{R}_g \to \mathcal{A}_{g-1}.$$

Let ι be the involution on $\tilde{C}$ over C. The Abel- Jacobi map

$$\varphi : \tilde{C} \to J(\tilde{C})$$

induces the Abel-Prym map

$$\psi : \tilde{C} \to \operatorname{Ker}(\mathrm{Nm})$$

$$x \longmapsto \varphi(x) - \varphi(\iota x).$$

The image actually lands in the wrong component, P^-, but at least ψ is well-defined up to translation (by a point of order 2). In particular, its derivative is well-defined; it factors through C, yielding the Prym-canonical map

$$\Psi : C \to \mathbf{P}^{g-2}$$

given by the complete linear system $|\omega_C \otimes \mu|$. Beauville computed the codifferential of the Prym map:

Theorem 1.6 [B1] The codifferential

$$d\mathcal{P} : T_P^* \mathcal{A}_{g-1} \to T_{(C,\mu)}^* \mathcal{R}_g$$

can be naturally identified with restriction

$$\Psi^* : S^2 H^0(\omega_C \otimes \mu) \to H^0(\omega_C^2).$$

In particular, $\mathrm{Ker}(d\mathcal{P})$ is given by quadrics through the Prym-canonical curve $\Psi(C) \subset \mathbf{P}^{g-2}$.

Let $\bar{\mathcal{A}}_g$ denote a toroidal compactification of $\mathcal{A}_g$. Its boundary $\partial \bar{\mathcal{A}}_g$ maps to $\bar{\mathcal{A}}_{g-1}$, and the fiber over generic $A \in \mathcal{A}_{g-1} \subset \bar{\mathcal{A}}_{g-1}$ is the Kummer variety $K(A) := A/(\pm 1)$. In codimension 1, this picture is independent of the toroidal compactification used.

Let $\mathcal{RA}_g$ denote the level moduli space parametrizing pairs (A, μ) with $A \in \mathcal{A}_g$, $\mu \in A_2\backslash(0)$, and let $\overline{\mathcal{RA}}_g$ be a toroidal compactification. In [D3] we noted that its boundary has 3 irreducible components, distinguished by the relation of the vanishing cycle (mod. 2), λ, to the semiperiod μ:

(1.7)
$$\begin{aligned} \partial^{\mathrm{I}} &: \lambda = \mu \\ \partial^{\mathrm{II}} &: \lambda \neq \mu, \ \langle \lambda, \mu \rangle = 0 \in \mathbf{F}_2 \\ \partial^{\mathrm{III}} &: \langle \lambda, \mu \rangle \neq 0. \end{aligned}$$

Let $\bar{\mathcal{M}}_g$, $\overline{\overline{\mathcal{R}}}_g$ denote the Deligne-Mumford stable-curve compactifications of $\mathcal{M}_g$ and $\mathcal{R}_g$. At least in codimension one, the Jacobi map extends:

$$\bar{\mathcal{M}}_g \to \bar{\mathcal{A}}_g \ , \ \overline{\overline{\mathcal{R}}}_g \to \overline{\mathcal{RA}}_g.$$

We use $\partial \bar{\mathcal{M}}_g$, $\partial^i \overline{\overline{\mathcal{R}}}_g$ $(i = \mathrm{I, II, III})$ to denote the intersections of $\bar{\mathcal{M}}_g$, $\overline{\overline{\mathcal{R}}}_g$ with the corresponding boundary divisors in $\bar{\mathcal{A}}_g$, $\overline{\mathcal{RA}}_g$.

In [B1], Beauville introduced the notion of an allowable double cover. This leads to the construction ([DS] I, 1.1) of a proper version of the Prym map,

$$\bar{\mathcal{P}} : \bar{\mathcal{R}}_g \to \mathcal{A}_{g-1}.$$

Roughly, one extends $\mathcal{P}$ to

$$\overline{\overline{\mathcal{P}}} : \overline{\overline{\mathcal{R}}}_g \to \bar{\mathcal{A}}_{g-1},$$

then restricts to the open subset $\bar{\mathcal{R}}_g \subset \overline{\overline{\mathcal{R}}}_g$ of covers which are allowable, in the sense that their Prym is in $\mathcal{A}_{g-1}$. This condition can be made more explicit:

Theorem 1.8 [B1] A stable curve $\tilde{C}$ with involution ι, quotient C, is allowable if and only if all the fixed points of ι are nodes of $\tilde{C}$ where the branches are not exchanged, and the number of nodes exchanged under ι equals the number of irreducible components exchanged under ι.

We illustrate the possibilities in codimension 1:

Examples 1.9

(I) $X \in \mathcal{M}_{g-1}$, $p, q \in X$, $p \neq q$; let X_0, X_1 be isomorphic copies of X. Then $C := X/(p \sim q)$ is a point of $\partial \mathcal{M}_g$. The Wirtinger cover

$$\tilde{C} := (X_0 \amalg X_1)/(p_0 \sim q_1, p_1 \sim q_0)$$

gives a point

$$(C, \tilde{C}) \in \partial^{\mathrm{I}} \bar{\mathcal{R}}_g$$

which is allowable. The Prym is

$$\mathcal{P}(C, \tilde{C}) \approx J(X) \in \mathcal{A}_{g-1}.$$

(II) Start with $(\widetilde{X} \to X) \in \mathcal{R}_{g-1}$, choose distinct points $p, q \in X$, let $p_i, q_i(i = 0, 1)$ be their inverse images in $\widetilde{X}$, and set

$$C := X/(p \sim q), \quad \tilde{C} := \widetilde{X}/(p_0 \sim q_0, p_1 \sim q_1).$$

Then

$$(C, \tilde{C}) \in \partial^{\mathrm{II}} \overline{\overline{\mathcal{R}}}_g$$

is an unallowable cover. Its Prym is a $\mathbf{C}^*$-extension of $\mathcal{P}(X, \widetilde{X})$; the extension datum defining this extension is given by

$$\psi(p_0) - \psi(q_0) \in \mathcal{P}(X, \widetilde{X}),$$

which is well defined modulo ± 1 (i.e. in the Kummer), as it should be.

(III) X, p, q as before, but now $\widetilde{X} \to X$ is a double cover branched at p, q; consider Beauville's cover

$$C := X/(p \sim q), \quad \tilde{C} := \widetilde{X}/(\tilde{p} \sim \tilde{q})$$

where $\tilde{p}, \tilde{q}$ are the ramification points in $\widetilde{X}$ above p, q. Then $(C, \tilde{C}) \in \partial^{\mathrm{III}} \mathcal{R}_g$ is allowable.

In [M1], Mumford lists all covers $(C, \tilde{C}) \in \mathcal{R}_g$ whose Pryms are in the Andreotti-Mayer locus (i.e. have theta divisors singular in codimension 4). A major result in [B1] (Theorem (4.10)) is the extension of this list to allowable covers in $\bar{\mathcal{R}}_g$. We do not copy Beauville's list here, but we will refer to it when needed.

§2 Polygonal constructions

§2.1 The n-gonal constructions

Let

$$f : C \to K$$

be a map of non singular algebraic curves, of degree n, and

$$\pi : \tilde{C} \to C$$

a branched double cover. These two determine a 2^n-sheeted branched cover of K,

$$f_* \tilde{C} \to K,$$

whose fiber over a general point $k \in K$ consists of the 2^n sections s of π over k:

$$s : f^{-1}(k) \to \pi^{-1} f^{-1}(k), \quad \pi \circ s = id.$$

The curve $f_* \tilde{C}$ can be realized, for instance, as sitting in $\mathrm{Pic}^n(\tilde{C})$ or $S^n \tilde{C}$:

(2.1) $\quad f_* \tilde{C} = \{D \in S^n \tilde{C} \mid \mathrm{Nm}(D) = f^{-1}(k), \text{ some } k \in K\}.$

(If we think of $\tilde{C}$ as a local system on an open subset of C, this is just the direct image local system on K, hence our notation $f_* \tilde{C}$.) On $f_* \tilde{C}$ we have two structures: an involution

$$\iota : f_* \tilde{C} \to f_* \tilde{C}$$

obtained by changing all n choices in the section s via the involution (also denoted ι) of $\tilde{C}$, and an equivalence relation

$$f_* \tilde{C} \to \widetilde{K} \to K$$

where $\widetilde{K}$ is a branched double cover of K: two sections

$$s_1, s_2 : f^{-1}(k) \to \pi^{-1} f^{-1}(k)$$

are equivalent if they differ by an even number of changes.

For even n, the involution ι respects equivalence, so we have a sequence of maps

(2.1.1) $$f_*\widetilde{C} \to f_*\widetilde{C}/\iota \to \widetilde{K} \to K$$

of degrees $2, 2^{n-2}, 2$ respectively. For odd n the equivalence classes are exchanged by ι, so we have instead a Cartesian diagram:

(2.1.2) $$\begin{array}{ccccc} & & f_*\widetilde{C} & & \\ & \swarrow & & \searrow & \\ f_*\widetilde{C}/\iota & & & & \widetilde{K} \\ & \searrow & & \swarrow & \\ & & K & & \end{array}$$

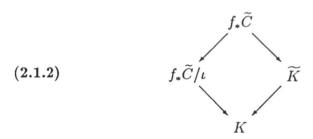

Remark 2.1.3 In prctice we will often want to allow C to acquire some nodes, over which π may be etale (as in (1.9 II)) or ramified (as in (1.9 III)). We will always consider this as a limiting case of the non-singular situation, and interpret the n-gonal construction in the limit so as to make it depend continuously on the parameters, whenever possible. We will see various examples of this below.

§2.2 Orientation

We observe that the branched cover $\widetilde{K} \to K$ depends on $f \circ \pi : \widetilde{C} \to K$, but not on f, π or C directly. More generally, to an m-sheeted branched cover

$$g : M \to K$$

we can associate an $m!$-sheeted branched cover (the Galois closure of M)

$$g! : M! \to K,$$

with an action of the symmetric group S_m; the quotient by the alternating group A_m gives a branched double cover

$$O(g) : O(M) \to K$$

which we call the orientation cover of M. We say M is orientable (over K) if the double cover $O(M)$ is trivial. One verifies easily that the double cover $\widetilde{K} \to K$ (obtained in §2.1 from the maps $\widetilde{C} \xrightarrow{\pi} C \xrightarrow{f} K$ as quotient of $f_*\widetilde{C}$) is the orientation cover $O(f \circ \pi)$ of $\widetilde{C}$.

Corollary 2.2 If $\widetilde{C}$ is orientable over K then $f_*\widetilde{C} = \widetilde{C}_0 \cup \widetilde{C}_1$ is reducible:

(i) For n even, the involution ι acts on each $\widetilde{C}_i$ with quotient C_i of degree 2^{n-2} over K, $i = 0, 1$.

(ii) For n odd, ι exchanges $\widetilde{C}_0, \widetilde{C}_1$. Each $\widetilde{C}_i$ has degree 2^{n-1} over K.

Lemma 2.3 Branch $(\widetilde{K}/K) = f_*(\text{Branch } (\widetilde{C}/C))$.

This means: if one point of $f^{-1}(k)$ is a branch point of $\widetilde{C} \to C$, then k is a branch point of $\widetilde{K} \to K$; if two points of $f^{-1}(k)$ are branch points of $\widetilde{C} \to C$, then k is not a branch point of (the normalization of) $\widetilde{K} \to K$, but the two sheets of $\widetilde{K}$ there intersect; etc. In particular, the ramification behavior of $f : C \to K$ does not affect the ramification of $\widetilde{K}$.

Corollary 2.4 Let $f : C \to \mathbf{P}^1$ be a branched cover, $\pi : \widetilde{C} \to C$ an (unramified) double cover. Then $\widetilde{C}$ is orientable over $\mathbf{P}^1$.

(More generally, the conclusion holds whenever

$$f_*(\text{Branch}(\pi)) = 2D$$

for some divisor D on $\mathbf{P}^1$, since the normalization of $O(\widetilde{C})$ is then an unramified double cover of the simply connected $\mathbf{P}^1$, by (2.3). In this situation we say that π has <u>cancelling ramification</u>.)

Remark 2.5 Assume $K = \mathbf{P}^1$ and π unramified. The image of $f_*\widetilde{C}$ in $\text{Pic}(\widetilde{C})$ is:

$$\{L \in \text{Pic}^n(\widetilde{C}) \mid \text{Nm}(L) = f^*\mathcal{O}_{\mathbf{P}^1}(1), \quad h^0(L) > 0\}.$$

This is contained in a translate of

$$\text{Nm}^{-1}(\omega_C) = P_0 \cup P_1,$$

and the splitting (2.2) of $f_*\widetilde{C}$ is "explained", in this case, by the splitting (1.1) of $\text{Ker}(\text{Nm})$, i.e. after translation:

$$\widetilde{C}_i \subset P_i, \quad i = 0, 1,$$

cf. [D1, §6], [B2].

Remark 2.6 The splitting of $f_*\widetilde{C}$ can also be explained group theoretically. Let WC_n be the group of signed permutations of n letters, i.e. the subgroup of S_{2n} centralizing a fixed-point-free involution of the $2n$ letters. Let WD_n be its subgroup of index 2 consisting of even signed permutations, i.e. permutations of n letters followed by an even number of sign changes. (These are the Weyl groups of the Dynkin diagrams C_n, D_n.) Over an arbitrary space X, we have equivalences:

$$\{ \quad n\text{-sheeted cover } Y \to X \quad \} \longleftrightarrow \{ \text{ Representation } \pi_1(X) \to \quad S_n \}$$

$$\left\{ \begin{array}{l} n\text{-sheeted cover } Y \to X \\ \text{with a double cover } \tilde{Y} \to Y \end{array} \right\} \longleftrightarrow \{ \text{ Representation } \pi_1(X) \to WC_n \}$$

$$\left\{ \begin{array}{l} n\text{-sheeted cover } Y \to X \\ \text{with an orientable double} \\ \text{cover } \tilde{Y} \to Y \end{array} \right\} \longleftrightarrow \{ \text{ Representation } \pi_1(X) \to WD_n \}$$

The basic construction of $f_*\tilde{C}$ then corresponds to the standard representation

$$\rho : WC_n \hookrightarrow S_{2^n}.$$

The existence of the involution ι on $f_*\tilde{C}$ corresponds to the factoring of ρ through

$$WC_{2^{n-1}} \subset S_{2^n}.$$

The restriction $\bar{\rho}$ of ρ to WD_n factors through

$$S_{2^{n-1}} \times S_{2^{n-1}},$$

explaining the splitting when $\tilde{C}$ is orientable.

§2.3   The bigonal construction

The case $n = 2$ of our construction ("bigonal") takes a pair of maps of degree 2:

$$\tilde{C} \xrightarrow{g} C \xrightarrow{f} K$$

and produces another such pair

$$f_*\tilde{C} \xrightarrow{g'} \widetilde{K} \xrightarrow{f'} K.$$

Above any given point $k \in K$, the possibilities are:

(i) If f, g are etale then so are f' and g'.

(ii) If f is etale and g is branched at one of the two points $f^{-1}(k)$, then f' is branched at k and g' is etale there.

(iii) Vise versa, if f is branched and g is etale then f' is etale and g' is branched at one point of $f'^{-1}(k)$.

(iv) If both f and g are branched over k then so are f', g'.

(v) If f is etale and g is branched at both points $f^{-1}(k)$, then $\widetilde{K}$ will have a node over k, and $g' : f_*\widetilde{C} \to \widetilde{K}$ will be a ∂^{III} degeneration, i.e. will look like (1.9 III).

(vi) Vice versa, we can extend the bigonal construction by continuity, as in (2.1.3), to allow $g : \widetilde{C} \to C$ to degenerate to a ∂^{III}-cover. This leads to f' which is etale and g' which is branched at both points of $f'^{-1}(k)$.

The following general properties are immediately verified:

Lemma 2.7

(1) The bigonal construction is symmetric, i.e. if it takes $\widetilde{C} \xrightarrow{g} C \xrightarrow{f} K$

to $\widetilde{C}' \xrightarrow{g'} C' \xrightarrow{f'} K$ then it takes $\widetilde{C}' \to C' \to K$ to $\widetilde{C} \to C \to K$.

(2) The bigonal construction exchanges branch loci:

$$\text{Branch}(g') = f_*(\text{Branch}(g)), \qquad \text{Branch}(f) = g'_*(\text{Branch}(f')).$$

(As in lemma (2.3), this requires the following convention in case (vi) above: the local contribution to $\text{Branch}(f)$ is $2k$, and the contribution to $\text{Branch}(g)$ is 0).

The symmetry group of this situation, WC_2, is the dihedral group of the square:

$$WC_2 = \langle r, f \mid f^2 = r^4 = (rf)^2 = 1 \rangle.$$

($r = 90°$ rotation, $f =$ flip around x-axis, in the 2-dimensional representation.) It has a non-trivial outer automorphism (=conjugation by a $45°$ rotation), which explains why conjugacy classes of representations (of $\pi_1(X)$) in WC_2 come in (bigonally related) pairs. We list all conjugacy classes of subgroups of WC_2 in the following diagram ($\sim$ denotes conjugate subgroups):

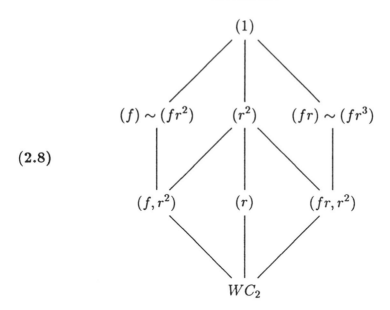

(2.8)

Correspondingly, we obtain the diagram of curves and maps of degree 2:

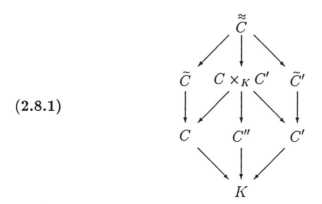

(2.8.1)

Here the two sides are bigonally related.

Note that C' is $O(\tilde{C})$; so if $\tilde{C}$ is orientable (e.g. if $K = \mathbf{P}^1$ and g is unramified) then everything splits:

$$\tilde{C}' = C_0 \amalg C_1 \to K \amalg K = C',$$

$\tilde{C}$ is Galois over K with group $(\mathbf{Z}/2\mathbf{Z})^2$ and quotients

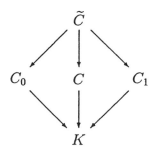

(cf. [M1]), and (2.8.1) simplifies to:

(2.8.2)

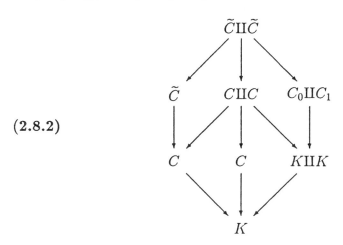

Given an arbitrary branched double cover $\widetilde{C} \to C$, we form its Prym variety

$$P(\widetilde{C}/C) := \mathrm{Ker}^0(\mathrm{Nm} : J(\widetilde{C}) \to J(C)).$$

It is an abelian variety (for $C, \widetilde{C}$ non-singular), but in general not a principally polarized one. Nevertheless, there is a simple relationship between the bigonally-related Pryms $P(\widetilde{C}/C)$ and $P(\widetilde{C}'/C')$: in the case $K = \mathbf{P}^1$, Pantazis [P] showed that these abelian varieties are dual to each other.

§2.4 The trigonal construction.

The case $n = 3$ of our construction was discovered by Recillas [R]. Start with a tower

$$\widetilde{C} \xrightarrow{\pi} C \xrightarrow{f} \mathbf{P}^1$$

where f has degree 3, and $\widetilde{C} \to C$ is an unramified double cover. By Corollaries (2.4) and (2.2), $f_*\widetilde{C}$ consists of two copies of a tetragonal curve $g : X \to \mathbf{P}^1$. Since f and g have the same branch locus by Lemma (2.3), we find from Hurwitz' formula:

$$\operatorname{genus}(X) = \operatorname{genus}(C) - 1.$$

All in all, we have constructed a map:

$$T : \left\{ \begin{array}{c} \text{trigonal curves } C \text{ of} \\ \text{genus } g \text{ with a double cover } \tilde{C} \end{array} \right\} \to \left\{ \begin{array}{c} \text{tetragonal curves} \\ X \text{ of genus } g - 1 \end{array} \right\}.$$

We claim that this map is a bijection (except that sometimes a nonsingular object on one side may correspond to a singular one on the other): given $g : X \to \mathbf{P}^1$, we recover $\tilde{C}$ as the relative second symmetric product of X over $\mathbf{P}^1$,

$$\tilde{C} := S^2_{\mathbf{P}^1} X \to \mathbf{P}^1,$$

whose fiber over $p \in \mathbf{P}^1$ consists of all unordered pairs in $g^{-1}(p)$; this has an involution ι (=complementation of pairs), giving the quotient $C := \tilde{C}/\iota$.

X $\qquad\qquad\qquad\qquad$ $S^2_{\mathbf{P}^1} X$ and its involution

In the group-theoretic setup of Remark (2.6), $\bar{\rho}$ induces an isomorphism

$$WD_3 \xrightarrow{\sim} S_4.$$

(This is the standard isomorphism, reflecting the isomorphism of the Dynkin diagrams D_3, A_3.) Recillas' map T then corresponds to composition of a representation with this isomorphism.

We list a few of the subgroups of S_4:

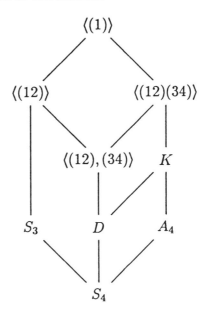

D: The dihedral group $\langle\!\langle (12), (1324) \rangle\!\rangle$
$K = D \cap A_4$: The Klein group $\langle\!\langle (12)(34), (13)(24) \rangle\!\rangle$.

The corresponding curves are:

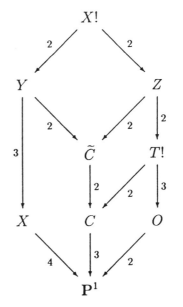

$O \approx O(X) \approx O(C)$: The orientation
$Y \approx (X \times_{\mathbf{P}^1} X) \setminus (\text{diagonal})$
$Z \approx \widetilde{C} \times_{\mathbf{P}^1} O$.

Using either of these constructions, we can easily describe the behavior of $X, C, \widetilde{C}$ around various types of branch points. Keeping X non-singular, there are the following five possible local pictures, cf. [DS, III 1.4].

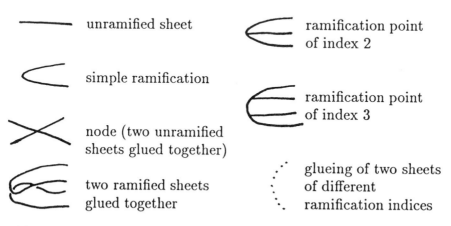

Legend

—— unramified sheet	⊏ ramification point of index 2
⟨ simple ramification	⊏ ramification point of index 3
✕ node (two unramified sheets glued together)	⋰ glueing of two sheets of different ramification indices
⟨ two ramified sheets glued together	

(i) f, π, g are étale.

(ii) f and g have simple ramification points, π is étale.

(iii) f and g each have a ramification point of index 2, π is étale.

(iv) g has two simple ramification points, π is a Beauville cover:
  $\bar{f} : N \to \mathbf{P}^1$ is trigonal, with a fiber $\{p, q, r\}$; $\bar{\pi} : \widetilde{N} \to N$ is

branched at p, q : $\bar{\pi}^{-1}(p) = \tilde{p}$, $\bar{\pi}^{-1}(q) = \tilde{q}$; and we have $C = N/(p \sim q)$, $\tilde{C} = \tilde{N}/(\tilde{p} \sim \tilde{q})$, and $\pi : \tilde{C} \to C$, $f : C \to \mathbf{P}^1$ are induced by $\bar{\pi}, \bar{f}$.

(v) g has a ramification point of index 3, π is Beauville, f is ramified at one of the two branches of the node of C.

Considering first the first three cases, then all five, we conclude:

Theorem 2.9 The trigonal construction gives isomorphisms

$$T^0 : \mathcal{R}_g^{\mathrm{Trig}} \xrightarrow{\sim} \mathcal{M}_{g-1}^{\mathrm{Tet},0}$$

and

$$T : \bar{\mathcal{R}}_g^{\mathrm{Trig}} \xrightarrow{\sim} \mathcal{M}_{g-1}^{\mathrm{Tet}},$$

where:

$\mathcal{M}_{g-1}^{\mathrm{Tet}}$ is the moduli space of (non-singular) curves of genus $g - 1$ with a tetragonal line bundle.

$\mathcal{M}_{g-1}^{\mathrm{Tet},0}$ is the open subset of tetragonal curves X with the property that above each point of $\mathbf{P}^1$ there is at least one etale point of X.

$\mathcal{R}_g^{\mathrm{Trig}}$ is the moduli space of etale double covers of non-singular curves of genus g with a trigonal bundle.

$\bar{\mathcal{R}}_g^{\mathrm{Trig}}$ is the partial compactification of $\mathcal{R}_g^{\mathrm{Trig}}$ using allowable covers in $\bar{\mathcal{R}}_g$ of type ∂^{III} (cf (1.9.III)).

Examples 2.10

(i) $\tilde{C}$ is the trivial cover, $\tilde{C} = C_0 \amalg C_1$, iff X is disconnected, $X = \mathbf{P}^1 \amalg C$, with $f = g|_C$, $id_{\mathbf{P}^1} = g|_{\mathbf{P}^1}$.

(ii) Wirtinger covers $(C_0 \amalg C_1) / (p_0 \sim q_1, q_0 \sim p_1) \to C/(p \sim q)$, where $\{p, q, r\}$ form a trigonal fiber in C, correspond to reducible $X = \mathbf{P}^1 \cup_r C$, the two components meeting at $r \in C$.

(iii) C is reducible: $C = H \cup \mathbf{P}^1$, with H hyperelliptic, and $\tilde{C} = \tilde{H} \cup \mathbf{P}^1$ with $\tilde{H} \to H$ and $\tilde{\mathbf{P}}^1 \to \mathbf{P}^1$ branched over $B := H \cap \mathbf{P}^1$. This situation corresponds to $g : X \to \mathbf{P}^1$ factoring through a hyperelliptic H'. Indeed, such a pair $(C, \tilde{C})$ is uniquely determined by the tower $\tilde{H} \to H \to \mathbf{P}^1$. The trigonal construction for C is reduced to the bigonal construction for H, which then gives $X = \tilde{H}' \to H' \to \mathbf{P}^1$. In particular:

(iv) $C = H \amalg \mathbf{P}^1$ is disconnected iff $X = H_0 \amalg H_1$ is disconnected with
    hyperelliptic pieces, and then $\tilde{C} = \tilde{H} \amalg \mathbf{P}^1 \amalg \mathbf{P}^1$, where $\tilde{H}$ is the
    Cartesian cover:

$$\tilde{H} = H_0 \times_{\mathbf{P}^1} H_1.$$

So far, we have only used the fact that $\tilde{C}$ is an orientable double
cover of a triple cover. We now use our two assumptions, that π is
unramified and that the base K equals $\mathbf{P}^1$, to obtain an identity of
abelian varieties. Namely, by Remark 2.5 we have a map, natural up
to translation.

$$\alpha : X \to P(\tilde{C}/C).$$

The result, due to S. Recillas, is:

Theorem 2.11 [R] If X is trigonally related to $(\tilde{C}, C)$, then the above
map α induces an isomorphism

$$\alpha_* : J(X) \overset{\sim}{\to} P(\tilde{C}/C).$$

Proof.

By naturality of α and irreducibility of $\mathcal{M}_{g-1}^{\text{Tet}}$, it suffices to prove
this for any one convenient X. We take $\tilde{C} \to C$ to be a Wirtinger cover
as in (2.10)(ii), so

$$X = \mathbf{P}^1 \cup_r C'.$$

where $p + q + r$ is a trigonal divisor on C', and $C = C'/(p \sim q)$. We
have natural identifications:

$$J(X) \approx J(C') \approx P(\tilde{C}/C),$$

in terms of which α becomes the Abel-Jacobi map φ on C', and collapses
$\mathbf{P}^1$ to a point. The induced α_* is therefore the identity.

                                                                QED

Corollary 2.12 All trigonal Pryms are Jacobians, and all tetragonal
Jacobians are Pryms.

§2.5   The tetragonal construction

Consider now a tower

$$\tilde{C} \to C \overset{f}{\to} \mathbf{P}^1$$

where f has degree 4 and $\tilde{C}$ is a double cover (unramified) of C. The
general construction yields a sequence of maps of degrees 2, 4, 2:

$$f_*\tilde{C} \to f_*\tilde{C}/\iota \to \tilde{\mathbf{P}}^1 \to \mathbf{P}^1.$$

By (2.2) and (2.4) again, $\tilde{\mathbf{P}}^1$ is unramified, hence we have splittings:

$$\tilde{\mathbf{P}}^1 = \mathbf{P}_0^1 \amalg \mathbf{P}_1^1$$
$$f_*\tilde{C} = \tilde{C}_0 \amalg \tilde{C}_1$$
$$f_*\tilde{C}/\iota = C_0 \amalg C_1.$$

The tetragonal construction thus associates to a tower

$$\tilde{C} \xrightarrow{2} C \xrightarrow{4} \mathbf{P}^1$$

two other towers

$$\tilde{C}_i \to C_i \to \mathbf{P}^1, \qquad\qquad i = 0, 1$$

of the same type.

Lemma 2.13 The tetragonal construction is a triality, i.e. starting with $\tilde{C}_0 \to C_0 \to \mathbf{P}^1$ it returns $\tilde{C} \to C \to \mathbf{P}^1$ and $\tilde{C}_1 \to C_1 \to \mathbf{P}^1$.

On the group level, the point is this: Our tower $\tilde{C} \to C \to \mathbf{P}^1$ corresponds to a representation (of $\pi_1(\mathbf{P}^1 \setminus (\text{branch locus}))$) in WD_4. Now the Dynkin diagram D_4 has an automorphism of order 3:

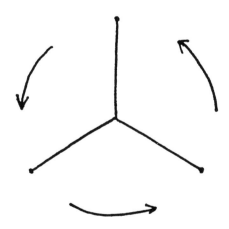

This corresponds to an outer automorphism of WD_4, of order 3. Hence representations in WD_4 come in packets of three. The various groups involved are described in some detail in the proof of Lemma (5.5), below.

Local pictures 2.14 Given the local behavior of C and $\widetilde{C}$ over a point of $\mathbf{P}^1$, it is quite straightforward to compute $f_*\widetilde{C}$ and hence $\widetilde{C}_i, C_i$ ($i = 0, 1$) over the same point. Since these local pictures are needed quite frequently, we record the simplest ones here.

(1) $C, \widetilde{C}$ unramified $\Rightarrow C_i, \widetilde{C}_i$ are also unramified.

(2) C has one simple ramification point (and two unramified sheets), $\widetilde{C} \to C$ unramified $\Rightarrow C_i, \widetilde{C}_i$ have the same local picture as $C, \widetilde{C}$ respectively.

(3) C has two distinct simple ramification points, $\widetilde{C} \to C$ unramified $\Rightarrow$ One pair, say $C_0, \widetilde{C}_0$, has the same local pictures as $C, \widetilde{C}$, while the other is a Beauville degeneration: C_1 is unramified but two of its four sheets are glued, $\widetilde{C}_1 \to C_1$ is ramified over these two sheets (and the ramification points are glued) while the other sheets are unramified.

(4) C is unramified but two of its sheets are glued, $\widetilde{C} \to C$ is ramified over these two sheets $\Rightarrow C_i$ has two distinct ramification points, $\widetilde{C}_i \to C_i$ is unramified ($i = 0, 1$). (This is the same triple as in (3).)

(5) C has a simple ramification point and the other two sheets are glued, $\widetilde{C}$ is ramified over the glued sheets $\Rightarrow C_i, \widetilde{C}_i$ have the same local pictures as $C, \widetilde{C}$.

(6) C has a ramification point of index 2 (i.e. 3 of its sheets are permuted by the local monodromy), $\widetilde{C} \to C$ unramified $\Rightarrow$ same local picture for $\widetilde{C}_i \to C_i$.

(7) C has a ramification point of index 3 (all 4 sheets permuted), $\widetilde{C} \to C$ unramified $\Rightarrow C_0, \widetilde{C}_0$ have the same local picture as $C, \widetilde{C}$, but C_1 has a simple ramification point glued to an unramified point, so $\widetilde{C}_1$ must be simply ramified over each. (I.e. $\widetilde{C}_1$ has a point which is simply ramified over $\mathbf{P}^1$, glued to a point which has ramification index 3 over $\mathbf{P}^1$!)

We note that in examples (3) and (7), the tetragonal construction applied to $(\widetilde{C} \to C) \in \mathcal{RM}_g$ produces an (allowable) degenerate cover, $(\widetilde{C}_1 \to C_1) \in \partial^{\mathrm{III}}(\mathcal{RM}_g)$.

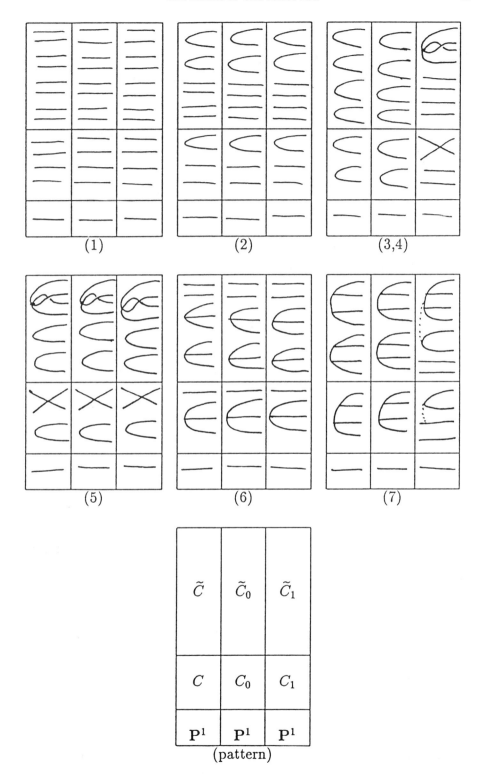

Examples 2.15

(1) It is perhaps not terribly surprising that the trigonal construction is a degenerate case of the tetragonal construction. Start with $\widetilde{C} \to C$ the split double cover of the curve C with the tetragonal map $f : C \xrightarrow{4} \mathbf{P}^1$. Then $f_* \widetilde{C}$ splits into 5 components, of degrees 1, 4, 6, 4, 1 respectively over $\mathbf{P}^1$. The components of degree 4 make up $\widetilde{C}_1 \to C_1$, which is isomorphic to $\widetilde{C} \to C$. The remaining components give

$$\mathbf{P}^1 \amalg \widetilde{T} \amalg \mathbf{P}^1 \to T \amalg \mathbf{P}^1$$

where $(\widetilde{T}, T)$ is associated to C by the trigonal construction. Vice versa, starting with an (unramified) double cover $\mathbf{P}^1 \amalg \widetilde{T} \amalg \mathbf{P}^1$ of $T \amalg \mathbf{P}^1$, the tetragonal construction produces $C \amalg C \to C$, twice.

(2) Let $p + q + r + s$ be a tetragonal divisor on C. Then $C/(p \sim q)$ is still tetragonal. Tacking a node onto the previous example, we see that the Wirtinger cover

$$(C' \amalg C'')/(p' \sim q'', q' \sim p'') \to C/(p \sim q)$$

is taken by the tetragonal construction to :

• Another Wirtinger Cover,

$$(C' \amalg C'')/(r' \sim s'', s' \sim r'') \to C/(r \sim s),$$

and to:

• $\mathbf{P}^1 \cup_{t'} \widetilde{T} \cup_{t''} \mathbf{P}^1 \to T \cup_t \mathbf{P}^1$, where $(\widetilde{T}, T)$ is associated by the trigonal construction to C. (Each copy of $\mathbf{P}^1$ meets $\widetilde{T}$ or T in the unique point indicated. $t \in T$ corresponds to the partition $\{\{p, q\}, \{r, s\}\}$.)

(3) We will see in Lemma (3.5) that if $C \to \mathbf{P}^1$ factors through a hyperelliptic curve, so do C_0, C_1. An interesting subcase occurs when $C = H^0 \cup H^1$ has two hyperelliptic components, cf. Proposition (3.6).

(4) Let X be a non-singular cubic hypersurface in $\mathbf{P}^4$, $\ell \subset X$ a line, and $\widetilde{X}$ the blowup of X along ℓ, with projection from ℓ:

$$\pi : \widetilde{X} \to \mathbf{P}^2.$$

This is a conic bundle [CG] whose discriminant is a plane quintic curve $Q \subset \mathbf{P}^2$. The set of lines $\ell' \subset X$ meeting ℓ is a double cover $\widetilde{Q}$ of Q. Now choose a plane $A \subset \mathbf{P}^4$ meeting X in 3 lines ℓ, ℓ', ℓ'';

we get 3 plane quinties Q, Q', Q'', with double covers $\tilde{Q}, \tilde{Q}', \tilde{Q}''$. Note that ℓ', ℓ'' map to a point $p \in Q$, hence determine a tetragonal map $f : Q \to \mathbf{P}^1$, given by $\mathcal{O}_Q(1)(-p)$, and similarly for Q', Q''. Our observation is that the 3 objects

$$(\tilde{Q}, Q, f) \ ; \ (\tilde{Q}', Q', f') \ ; \ (\tilde{Q}'', Q'', f'')$$

are tetragonally related. Indeed, the 3 maps can be realized simultaneously via the pencil of hyperplanes $S \subset \mathbf{P}^4$ containing A. Such an S meets X in a (generally non-singular) cubic surface Y. A line in Y (and not in A) which meets ℓ', also meets 4 of the 8 lines (in Y, not in A) meeting ℓ, one in each of 4 coplanar pairs. this gives the desired injection $\tilde{Q}' \hookrightarrow f_* \tilde{Q}$.

Our main interest in the tetragonal construction stems from:

Theorem 2.16 The tetragonal construction commutes with the Prym map,

$$P(\tilde{C}/C) \approx P(\tilde{C}_0/C_0) \approx P(\tilde{C}_1/C_1).$$

Proof

As in Remark (2.5), we have a map

$$\alpha : \tilde{C}_i \hookrightarrow f_* \tilde{C} \to \operatorname{Pic}(\tilde{C}), \qquad i = 0, 1.$$

The image sits in a translate of $P(\tilde{C}/C)$, so we get induced maps

$$\alpha_* : J(\tilde{C}_i) \to P(\tilde{C}/C)$$

and by restriction

$$\beta : P(\tilde{C}_i/C_i) \to P(\tilde{C}/C).$$

By Masiewicki's criterion [Ma], β will be an isomorphism if we can show:

(1) The image $\alpha(\tilde{C}_i)$ of $\tilde{C}_i$ in $P(\tilde{C}/C)$ is symmetric;

(2) The fundamental class in $P(\tilde{C}/C)$ of $\alpha(\tilde{C}_i)$ is twice the minimal class, $\frac{2}{(g-1)!}[\Theta]^{g-1}$.

Now (1) is clear, since the involution on $\tilde{C}_i$ commutes with -1 in $P(\tilde{C}/C)$. The fundamental class in (2) can be computed directly, as is done in [B2]. Instead, we find it here by a degeneration argument: it varies continuously with $(C, \tilde{C}) \in \mathcal{RM}_g^{\text{Tet}}$, which is an irreducible parameter space, so it suffices to do the computation for a single $(C, \tilde{C})$. We take

$$C = T \cup_t \mathbf{P}^1, \quad \tilde{C} = \mathbf{P}^1 \cup_{t'} \tilde{T} \cup_{t''} \mathbf{P}^1,$$

as in Example (2.15)(2). Then $(C_i, \tilde{C}_i)$ is a Wirtinger cover, $i = 0, 1$, and the normalization of C_i is the tetragonal curve N associated to $(T, \tilde{T})$ by the trigonal construction. We have identifications

$$J(N) \approx P(\tilde{T}/T) \approx P(\tilde{C}/C)$$

(Theorem (2.11)), in terms of which $\alpha(\tilde{C}_i)$ consists of the Abel-Jacobi image $\varphi(N) \subset J(N)$ and of its image under the involution. Thus the fundamental class is twice that of $\varphi(N)$, as required.

(Note: since this argument works for any double cover $\tilde{T} \to T$, and since any semiperiod on a nearby non-singular C specializes to a semiperiod on $T \cup_t \mathbf{P}^1$ which is supported on T, we need only the irreducibility of $\mathcal{M}_g^{\mathrm{Tet}}$, instead of $\mathcal{RM}_g^{\mathrm{Tet}}$.)

QED

§3 Bielliptic Pryms.

As a first application of the tetragonal construction, we show that some remarkable coincidences occur among the various loci in Beauville's list [B1]. The central role here is played by Pryms of bielliptic curves. We see in (3.7), (3.8) that the bielliptic loci can be tetragonally related to various other components in Beauville's list, and therefore give the same Pryms. As suggested in [D1], this leads to a complete, short list of the irreducible components of the Andreotti-Mayer locus in genus ≤ 5, and of its intersection with the image of the proper Prym map for arbitrary g. We do not include here the complete analysis of the Andreotti-Mayer locus itself, since this has already appeared in [Deb1] and [D5] (together with some corrections to the original list in [D1]). Nevertheless, we could not resist describing explicitly the operation of the tetragonal construction on Beauville's list, as it is such a pretty and straightforward application of the results of §2.

We recall Mumford's results on hyperelliptic Pryms. Let

$$f_i : C^i \to K, \qquad i = 0, 1$$

be two ramified double covers of a curve K. The fiber product

$$\tilde{C} := C^0 \times_K C^1$$

has 3 natural involutions: $\tau_i (i = 0, 1)$, with quotient C^i, and $\tau := \tau_0 \circ \tau_1$, with a new quotient, C. This all fits in a Cartesian diagram:

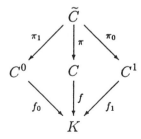

If the branch loci of f_0, f_1 are disjoint, then

$$\pi : \widetilde{C} \to C$$

is unramified. We say that a double cover obtained this way is <u>Cartesian</u>.

Lemma 3.1 Let $f : C \to K$ be a ramified double cover. A double cover

$$\pi : \widetilde{C} \to C,$$

given by a semiperiod $\eta \in J_2(C)$, is Cartesian if and only if $f_*(\eta) = 0 \in J_2(K)$.

Proof: apply the bigonal construction.

QED

Proposition 3.2 [M1]

(1) Any double cover $\widetilde{C}$ of a hyperelliptic C is Cartesian.

(2) Any hyperelliptic Prym is a product of 2 hyperelliptic Jacobians (one of which may vanish): If $\widetilde{C}$ arises as $C^0 \times_{\mathbf{P}^1} C^1$ then

$$P(\widetilde{C}/C) \approx J(C^0) \times J(C^1).$$

Proof: (2) follows from (1), (1) follows from lemma (3.1) with $K = \mathbf{P}^1$.

QED

A bielliptic curve (aka elliptic-hyperelliptic, superelliptic, ...) is a branched double cover of an elliptic curve. In this section we apply the tetragonal construction to find various identities between bielliptic Pryms and Pryms of other, usually degenerate, curves. Some of the results extend to bihyperelliptic curves, i.e. branched double covers of hyperelliptic curves. To warm up, we consider Jacobians of bihyperelliptic curves. Example (2.10)(iii) can be restated:

Lemma 3.3 The trigonal construction gives a bijection between:

- Bihyperelliptic, non singular curves C:

$$C \xrightarrow{f} H \xrightarrow{g} \mathbf{P}^1;$$

- Reducible trigonal double covers $\widetilde{X} \to X$:

$$
\begin{array}{ccccc}
\widetilde{X} & = & C' & \cup & H \\
\downarrow & & \downarrow & & \downarrow \\
X & = & H' & \cup & \mathbf{P}^1
\end{array}
$$

where

$X = H' \cup \mathbf{P}^1$ is reducible

$\tau : X \to \mathbf{P}^1$, the trigonal map, has degree 2 on H' and 1 on $\mathbf{P}^1$.

$\tau(H' \cap \mathbf{P}^1) = \mathrm{Branch}(g)$

$\widetilde{X} \to X$ is allowable of type ∂^{III} at each point of $H' \cap \mathbf{P}^1$.

We note that $C' \to H' \to \mathbf{P}^1$ is bigonally related to $C \to H \to \mathbf{P}^1$.

Corollary 3.4 The Jacobian of a bihyperelliptic curve C,

$$C \xrightarrow{f} H \xrightarrow{g} \mathbf{P}^1,$$

is isogenous to the product

$$J(H) \times P(g_*C, \iota)$$

of a hyperelliptic Jacobian and a bihyperelliptic (branched) Prym.

We move to the Pryms of bihyperelliptic curves. First we note that this class is closed under the tetragonal construction:

Lemma 3.5 Let $(\widetilde{C}_i, C_i)$ be tetragonally related to $(\widetilde{C}, C)$, with C non-singular. If $C \to \mathbf{P}^1$ factors through a (possibly reducible) hyperelliptic H, so do the C_i:

$$C_i \xrightarrow{f_i} H_i \xrightarrow{g_i} \mathbf{P}^1, \qquad i = 0, 1.$$

Proof.

The bigonal construction applied to

$$\widetilde{C} \xrightarrow{\pi} C \xrightarrow{f} H$$

yields

$$f_*\widetilde{C} \to \widetilde{H} \to H,$$

and when applied again to

$$\widetilde{H} \to H \xrightarrow{g} \mathbf{P}^1$$

yields

$$g_*\widetilde{H} \to \widetilde{\mathbf{P}}^1 \to \mathbf{P}^1.$$

Since π is unramified, so are $\widetilde{H} \to H$ and $\widetilde{\mathbf{P}}^1 \to \mathbf{P}^1$. Hence $\widetilde{\mathbf{P}}^1$ splits:

$$\widetilde{\mathbf{P}}^1 = \mathbf{P}^1_0 \amalg \mathbf{P}^1_1,$$

and this splitting climbs its way up the tower:

$$
\begin{array}{rcl}
(g \circ f)_*\widetilde{C} & = & \widetilde{C}_0 \amalg \widetilde{C}_1 \\
& & \downarrow \\
& & C_0 \amalg C_1 \\
& & \downarrow \\
g_*\widetilde{H} & = & H_0 \amalg H_1 \\
& & \downarrow \\
\widetilde{\mathbf{P}}^1 & = & \mathbf{P}^1_0 \amalg \mathbf{P}^1_1 \\
& & \downarrow \\
& & \mathbf{P}^1
\end{array}
$$

QED

Remark 3.5.1 The rational map $f_i : C_i \to H_i$ can, in a couple of cases, fail to be a morphism; this is easily remedied by identifying a pair of points in H_i. Among the local pictures (2.14), the ones that can occur here are (1), (2), (7) and (3) :

- In cases (1), (2), the hyperelliptic maps g, g_0, g_1 are all unramified at the relevant point, and the f_i are morphisms.

- In case (7), g and g_0 are ramified, g_1 is not, f and f_0 are (ramified) morphisms, but f_1 is not, since C_1 is singular above a point where H_1, as constructed above, is nonsingular. To make f_1 into a morphism, we must glue the two points of $g_1^{-1}(k)$.

- In case (3) we find two possibilities:

  (3a) g is etale, f is ramified at both points of $g^{-1}(k)$; then g_0, g_1 are also etale, f_0 is ramified at both points of $g_0^{-1}(k)$, C_1 has a node but f_1 is still a morphism.

  (3b) g is ramified, f is etale; then g_0 is ramified, f_0 is etale, g_1 is etale, but the two branches of the node of C_1 are sent by f_1 to opposite sheets of H_1, so f_1 is again not a morphism.

Proposition 3.6 Let $\tilde{C} \to C$ be a Cartesian double cover of a bihyperelliptic C:

$$C \xrightarrow{f} H \xrightarrow{g} \mathbf{P}^1, \ \tilde{C} = C^0 \times_H C^1.$$

The tetragonal construction applied to $\tilde{C} \to C \to \mathbf{P}^1$ yields:

- A similar Cartesian tower $\tilde{C}_0 \to C_0 \xrightarrow{f_0} H \xrightarrow{g_0} \mathbf{P}^1$, same H.
- A tower $\tilde{C}_1 \to C_1 \to \mathbf{P}^1$ where:

  C_1 is reducible, $C_1 = H^0 \cup H^1$,
  H^0, H^1 are hyperelliptic,
  $H^0 \cap H^1$ maps onto $B := \mathrm{Branch}(g) \subset \mathbf{F}^1$,
  $\tilde{C}_1 = \widetilde{H}^0 \cup \widetilde{H}^1$ is allowable over C_1,
  $C^i \to H \to \mathbf{P}^1$ is bigonally related to $\widetilde{H}^i \to H^i \to \mathbf{P}^1$, $i = 1, 2$.

Vice versa, the tetragonal construction takes any tower $\tilde{C}_1 \to C_1 \to \mathbf{P}^1$ as above to two Cartesian bihyperelliptic towers

$$\tilde{C} \to C \to H \to \mathbf{P}^1 \quad \text{and} \quad \tilde{C}_0 \to C_0 \to H \to \mathbf{P}^1.$$

The proof is quite straightforward, and we will simply write down a few of the relationships involved, using the notation of the previous proof:

- $\widetilde{H}$ splits into two copies of H, by (3.1). Hence:
- $g_* \widetilde{H} \approx H \cup \mathbf{P}^1 \cup \mathbf{P}^1$, say $H_0 \approx H$, $H_1 = R^0 \cup R^1$, $R^i \approx \mathbf{P}^1$, $i = 0, 1$.

- Let H^i, $\widetilde{H}^i$ be the inverse image of R^i in C_1, $\tilde{C}_1$ respectively. Then $\widetilde{H}^i \to H^i \to \mathbf{P}^1$ is bigonally related to $C^i \to H \to \mathbf{P}^1$.

- The intersection properties of the H^i (or $\widetilde{H}^i$) can be read off the local pictures (2.14.3).

- Finally, let $\varepsilon : H \to H$ be the hyperelliptic involution. A cover $C^1 \to H$ determines a mirror-image $\varepsilon^* C^1$. The remaining tower $\tilde{C}_0 \to C_0 \to H \to \mathbf{P}^1$ is given by the Cartesian diagram:

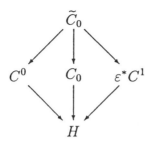

QED

Remarks

(3.6.1) Since the branch points of $C^i \to H$ map to the branch points of $H^i \to \mathbf{P}^1$, we have the relation between the genera:

$$g(H^i) = g(C^i) - 2 \cdot g(H).$$

(3.6.2) The possible local pictures are exactly the same as in (3.5.1). (The use of $C_0,\, C_1$ in (3.6) is consistent with that of (2.14).)

(3.6.3) Another way of proving both lemma (3.5) and proposition (3.6) is based on lemma (5.5), which says that the three tetragonal curves C, C_0, C_1 which are tetragonally related are obtained, via the trigonal construction, from one and the same trigonal curve X (with three distinct double covers). Lemma (3.3) characterizes the possible curves X, hence proves that the locus of bihyperelliptics is closed under the tetragonal construction, lemma (3.5). To complete the proof of proposition (3.6), one simply needs to characterize the double covers $\tilde{X}$ which correspond to Cartesian covers of C.

For the rest of this section, we specialize to the case where the hyperelliptic H is an elliptic curve E, i.e. C is bielliptic. First, we write out explicitly the content of Proposition (3.6) in this case:

Corollary 3.7 The Pryms of double covers $\pi : \tilde{C} \to C$ where

- C is bielliptic, $C \xrightarrow{f} E \xrightarrow{g} \mathbf{P}^1$,
- $\tilde{C} \to C$ is Cartesian, $\tilde{C} = C^0 \times_E C^1$, C^0 is of genus n,

are precisely (via the tetragonal construction) the Pryms of the following allowable double covers $\tilde{X} \to X$:

$n = 1$: X is obtained from a hyperelliptic curve by identifying
two pairs of points, $X = H/(p \sim q, \ r \sim s)$.

$n = 2$: $X = X_0 \cup X_1$, X_0 rational, X_1 hyperelliptic,
$\#(X_0 \cap X_1) = 4$.

$n \geq 3$: $X = X_0 \cup X_1$, each X_i hyperelliptic, $g(X_0) = n - 2$,
$g(X_1) = g(C) - n - 1$, $\#(X_0 \cap X_1) = 4$, and both
hyperelliptic maps are restrictions of the same tetragonal
map on X (i.e. they agree on $X_0 \cap X_1$).

Everything here follows directly from the proposition, except that
for $n = 1$ we need to use (twice) the following observation of Beauville.
Let $\pi : \widetilde{X} \to X$ be an allowable double cover where

$$X = Y \cup R, \qquad R \text{ rational}, \qquad Y \cap R = \{a, b\}$$

$$\widetilde{X} = \widetilde{Y} \cup \widetilde{R}, \ \widetilde{R} = \pi^{-1}(R) \text{ rational}, \ \widetilde{Y} \cap \widetilde{R} = \{\tilde{a}, \tilde{b}\}$$

and π is ramified at $\tilde{a}, \tilde{b}$, which map to a, b. Construct a new cover
$\widetilde{Z} \to Z$ where

$$\widetilde{Z} := \widetilde{Y}/(\tilde{a} \sim \tilde{b})$$

$$Z := Y/(a \sim b).$$

Then this is still allowable, and

$$P(\widetilde{Z}/Z) \approx P(\widetilde{X}/X).$$

(Indeed, there are natural isomorphisms of generalized Jacobians

$$J(\widetilde{Z}) \approx J(\widetilde{X}), \ \ J(Z) \approx J(X)$$

commuting with π_* and inducing the desired isomorphisms.)

QED

We are left with the Pryms of non-Cartesian double covers of
bielliptic curves. The result here may be somewhat surprising:

Proposition 3.8 Pryms of non-Cartesian double covers of bielliptic
curves are precisely the Pryms of Cartesian covers (of bielliptic curves)
with $n(:= g(C_0)) = 1$. (The isomorphism is obtained through a sequence of 2 tetragonal moves.)

The point is that if $X = H/(p \sim q, \ r \sim s)$ with H hyperelliptic,
and $\widetilde{X} \to X$ is an allowable double cover, then $P(\widetilde{X}/X)$ is the Prym of
a Cartesian cover (with $n = 1$) of a bielliptic curve, as we've just seen;

but X has another g_4^1, and applying the tetragonal construction to it yields a non-Cartesian double cover of a bielliptic curve.

The g_4^1 is obtained as follows: map H to a conic in $\mathbf{P}^2$ (by the hyperelliptic map), then project the conic to $\mathbf{P}^1$ from the unique point x in $\mathbf{P}^2$ (and not on the conic) on the intersection of the lines $\overline{pq}$ and $\overline{rs}$.

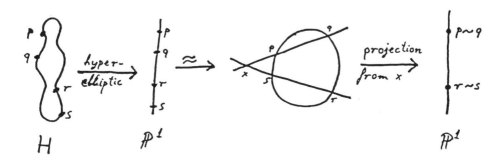

We should now check that the tetragonal construction yields a non-Cartesian cover of a bielliptic curve, and that all covers arise this way. We leave the former to the reader, and do the latter.

Let $\widetilde{C} \to C$ be a non-Cartesian cover of C, which is bielliptic:

$$C \xrightarrow{f} E \xrightarrow{g} \mathbf{P}^1.$$

Let $(\widetilde{C}_i, C_i)$, $i = 0, 1$, be the tetragonally related covers. By lemma (3.5), C_i is bihyperelliptic:

$$C_i \xrightarrow{f_i} H_i \xrightarrow{g_i} \mathbf{P}^1.$$

By the local pictures (2.14),

$$B := \operatorname{Branch}(g) = B_0 \amalg B_1, \quad B_i := \operatorname{Branch}(g_i).$$

(As we saw in Remark (3.5.1), the possible pictures are (1), (2), (7), (3a) and (3b). Of these, (7) and (3b) contribute to B, and each contributes also to one of the B_i.) Since $\#B = 4$ (E is elliptic), and $\#B_i$ is even and > 0 (non-Cartesian!), we find

$$\#B_i = 2, \quad i = 0, 1,$$

hence H_i is rational and C_i is hyperelliptic. Again by the local pictures, C_i will have two nodes, at points lying over B_{1-i}.

$$\text{QED}$$

We observe that the last argument works not only for bielliptics but also for branched double covers of hyperelliptic curves of genus 2, since now

$$\#B_0 > 0, \quad \#B_1 > 0, \quad \#B_0 + \#B_1 = 6, \quad \#B_i \text{ even} \Rightarrow$$

$$\text{either } \#B_0 = 2 \text{ or } \#B_1 = 2.$$

However, the resulting hyperelliptic curve with 4 nodes does not carry other g_4^1's and is not necessarily related to any other covers.

We leave one more corollary of proposition (3.6) to the reader.

Corollary 3.9 Let K be hyperelliptic, $\widetilde{K} \to K$ a double cover with 2 branch points. Then $P(\widetilde{K}/K)$ is a hyperelliptic Jacobian.

(Hint: take both H and C^0 in proposition (3.6) to be rational, show $P(\widetilde{C}_1/C_1) \approx J(C^1)$ and $C_1 = K \cup_{(2 \text{ points})} \mathbf{P}^1$, K hyperelliptic.)

§4   Fibers of $\mathcal{P} : \mathcal{R}_6 \to \mathcal{A}_5$.

§4.1   The structure

We recall the main result of [DS]:

Theorem 4.1 [DS] $\mathcal{P} : \mathcal{R}_6 \to \mathcal{A}_5$ is generically finite, of degree 27.

Recall that $\mathcal{M}_6^{\mathrm{Tet}}$ denotes the moduli space of curves of genus 6 with a g_4^1. The forgetful map $\mathcal{M}_6^{\mathrm{Tet}} \to \mathcal{M}_6$ is generically finite, of degree 5 [ACGH]. By base change we get a corresponding object $\mathcal{R}_6^{\mathrm{Tet}}$, with map

$$\mathcal{R}_6^{\mathrm{Tet}} \to \mathcal{R}_6$$

of degree 5. The tetragonal construction gives a triality, or (2,2)-correspondence, on $\mathcal{R}_6^{\mathrm{Tet}}$. The image in $\mathcal{R}_6$ is then a (10,10)-correspondence:

(4.1.1) $\mathrm{Tet} \subset \mathcal{R}_6 \times \mathcal{R}_6.$

Theorem 4.2 The correspondence Tet induced by the tetragonal construction on the fiber $\mathcal{P}^{-1}(A)$, for generic $A \in \mathcal{A}_4$, is isomorphic to the incidence correspondence on the 27 lines on a non-singular cubic surface. The monodromy group of $\mathcal{R}_6$ over $\mathcal{A}_5$ (i.e. the Galois group of its Galois closure) is the Weyl group WE_6, the symmetry group of the incidence of the 27 lines on the cubic surface.

This was conjectured in [DS] and announced in [D1]. The proof will be given below. For the symmetry group of the line incidence on a cubic surface, or other del Pezzo surfaces, we refer to [Dem].

(4.3) The blownup map

Let $\mathcal{Q} \subset \mathcal{M}_6$ denote the moduli space of non-singular plane quintic curves, $\mathcal{R}\mathcal{Q}$ its inverse image in $\mathcal{R}_6$. By Theorem (1.2), it splits:

$$\mathcal{R}\mathcal{Q} = \mathcal{R}\mathcal{Q}^+ \cup \mathcal{R}\mathcal{Q}^-$$

with $(Q, \mu) \in \mathcal{R}\mathcal{Q}^+$ (respectively, $\mathcal{R}\mathcal{Q}^-$) iff $h^0(\mu \otimes \mathcal{O}_Q(1))$ is even (respectively, odd). The point is that $\mathcal{O}_Q(1)$ gives a uniform choice of theta characteristics over $\mathcal{Q}$, hence the spaces of theta characteristics and semiperiods over $\mathcal{Q}$ are identified.

Let $\mathcal{J}$ be the closure in $\mathcal{A}_5$ of the locus of Jacobians of curves, and let $\mathcal{C}$ denote the moduli space of non-singular cubic threefolds. Via the intermediate Jacobian map, we identify $\mathcal{C}$ with its image in $\mathcal{A}_5$.

The Prym map sends $\mathcal{R}\mathcal{Q}^+$ to $\mathcal{J}$ and $\mathcal{R}\mathcal{Q}^-$ to $\mathcal{C}$. Since the fiber dimensions can be positive, it is useful to consider the blownup Prym map

$$\widetilde{\mathcal{P}} : \widetilde{\mathcal{R}}_6 \to \widetilde{\mathcal{A}}_5$$

where $\mathcal{J}, \mathcal{C}$ on the right are blown up to divisors $\widetilde{\mathcal{J}}, \widetilde{\mathcal{C}}$, while on the left we blow up $\mathcal{R}\mathcal{Q}^+, \mathcal{R}\mathcal{Q}^-$, as well as the locus $\mathcal{R}_6^{Trig}$ of double covers of trigonal curves. The result is a morphism which is generically finite over $\widetilde{\mathcal{J}}$ and $\widetilde{\mathcal{C}}$. We recall the geometric description of points of the various loci, and give the map in these geometric terms. This is taken from [CG], [T] and [DS].

(4.3.1) A point of $\mathcal{C}$ is given by a non-singular cubic threefold $X \subset \mathbf{P}^4$. A point of $\widetilde{\mathcal{C}}$ is given by a pair (X, H), $H \in (\mathbf{P}^4)^*$ a hyperplane.

(4.3.2) A point of $\widetilde{\mathcal{R}\mathcal{Q}}$ is given by (Q, μ, L), or $(Q, \widetilde{Q}, L)$, where $Q \subset \mathbf{P}^2$ is a plane quintic, $L \in (\mathbf{P}^2)^*$ a line, and μ a semiperiod on Q (or $\widetilde{Q}$ the corresponding double cover).

(4.3.3) The fiber $\mathcal{P}^{-1}(J(X)) \subset \mathcal{R}\mathcal{Q}^-$ over a cubic threefold X can be identified with the Fano surface $F(X)$ of lines $\ell \subset X$. (Projection from ℓ puts a conic bundle structure $\pi : X -- \to \mathbf{P}^2 = \mathbf{P}^4/\ell$ on X; the corresponding point of $\mathcal{R}\mathcal{Q}^-$ is $(Q, \widetilde{Q})$, where the plane quintic Q is the discriminant locus of π, and its double cover $\widetilde{Q}$ parametrizes lines $\ell' \in F(X)$ meeting ℓ.)

(4.3.4) The fiber $\widetilde{\mathcal{P}}^{-1}(X, H)$ corresponds to the lines ℓ in the cubic surface $X \cap H$. For general X, H, there are 27 of these. The corresponding objects are of the form $(Q, \widetilde{Q}, L)$ where $(Q, \widetilde{Q})$ are as above, and $L \subset \mathbf{P}^2$ is the projection, $L = \pi(H)$.

(4.3.5) A point of $\mathcal{R}_6^{Trig}$ is given by a curve $T \in \mathcal{M}_6$ with a trigonal line bundle $\mathcal{L} \in \operatorname{Pic}^3(T), h^0(\mathcal{L}) = 2$, and a double cover $\tilde{T} \to T$. The fiber of $\widehat{\mathcal{R}}_6^{Trig}$ above it is given by the linear system $|\omega_T \otimes \mathcal{L}^{-2}|$, a $\mathbf{P}^1$.

(4.3.6) A point of $\mathcal{J}$ is given by the Jacobian of a curve $C \in \mathcal{M}_5$. The canonical curve $\Phi(C) \subset \mathbf{P}^4$, for general C, is the base locus of a net of quadrics:

$$A_p \subset \mathbf{P}^4, \quad p \in \mathbf{P}^2 = \mathbf{P}^2(C).$$

A point of $\widetilde{\mathcal{J}}$ above C is then given by a pair (C, L), where L is a line in $\mathbf{P}^2(C)$. (Choosing such a line is the same as choosing a quartic del Pezzo surface

$$S = S_L = \cap_{p \in L} A_p$$

containing $\Phi(C)$.)

(4.3.7) Consider the map

$$\alpha : \mathcal{M}_5 \to \mathcal{RQ}^+$$

sending $C \in \mathcal{M}_5$ to $\alpha(C) = (Q, \tilde{Q})$, where:

$$Q := \{p \in \mathbf{P}^2(C) \mid A_p \text{ is singular}\} \subset \mathbf{P}^2(C),$$

and $\tilde{Q}$ is the double cover whose fiber over a general $p \in Q$ corresponds to the two rulings on the rank-4 quadric A_p. This α is a birational isomorphism; its inverse is the restriction to $\mathcal{RQ}^+$ of $\mathcal{P}$.

The fiber $\widetilde{\mathcal{P}}^{-1}(C, L)$ over generic $(C, L) \in \widetilde{\mathcal{J}}$ is given by the following 27 objects:

• The quintic object $(Q, \tilde{Q}, L) \in \widetilde{\mathcal{RQ}}^+$, where $(Q, \tilde{Q}) = \alpha(C)$ and L is the given line in $\mathbf{P}^2(C)$.

• Ten trigonals T_i^ε, $1 \leq i \leq 5$, $\varepsilon = 0, 1$, each with a double cover $\tilde{T}_i^\varepsilon$: each of the 5 points $p_i \in Q \cap L$ determines two g_4^1's on C, cut out by the rulings R_i^ε on A_{p_i}, and the $(T_i^\varepsilon, \tilde{T}_i^\varepsilon)$ are associated to these by the trigonal construction.

• Sixteen Wirtinger covers $(X_j, \widetilde{X}_j) \in \partial^I \mathcal{R}_6$: the quartic del Pezzo surface S_L contains 16 lines ℓ_j [Dem], each meeting $\Phi(C)$ in two points, say p_j, q_j, and then

$$X_j = C/(p_j \sim q_j)$$

and $\widetilde{X}_j$ is its unique Wirtinger cover (1.9.I).

(4.3.8) We observe that the generically finite map $\mathcal{R}_6^{Tet} \to \mathcal{R}_6$ has 1-dimensional fibers over both $\mathcal{RQ}$ and $\mathcal{R}_6^{Trig}$. After blowing up and

normalizing, we obtain finite fibers generically over the exceptional loci.
In the limit:

- Over (Q, L), the 5 g_4^1's correspond to projections of the plane quintic
Q from one of the 5 points $p_i \in Q \cap L$.

- Over (T, D), with T a trigonal curve, $\mathcal{L}$ the trigonal bundle, and
$D \in |\omega_T \otimes \mathcal{L}^{-2}|$, four of the g_4^1's are of the form $\mathcal{L}(q)$ with $q \in D$ (i.e.
they are the trigonal $\mathcal{L}$ with base point q); the fifth g_4^1 is $\omega_T \otimes \mathcal{L}^{-2}$.

- Given $X = C/(p \sim q) \in \partial \bar{\mathcal{M}}_5$, there is a pencil $L \subset \mathbf{P}^2(C)$ of
quadrics A_p, $p \in L$, which contain both $\Phi(C)$ and its chord $\overline{pq}$. Among
these there are 5 quadrics A_{p_i} which are singular, generically of rank 4.
Each of these has a single ruling R_i containing a plane containing $\overline{pq}$.
These R_i cut the 5 g_4^1's on X.

We conclude that the tetragonal correspondence Tet of (4.1.1) lifts
to

$$\widetilde{\text{Tet}} \subset \widetilde{\mathcal{R}}_6 \times \widetilde{\mathcal{R}}_6$$

which is generically finite, of type (10,10), over each of our special loci.

Theorem 4.4 Structure of the blownup Prym map.

Over each of the following loci, the blownup Prym map $\widetilde{\mathcal{P}}$ has the
listed monodromy group, and the lifted tetragonal correspondence $\widetilde{\text{Tet}}$
induces the listed structure.

(1) $\widetilde{\mathcal{C}}$: The group is WE_6, the structure is that of lines on a general
non-singular cubic surface.

(2) $\widetilde{\mathcal{J}}$: The group is WD_5, the symmetry group of the incidence of
lines on a quartic del Pezzo surface, or stabilizer in WE_6 of a line.
The structure is that of lines on a non-singular cubic surface, one
of which is marked.

(3) $\mathcal{B}=$ the locus of intermediate Jacobians of Clemens' quartic double
solids of genus 5 [C1]: The group is $WA_5 = S_6$, the structure is
that of lines on a nodal cubic surface.

[Note: $\mathcal{B}$ is contained in the branch locus of $\mathcal{P}$ [DS, V.4] and in
fact ([D6], and compare also [SV], [I]) equals the branch locus.
The monodromy along $\mathcal{B}$ acting on a nearby, unramified fiber is
$(\mathbf{Z}/2\mathbf{Z}) \times S_6$, or the symmetry group of a double-six, which is a
subgroup of WE_6. The group S_6 thus occurs as a subquotient of
WE_6.]

(4) (cf. [I]) $\widetilde{\mathcal{P}}$ extends naturally to the boundary $\partial = \partial \mathcal{A}_5$; the monodromy is WE_6 and the structure is that of lines on a general cubic
surface.

We will prove parts (1), (2) and (3) in §4.2. In the rest of this section we show that theorems (4.1) and (4.2) follow from (4.4).

(4.5) Proofs of Theorem (4.2).

By Theorem (2.16), Tet commutes with $\mathcal{P}$, therefore $\widetilde{\mathrm{Tet}}$ commutes with $\widetilde{\mathcal{P}}$. To identify this structure over a generic point, it suffices to do so over any point over which $\widetilde{\mathcal{P}}$ and $\widetilde{\mathrm{Tet}}$ are etale. These conditions hold, e.g., over a generic $(X, H) \in \widetilde{\mathcal{C}}$, where (4.4.1) identifies the structure. This implies that the monodromy is contained in WE_6, but we get all of WE_6 already over $\widetilde{\mathcal{C}}$ (by (4.4.1) again), so we are done.

We can work instead over $\widetilde{\mathcal{J}}$: again, $\widetilde{\mathcal{P}}$ and $\widetilde{\mathrm{Tet}}$ are etale over generic $(C, L) \in \widetilde{\mathcal{J}}$, and $\widetilde{\mathrm{Tet}}$ has the right structure there by (4.4.2). This shows

$$WD_5 \subset \text{Monodromy} \subset WE_6.$$

As there are no intermediate groups, the monodromy must equal WD_5 or WE_6. But if it were the former, $\mathcal{R}_6$ would be reducible (since WD_5 is the stabilizer in WE_6 of one of the 27 lines), contradiction.

QED

Remark 4.5.1 Along the same lines, we can also reprove Theorem (4.1). Let $\widetilde{\mathrm{Tet}}^i$ denote the i-th iterate of the correspondence $\widetilde{\mathrm{Tet}}$. On $\mathcal{RQ}^-$ we have:

$$\widetilde{\mathrm{Tet}}^2 \text{ has degree } 27,$$
$$\widetilde{\mathrm{Tet}}^i = \widetilde{\mathrm{Tet}}^2 \text{ for } i \geq 2.$$

Since $\widetilde{\mathrm{Tet}}$ is etale there, these properties persist generically on $\widetilde{\mathcal{R}}_6$. Let $\sim$ be the equivalence relation generated by $\widetilde{\mathrm{Tet}}$. We conclude that $\sim$ has degree 27, and that $\widetilde{\mathcal{P}}$ factors through a proper quotient:

$$\mathcal{P}' : \widetilde{\mathcal{R}}_6 / \sim \longrightarrow \tilde{\mathcal{A}}_5.$$

We still need to verify that $\deg(\mathcal{P}') = 1$. There are several possibilities:

• We can work over $\widetilde{\mathcal{J}}$; as we will see in (4.7), the fiber of $\widetilde{\mathcal{P}}$ there consists of a unique $\sim$-equivalence class; so we need to check that $\mathcal{P}'$ is unramified at that equivalence class. This reduces to seeing that $\widetilde{\mathcal{P}}$ is unramified at least at one point of the fiber; this is trivial at the plane-quintic point. (This argument avoids some of the detailed computations of the codifferential on the boundary, [DS, Ch., IV], but is still very close in spirit to [DS].)

• We could instead work over any other point of $\tilde{\mathcal{A}}_5$ over which we know the complete fiber, e.g. over Andreotti-Mayer points, coming from bielliptic Pryms, as in §3. (This was proposed in [D1], as a way to avoid the boundary computations.)

• Izadi [I] applies a similar argument over boundary points, in $\partial \mathcal{A}_5$. This lets her reduce the degree computation over $\mathcal{A}_5$ to her results on $\mathcal{A}_4$, cf. (4.9).

§4.2 Special Fibers

In this section we exhibit the cubic surface of theorem (4.2) explicitly over three special loci in $\mathcal{A}_5$. We do not know how to do this at the generic point of $\mathcal{A}_5$.

(4.6) Cubic threefolds

From (4.3.4) we have an identification of $\tilde{\mathcal{P}}^{-1}(X, H)$, where $X \subset \mathbf{P}^4$ is a cubic threefold and $H \subset \mathbf{P}^4$ a hyperplane, with the set of lines ℓ on the cubic surface $X \cap H$. For Theorem (4.4.1) we need to check that two of these, say $\ell, \ell' \in F(X)$, intersect each other if and only if the corresponding objects $(Q, \tilde{Q}, L), (Q', \tilde{Q}', L')$ correspond under $\widetilde{\text{Tet}}$. If the lines ℓ, ℓ' intersect, we are in the situation of (2.15.4), so the corresponding objects

$$(Q, \tilde{Q}, f), (Q', \tilde{Q}', f')$$

(notation of (2.15.4)) are tetragonally related. Since f, f' are both cut out by hyperplanes through the span A of ℓ, ℓ', we find points

$$p \in Q \cap L, \quad p' \in Q' \cap L'$$

(namely, the projection of A from ℓ, ℓ' respectively) such that f, f' are the projections of Q from p and of Q' from p', respectively. The description of $\widetilde{\mathcal{R}}_6^{\text{Tet}}$ in (4.3.8) then shows that

$$((Q, \tilde{Q}, L), (Q', \tilde{Q}', L')) \in \widetilde{\text{Tet}},$$

as required. Since both the line incidence and $\widetilde{\text{Tet}}$ are of bidegree (10,10), and we have an inclusion, it must be an equality.

This shows that $\widetilde{\text{Tet}}$ induces on $\tilde{\mathcal{P}}^{-1}(X, H)$ the structure of line incidence on the cubic surface $X \cap H$. Fix the ambient $\mathbf{P}^4$ and the hyperplane H, and let the cubic X vary. We clearly get all cubic surfaces in H as intersections $X \cap H$; therefore the monodromy group is the full symmetry group of the line configuration. This completes the proof of (4.4.1), hence also of Theorem (4.2).

(4.7) Jacobians

Start with $(C, L) \in \widetilde{\mathcal{J}}$. The fiber $\widetilde{\mathcal{P}}^{-1}(C, L)$ consists of the 27 objects listed in (4.3.7). Each of these comes with the 5 g_4^1's given in (4.3.8). These give the correspondence $\widetilde{\text{Tet}}$, which we claim is equivalent to the line incidence on a cubic surface.

Let $S = S_L$ be the quartic del Pezzo surface determined by (C, L), as in (4.3.6). Let S' be its blowup at a generic point $r \in S$. Then S' is a cubic surface; its lines correspond to:

- ℓ_Q, the exceptional divisor over r.

- 10 conics through r in S; these correspond naturally to the 10 rulings $\mathcal{R}_i^\varepsilon$ (as in (4.3.7)). [Each $\mathcal{R}_i^\varepsilon$ contains a unique plane through r, which meets S in a conic through r.]

- The 16 lines ℓ_j in S.

There is thus a natural bijection between the lines of S' and $\widetilde{\mathcal{P}}^{-1}(C, L)$. We need to check that this correspondence takes incident lines to covers which are tetragonally related to each other through the g_4^1's of (4.3.8). To that end, we list the effects of the tetragonal constructions on our curves. The details are straightforward, and are omitted.

(4.7.1) The quintic $(Q, \widetilde{Q})$, with the $g_4^1 : \mathcal{O}_Q(1)(-p_i)$, $p_i \in Q \cap L$, is taken to the two trigonals

$$(T_i^\varepsilon, \widetilde{T}_i^\varepsilon), \quad \varepsilon = 0, 1,$$

each with its unique base-point-free g_4^1, $\omega_T \otimes \mathcal{L}^{-2}$.

(4.7.2) The trigonal $(T_i^\varepsilon, \widetilde{T}_i^\varepsilon)$ with its base-point-free g_4^1 goes to $(Q, \widetilde{Q})$ with $\mathcal{O}_Q(1)(-p_i)$, and to $(T_i^{1-\varepsilon}, \widetilde{T}_i^{1-\varepsilon})$ with the base-point-free g_4^1.

Consider $(T_i^\varepsilon, \widetilde{T}_i^\varepsilon)$ with the g_4^1 $\mathcal{L}_i^\varepsilon(p)$. The actual 4-sheeted cover of $\mathbf{P}^1$ in this case is reducible, consisting of the trigonal T_i^ε together with a copy of $\mathbf{P}^1$ glued to it at p. We are thus precisely in the situation of Example (2.15.2): both tetragonally related objects are Wirtinger covers $(X_j, \widetilde{X}_j)$.

(4.7.3) A Wirtinger cover $(X_j, \widetilde{X}_j)$ with the g_4^1 cut out by the ruling $\mathcal{R}_i^\varepsilon$ on the singular quadric A_{p_i}, is taken to the trigonal $(T_i^\varepsilon, \widetilde{T}_i^\varepsilon)$ and to another Wirtinger cover.

(4.8) Quartic double solids and the branch locus of $\mathcal{P}$.

The fiber of $\mathcal{P}$ over the Jacobian $J(X) \in \mathcal{B}$ of a quartic double solid X of genus 5 is described in [DS, V.4], following ideas of Clemens. It consists of 6 objects $(C_i, \widetilde{C}_i)$, $0 \leq i \leq 5$, each with multiplicity 2, and 15 objects $(C_{ij}, \widetilde{C}_{ij})$, $0 \leq i < j \leq 5$. The monodromy group S_6 permutes the six values of i: clearly the two sets $\{C_i\}$ and $\{C_{ij}\}$ must be separately permuted, and any permutation of the C_i induces a unique permutation of the C_{ij}. The situation is precisely that of lines on a nodal cubic surface: the C_i correspond to lines ℓ_i through the node; and the plane through ℓ_i, ℓ_j meets the cubic residually in a line $\ell_{i,j}$.

The best way to see the symmetry is to consider Segre's cubic three-fold $Y \subset \mathbf{P}^4$, image of $\mathbf{P}^3$ by the linear system of quadrics through 5 points p_i, $1 \leq i \leq 5$, in general position in $\mathbf{P}^3$. (cf. [SR] for the details.) Y contains six irreducible, two-dimensional families of lines, which we call the "rulings" R_i, $0 \leq i \leq 5$: For $1 \leq i \leq 5$, R_i consists of proper transforms of lines through p_i; while R_0 parametrizes twisted cubics through $p_1, \cdots, p_5$. Y also contains 15 planes Π_{ij}, $0 \leq i < j \leq 5$ (= the 5 exceptional divisors and the proper transforms of the 10 planes $\overline{p_i p_j p_k}$); the ruling R_i is characterized as the set of lines in $\mathbf{P}^4$ meeting the 5 planes Π_{ij}, $j \neq i$. The symmetric group S_6 acts linearly on $\mathbf{P}^4$, preserving Y, permuting the R_i and correspondingly the Π_{ij}.

The quartic double solids in question are essentially the double covers

$$\zeta : X \to Y$$

branched along the intersection of Y with a quadric $Q \subset \mathbf{P}^4$. The Prym fiber is obtained as follows:

- $C_i := \{$ lines $\ell \in R_i$, tangent to $Q\}$
  $\widetilde{C}_i := \{$ irreducible curves $\ell' \subset X$ such that $\zeta(\ell') = \ell \in C_i\}$

Thus $(C_i, \widetilde{C}_i)$ is the discriminant of a conic-bundle structure on X given by $\zeta^{-1}(R_i)$. The Prym canonical curve $\Psi(C_i) \subset \mathbf{P}^4$ is traced by the tangency points of ℓ and Q; in particular, $\Psi(C_i) \subset Q$, so $(C_i, \widetilde{C}_i)$ is a ramification point of $\mathcal{P}$, by (1.6).

- $(C_{ij}, \widetilde{C}_{ij})$ is similarly obtained as discriminant of a conic bundle structure on X given by projection from Π_{ij}, cf. [DS, V4.5].

(4.9) Boundary behavior

In [I], Izadi uses results on the structure of $\mathcal{P} : \mathcal{R}_5 \to \mathcal{A}_4$ to find the incidence structure on the fibers of the compactified map $\overline{\overline{\mathcal{P}}} : \overline{\mathcal{R}}_6 \to \bar{\mathcal{A}}_5$

over boundary points of the toroidal compactification $\bar{\mathcal{A}}_5$. The picture is as follows:

$$
\begin{array}{ccc}
\bar{\bar{\mathcal{P}}} \ : & \bar{\bar{\mathcal{R}}}_6 \longrightarrow \bar{\mathcal{A}}_5 \\
 & \cup \qquad\quad \cup \\
\partial\mathcal{P} \ : & \partial^{\mathrm{II}}\bar{\bar{\mathcal{R}}}_6 \longrightarrow \partial\bar{\mathcal{A}}_5 \\
 & \alpha\Big\downarrow \qquad\quad \Big\downarrow\beta \\
\mathcal{P} \ : & \mathcal{R}_5 \longrightarrow \mathcal{A}_4
\end{array}
$$

Over general $A \in \mathcal{A}_4$, the fiber $\beta^{-1}(A)$ is isomorphic to the Kummer variety $A/(\pm 1)$. Over $(\tilde{C}, C) \in \mathcal{R}_5$, the fiber of α is $S^2\tilde{C}/\iota$, and $\partial\mathcal{P}$ becomes (cf. [D3, (4.6)]) the map

$$
\begin{array}{ccc}
x + y \mapsto \psi(x) + \psi(y) \\
S^2\tilde{C} \longrightarrow A \\
\Big\downarrow \qquad\quad \Big\downarrow \\
S^2\tilde{C}/\iota \to A/(\pm 1)
\end{array}
$$

where ψ is the Abel-Prym map $\tilde{C} \to A$. All in all then, we are considering the map

$$
\partial\mathcal{P} : \cup_{(\tilde{C},C)\in\mathcal{P}^{-1}(A)} S^2\tilde{C} =: E \longrightarrow A.
$$

Theorem (4.1) says that its degree is 27, and Theorem (4.2) predicts an incidence structure on its fibers, i.e. a way of associating a cubic surface to each point $a \in A$.

In §5 we associate to $A \in \mathcal{A}_4$ a cubic threefold $X = \kappa(A) \subset \mathbf{P}^4$ such that $\mathcal{P}^{-1}(A)$ is a double cover of the Fano surface $F(X)$ of lines in X. For generic $a \in A$, we are looking for a cubic surface; it is reasonable to hope that this should be of the form $H(a) \cap X$, where $H(a)$ is an appropriate hyperplane in $\mathbf{P}^4$. We thus want a map

$$
H : A \to (\mathbf{P}^4)^*
$$

such that

$$
pr((\partial\mathcal{P})^{-1}(a)) = \{\text{lines in } H(a) \cap X\}.
$$

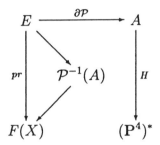

Izadi's beautiful observation is that such an H is given by the linear system Γ_{00} (sections of $|2\Theta|$ vanishing to order ≥ 4 at 0). The identification of Γ_{00} with the ambient $\mathbf{P}^4$ of X uses a construction of Clemens relating his double solids to Γ_{00}, and the interpretaton of (a cover of) X as parametrizing double solids with intermediate Jacobians isomorphic to A, cf. [D6] or [I].

§5   Fibers of $P : \mathcal{R}_5 \to A_4$.

§5.1   The general fiber.

Our main result in this section is:

Theorem 5.1 For generic $A \in A_4$, the fiber $\overline{\mathcal{P}}^{-1}(A)$ is isomorphic to a double cover of the Fano surface $F = F(X)$ of lines on some cubic threefold X.

Let $\mathcal{RC}$ denote the inverse image in $\mathcal{RA}_5$ of the locus $\mathcal{C}$ of (intermediate Jacobians $J(X)$ of) cubic threefolds X. We recall from [D4] that it splits into even and odd components:

(5.1.1) $$\mathcal{RC} = \mathcal{RC}^+ \amalg \mathcal{RC}^-,$$

distinguished by a parity funciton. This follows from the existence of a natural theta divisor $\Xi \subset J(X)$, characterized (cf. [CG]) by having a triple point at $0 : \Xi$ translates the parity function q of (1.2), on theta characteristics, to a parity on semiperiods. More explicitly, pick $(Q,\sigma) \in \mathcal{P}^{-1}(J(X)) \subset \mathcal{RQ}^-$; Mumford's exact sequence (Theorem (1.4)(2)) says that any $\delta \in J_2(X)$ is $\pi^*\nu$ for some $\nu \in (\sigma)^\perp \subset J_2(Q)$. The compatibility result, theorem (1.5), then gives (cf. [D4], Proposition (5.1)):

(5.1.2) $$q_X(\delta) = q_Q(\nu) = q_Q(\nu\sigma).$$

In case δ is even, we end up with an isotropic subgroup $(\nu,\sigma) \subset J_2(Q)$, with σ odd and $\nu, \nu\sigma$ even. The Pryms of the latter are therefore Jacobians of curves:

(5.1.3) $P(Q, \nu) \approx J(C), \quad P(Q, \nu\sigma) \approx J(C'),$

and the image of σ gives semiperiods $\mu \in J_2(C), \mu' \in J_2(C')$.

Reversing direction, we can construct an involution

$$\lambda : \mathcal{R}_5 \longrightarrow \mathcal{R}_5$$

and a map

$$\kappa : \mathcal{R}_5 \longrightarrow \mathcal{R}\mathcal{C}^+,$$

as follows: Start with $(C, \mu) \in \mathcal{R}_5$, pick the unique (Q, ν) in $\mathcal{P}^{-1}(C) \cap \mathcal{R}\mathcal{Q}^+$, and let $\sigma, \nu\sigma \in J_2(Q)$ map to $\mu \in J_2(C)$. Then formula (1.3) reads:

(5.1.4) $0 \equiv 3 + \text{ even} + q(\sigma) + q(\nu\sigma) \quad (\text{mod. } 2),$

so after possibly relabeling, we may assume

$$(Q, \sigma) \in \mathcal{R}\mathcal{Q}^-, \quad (Q, \nu\sigma) \in \mathcal{R}\mathcal{Q}^+$$

so that there is a well-defined curve $C' \in \mathcal{M}_5$ and a cubic threefold $X \in \mathcal{C}$ such that

$$P(Q, \sigma) \approx J(X)$$

(5.1.5)

$$P(Q, \nu\sigma) \approx J(C').$$

We can thus define λ and κ by:

$$\lambda(C, \mu) := (C', \mu')$$

(5.1.6)

$$\kappa(C, \mu) := (X, \delta),$$

where $\mu' \in J_2(C'), \quad \delta \in J_2(X)$ are the images of $\nu \in J_2(Q)$. The precise version of our results is in terms of λ and κ:

Theorem 5.2

(1) (C, μ) is related to $\lambda(C, \mu)$ by a sequence of two tetragonal constructions. Hence λ commutes with the Prym map:

$$\mathcal{P} \circ \lambda = \mathcal{P}, \quad \lambda \circ \lambda = id.$$

(2) κ factors through the Prym map:

$$\kappa : \mathcal{R}_5 \xrightarrow{\mathcal{P}} \mathcal{A}_4 \xrightarrow{\chi} \mathcal{R}\mathcal{C}^+,$$

where χ is a birational map.

Recall the Abel-Jacobi map [CG],

$$AJ : F(X) \longrightarrow J(X),$$

which is well-defined up to translation in $J(X)$. (It can be identified with the Albanese map of the Fano Surface $F(X)$.) A point $\delta \in J_2(X)$ determines a double cover of $J(X)$, hence of $F(X)$.

Theorem 5.3 For generic $A \in \mathcal{A}_4$, set

$$(X, \delta) := \chi(A) = \kappa(\mathcal{P}^{-1}(A)) \in \mathcal{RC}^+.$$

Let $F(X)$ be the Fano surface of X, $\widetilde{F(X)}$ its double cover determined by δ via the Abel-Jacobi map.

(1) There is a natural isomorphism

$$P^{-1}(A) \approx \widetilde{F(X)}.$$

(2) The action of λ on the left corresponds to the sheet interchange on the right.

(3) Two objects $(C, \mu), (C', \mu) \in \mathcal{P}^{-1}(A)$ are tetragonally related if and only of the lines $\ell, \ell' \in F(X)$ which they determine intersect.

Remark 5.4 Izadi has recently analyzed the birational map χ, in [I]. In particular, she shows that χ is an isomorphism on an explicitly described, large open subset of $\mathcal{A}_4$.

§5.2   Isotropic subgroups.

By isotropic subgroup of rank r on a curve C we mean an r-dimensional $\mathbf{F}_2$-subspace of $J_2(C)$ on which the intersection pairing $\langle \, , \, \rangle$ is identically zero. Choosing an isotropic subgroup of rank 1 is the same as choosing a non-zero semiperiod.

Start with a trigonal curve $T \in \mathcal{M}_{g+1}$, with a rank-2 isotropic subgroup $W \subset J_2(T)$ whose non-zero elements we denote ν_i, $i = 0, 1, 2$. The trigonal construction associates to (T, ν_i) the tetragonal curve $X_i \in \mathcal{M}_g$. Mumford's sequence (1.4)(2) sends W to an isotropic subgroup of rank 1 on X_i, whose non-zero element we denote μ_i.

Lemma 5.5 The construction above sets up a bijection between the following data:

- A trigonal curve $T \in \mathcal{M}_{g+1}$ with rank-2 isotropic subgroup.
- A tetragonally related triple $(X_i, \mu_i) \in \mathcal{R}_g$, $\quad i = 0, 1, 2$.

Proof.

We think of WD_4 as the group of signed permutations of the 8 objects $\{x_i^{\pm}\}$, $1 \le i \le 4$.   Start with a tetragonal double cover $\widetilde{X}_0 \longrightarrow X_0 \longrightarrow \mathbf{P}^1$. It determines a principal WD_4-bundle over $\mathbf{P}^1 \backslash (\text{Branch})$. The original covers $\widetilde{X}_0, X_0$ are recovered as quotients by the following subgroups of WD_4:

$$\widetilde{H}_0 := \text{Stab}(x_1^+),$$
$$H_0 := \text{Stab}(x_1^{\pm}),$$

Consider also the subgroup

$$G := \text{Stab}\{\{x_1^+, x_2^+\}, \{x_1^-, x_2^-\}\}.$$

It has index 12 in WD_4. Its normalizer is:

$$N(G) = \text{Stab}\{\{x_1^{\pm}, x_2^{\pm}\}, \{x_3^{\pm}, x_4^{\pm}\}\},$$

of index 3. The quotient is

$$N(G)/G \approx (\mathbf{Z}/2\mathbf{Z})^2,$$

so there are 3 intermediate groups $\widetilde{G}_i$, $\quad i = 0, 1, 2$. We single out one of them:

$$\widetilde{G}_0 := \text{Stab}\{x_1^{\pm}, x_2^{\pm}\}.$$

The three subgroups $\widetilde{G}_i$ are not conjugate to each other, but can be taken to each other by outer automorphisms of WD_4. In fact, the action of $\text{Out}(WD_4) \approx S_3$ sends G, and hence also $N(G)$, to conjugate subgroups; it permutes the $\widetilde{G}_i$ transitively, modulo conjugation; and it also takes $H_0, \widetilde{H}_0$ to non-conjugate subgroups $H_i, \widetilde{H}_i$, $\quad i = 1, 2$. We illustrate each of these subgroups as the stabilizer in WD_4 of a corresponding partition of $\begin{pmatrix} x_1^+ & x_2^+ & x_3^+ & x_4^+ \\ x_1^- & x_2^- & x_3^- & x_4^- \end{pmatrix}$:

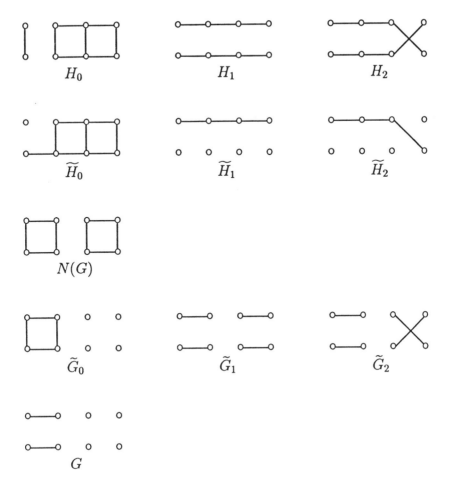

Let $X_i, \widetilde{X}_i, T, \widetilde{\widetilde{T}}, \widetilde{T}_i$ $\quad (i = 0, 1, 2)$ be the quotients of the principal WD_4-bundle by the subgroups $H_i, \widetilde{H}_i, N(G), G, \widetilde{G}_i$ respectively, compactified to branched covers of $\mathbf{P}^1$. We see immediately that:

• The trigonal construction takes $X_0 \to \mathbf{P}^1$ to $\widetilde{T}_0 \to T \to \mathbf{P}^1$.

• The double cover $\widetilde{X}_0 \to X_0$ corresponds via (1.4)(2) to the double cover $\widetilde{\widetilde{T}} \to \widetilde{T}_0$.

• The tetragonal construction acts by outer antomorphisms, hence exchanges the three tetragonal double covers $\widetilde{X}_i \to X_i \to \mathbf{P}^1$.

Applying the same outer automorphisms, we see that the trigonal construction also takes $X_i \to \mathbf{P}^1$ to $\widetilde{T}_i \to T \to \mathbf{P}^1$, $i = 1, 2$. To a tetragonally related triple $(\widetilde{X}_i \to X_i \to \mathbf{P}^1)$ we can thus unambiguously associate the trigonal $T \to \mathbf{P}^1$ together with the rank-2, isotropic subgroup corresponding to the covers $\tilde{T}_i$. This inverts the construction predecing the lemma.

QED.

Note 5.5.1 The basic fact in the above proof is that the 3 tetragonals X_i yield the same trigonal T. This can be explained more succinctly: outer automorphisms take the natural surjection $\alpha_0 : WD_4 \twoheadrightarrow S_4$ to homomorphisms α_1, α_2 which are not conjugate to it. But the composition $\beta \circ \alpha_i : WD_4 \twoheadrightarrow S_3$, where $\beta : S_4 \twoheadrightarrow S_3$ is the Klein map, are conjugate to each other.

Construction 5.6 Now let $T \in \mathcal{M}_{g+1}$ be a trigonal curve, together with an isotropic subgroup of rank 3,

$$V \subset J_2(T).$$

We think of V as a vector space over $\mathbf{F}_2$; the projective plane $\mathbf{P}(V)$ is identified with $V \backslash (0)$. For each $i \in \mathbf{P}(V)$, the trigonal construction gives a tetragonal curve $Y_i \in \mathcal{M}_g$. Mumford's sequence (1.4)(2) gives an isotropic subgroup of rank 2,

$$W_i \subset J_2(Y_i),$$

with a natural identification $W_i \approx V/(i)$.

Let $U \subset V$ be a rank-2 subgroup, so $\mathbf{P}(U) \subset \mathbf{P}(V)$ is a projective line. Lemma (5.5) shows that the 3 objects

$$(Y_i, U/(i)) \in \mathbf{R}_g, \qquad i \in \mathbf{P}(U)$$

are tetragonally related. In particular, they have a common Prym variety

$$P_U \approx \mathcal{P}(Y_i, U/(i)) \in \mathcal{A}_{g-1}, \qquad \forall i \in \mathbf{P}(U).$$

Applying (1.4) twice, we see that the original rank-3 subgroup V determines a rank-1 subgroup

$$V/U \subset (P_U)_2,$$

so we let $\mu_U \in (P_U)_2$ be its non-zero element. Altogether then, we have a map

$$\mathbf{P}(V)^* \longrightarrow \mathcal{R}\mathcal{A}_{g-1}$$
$$U \longmapsto (P_U, \mu_U).$$

(5.6.1) Assume now that one of the Y_i happens to be trigonal. (This can only happen if $g \leq 6$.) Whenever $U \ni i$, we find a tetragonal curve

$C_U \in \mathcal{M}_{g-1}$ such that $P_U \approx J(C_U)$. Lemma (5.5), applied to (Y_i, W_i), shows that the 3 objects

$$(C_U, \mu_U) \in \mathcal{R}_{g-1}, \qquad U \ni i$$

are tetragonally related, so they have a common Prym variety $A = P_V \in \mathcal{A}_{g-2}$.

(5.6.2) Assume instead that $g = 6$ and that P_U happens to be a Jacobian $J(C_U) \in \mathcal{J}_5$, for some $U \in \mathbf{P}(V)^*$. Of the three Y_i, $i \in U$, we claim two are trigonal and the third, a plane quintic. Indeed, by (4.7), the tetragonal triples above $J(C_U)$ consist either of a plane quintic and two trigonals, as claimed, or of a trigonal and two Wirtingers. The latter is excluded since the isomorphism

$$J(Y_i) \approx P(T, i)$$

implies that Y_i is non-singular for each $i \in \mathbf{P}(V)$.

Assume from now on that $g = 6$. Our data consists of:

- $T \in \mathcal{M}_7$, trigonal, with $V \subset J_2(T)$ isotropic of rank 3.
- For each $i \in \mathbf{P}(V)$, a curve $Y_i \in \mathcal{M}_6$, with a rank-2 isotropic subgroup $W_i \subset J_2(Y_i)$.
- For each $U \in \mathbf{P}(V)^*$, an object $(P_U, \mu_U) \in \mathcal{R}\mathcal{A}_5$
- An abelian variety $A = P_V \in \mathcal{A}_4$.

We display $\mathbf{P}(V)$ as a graph with 7 vertices $i \in \mathbf{P}(V)$ and 7 edges $U \in \mathbf{P}(V)^*$, in (3,3)-correspondence. We write T (or Q) on a vertex corresponding to a trigonal (or quintic) curve, and C on an edge corresponding to a Jacobian. We restate our observations:

(5.6.3): Edges through a T-vertex are C-edges.

(5.6.4): On a C-edge, the vertices are T, T, Q.

It follows that only one configuration is possible:

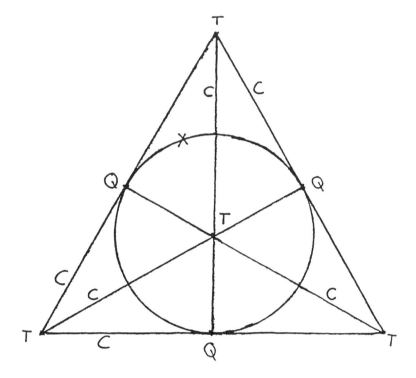

Figure 5.7

Thus four of the Y_i are trigonal, the other three are quintics, and six of the P_U, corresponding to the straight lines, are Jacobians of curves. Let $U_0 \in \mathbf{P}(V)^*$ correspond to the circle. For $i \in U_0$, Y_i is a quintic Q. Through Q pass two C edges and U_0, and the semiperiods corresponding to the C-edges are even; by (1.3), the semiperiod $U_0/(i)$ corresponding to U_0 must be <u>odd</u>, so there is a cubic threefold $X \in C$ such that

$$\mathbf{P}_{U_0} \approx J(X).$$

Finally, theorem (1.5), or formula (5.1.2), shows that the semiperiod $\delta := \mu_{U_0} \in J_2(X)$ is <u>even</u>.

We observe that the three tetragonally related quintics correspond to 3 lines on the cubic threefold which meet each other and thus form the intersection of X with a (tritangent) plane. We are thus exactly in the situation of (2.15.4).

§5.3   Proofs.

(5.8) Theorems (5.1),(5.2) and (5.3) all follow from the following statements:

(1) (C, μ) is related to $\lambda(C, \mu)$ by a sequence of two tetragonal constructions.

(2) κ is invariant under the tetragonal construction

(3) For $(X, \delta) \in \mathcal{RC}^+$, $\kappa^{-1}(X, \delta) \approx \widetilde{F(X)}$, the isomorphism takes λ to the involution on $\widetilde{F(X)}$ over $F(X)$, and two objects on the left are tetragonally related iff the corresponding lines intersect.

(4) Any two objects in $\mathcal{P}^{-1}(A)$, generic $A \in \mathcal{A}_4$, are connected by a sequence of (two) tetragonal constructions.

Indeed, (1) is (5.2)(1); (2) and (4) imply the existence of $\chi : \mathcal{A}_4 \longrightarrow \mathcal{RC}^+$ such that $\kappa = \chi \circ \mathcal{P}$, while (3) shows that any two objects in a κ-fiber are also connected by a sequence of two tetragonal constructions, so χ must be birational, giving (5.2)(2). This gives an isomorphism $\mathcal{P}^{-1}(A) \approx \kappa^{-1}(X, \delta)$, so (5.3) follows.

(5.9) We let $\mathcal{R}^2\mathcal{Q}^+, \mathcal{R}^2\mathcal{Q}^-$ denote the moduli spaces of plane quintic curves Q together with:

• A rank-2, isotropic subgroup $W \subset J_2(Q)$, containing one odd and two even semiperiods, and

• a marked even (respectively odd) semiperiod in $W \backslash (0)$.

Exchanging the two even semiperiods gives an involution on $\mathcal{R}^2\mathcal{Q}^+$, with quotient $\mathcal{R}^2\mathcal{Q}^-$. The birational map

$$\alpha : \mathcal{M}_5 \overset{\sim}{\longrightarrow} \mathcal{RQ}^+,$$

of (4.3.7), lifts to a birational map

(5.9.1) $\qquad\qquad \mathcal{R}\alpha : \mathcal{R}_5 \overset{\sim}{\longrightarrow} \mathcal{R}^2\mathcal{Q}^+.$

From the construction of λ in (5.1.6) it follows that the involution on the right hand side corresponds to λ on the left, so we have a commutative diagram:

(5.9.2)

$$
\begin{array}{ccc}
\mathcal{R}_5 & \overset{\mathcal{R}\alpha}{\underset{\sim}{\longrightarrow}} & \mathcal{R}^2\mathcal{Q}^+ \\
\downarrow & & {\scriptstyle \pi}\downarrow{\scriptstyle 2:1} \\
\mathcal{R}_5/\lambda & \overset{\sim}{\longrightarrow} & \mathcal{R}^2\mathcal{Q}^-
\end{array}
$$

Start with $(C, \mu) \in \mathcal{R}_5$ and any g_4^1 on C. The trigonal construction produces a trigonal $Y \in \mathcal{M}_6$ with rank-2, isotropic subgroup W_Y. On Y we have a natural g_4^1, namely $w_Y \otimes L^{-2}$, where L is the trigonal bundle; so we bootstrap again, to a trigonal $T \in \mathcal{M}_7$ with rank-3 isotropic subgroup V. Applying construction (4.6) we obtain a diagram like (5.7), including an edge for (C, μ) and on it a vertex for $(Q, W_Q) := \pi \mathcal{R}\alpha(C, \mu)$. But then $\lambda(C, \mu)$ and $\kappa(C, \mu)$ also appear in the same diagram, as the two other edges (the line, respectively the circle) through Q! Statement (5.8.1) now follows, since any two edges of (5.7) which meet in a trigonal vertex are tetragonally related. (5.8.2) also follows, since any (C', μ') tetragonally related to (C, μ) will appear in the same diagram with (C, μ) (for the obvious initial choice of g_4^1 on C), so they have the same κ.

From the restriction to $\mathcal{R}Q^-$ of the Prym map we obtain, by base change:

$$
\begin{array}{ccc}
\mathcal{R}^2 Q^- & \longrightarrow & \mathcal{R}Q^- \\
{\scriptstyle \mathcal{R}P}\big\downarrow & & \big\downarrow{\scriptstyle P} \\
\mathcal{R}C^+ & \longrightarrow & C
\end{array}
$$

(5.9.3)

Combining with (5.8)(1),(2) and (5.9.2), we find that κ factors

$$
\begin{array}{ccc}
\mathcal{R}_5 & \xrightarrow[\sim]{\mathcal{R}\alpha} & \mathcal{R}^2 Q^+ \\
\big\downarrow & & \big\downarrow{\scriptstyle \pi} \\
\mathcal{R}_5/\lambda & \xrightarrow[\sim]{} & \mathcal{R}^2 Q^- \\
& & \big\downarrow{\scriptstyle \mathcal{R}P} \\
& & \mathcal{R}C^+
\end{array}
$$

(5.9.4)

We know $\mathcal{P}^{-1}(X)$ from (4.6), so by (5.9.3):

(5.9.5) $\mathcal{R}P^{-1}(X, \delta) \approx \mathcal{P}^{-1}(X) \approx F(X),$

and $\kappa^{-1}(X, \delta)$ is a double cover, which by the following lemma is identified with $\widetilde{F(X)}$. (The compatibility with λ follows from (5.9.4); line incidence in $F(X)$ corresponds by (4.6) to the tetragonal relation among the quintics, which by figure (5.7) corresponds, in turn, to the tetragonal relation in $\mathcal{R}_5$, so the proof of (5.8)(3) is complete.)

Lemma 5.10 The Albanese double cover $\widetilde{F(X)}$ determined by $\delta \in J_2(X)$ is isomorphic to $\pi^{-1}\mathcal{R}P^{-1}(X, \delta)$ (notation of (5.9.4)).

Proof.

The second isomorphism in (5.9.5) sends a line $\ell \in F(X)$ to the object $(\tilde{Q}_\ell, Q_\ell) \in \mathcal{P}^{-1}(X)$, where the curves $\tilde{Q}_\ell, Q_\ell$ parametrize ordered (respectively, unordered) pairs $\ell', \ell'' \in F(X)$ satisfying:

$$\ell + \ell' + \ell'' = 0 \qquad (\text{sum in} \quad J(X)).$$

We may of course think of $\tilde{Q}_\ell$ as sitting in $F(X)$, since ℓ' uniquely determines ℓ'': $\tilde{Q}_\ell$ is the closure in $F(X)$ of

(5.10.1) $\qquad\qquad\qquad \{\ell' \in F(X) \mid \ell' \cap \ell \neq \phi, \ell' \neq \ell\}.$

The corresponding object of $\mathcal{RP}^{-1}(X, \delta)$ is $(\widetilde{\tilde{Q}}_\ell, \tilde{Q}_\ell, Q_\ell)$, where $\widetilde{\tilde{Q}}_\ell$ is the inverse image in $\widetilde{F(X)}$ of $\tilde{Q}_\ell$ embedded in $F(X)$ via (5.10.1). Now to specify a point in $\pi^{-1}\mathcal{RP}^{-1}(X, \delta)$ we need, additionally, a double cover $\tilde{Q}'_\ell \to Q_\ell$ satisfying:

(5.10.2) $\qquad\qquad\qquad \tilde{Q}_\ell \times_{Q_\ell} \tilde{Q}'_\ell \approx \widetilde{\tilde{Q}}_\ell .$

We need to show that a choice of $\tilde{\ell} \in \widetilde{F(X)}$ over $\ell \in F(X)$ determines such a $\tilde{Q}'_\ell$. Recall that $\widetilde{F(X)} \to F(X)$ is obtained by base change, via the Albanese map, from the double cover $\widetilde{J(X)} \to J(X)$ determined by δ. $\tilde{Q}'_\ell$ can thus be taken to parametrize unordered pairs $\tilde{\ell}', \tilde{\ell}'' \in \widetilde{F(X)}$ satisfying:

$$\tilde{\ell} + \tilde{\ell}' + \tilde{\ell}'' = 0 \qquad (\text{sum in} \quad \widetilde{J(X)}) .$$

The fiber product in (5.10.2) then parametrizes such ordered pairs, so the required isomorphism to $\widetilde{\tilde{Q}}_\ell$ simply sends

$$(\tilde{\ell}', \tilde{\ell}'') \mapsto \tilde{\ell}'.$$

Q.E.D.

Finally, we prove (5.8)(4). Let $\overline{\mathcal{P}} : \overline{\mathcal{R}}_5 \to \mathcal{A}_4$ be the proper Prym map. By (5.8)(3) it factors

$$\overline{\mathcal{P}} = \iota \circ \kappa$$

where $\iota : \mathcal{RC}^+ \to \mathcal{A}_4$ is a rational map, which we are trying to show is birational. It suffices to find some $A \in \mathcal{A}_4$ such that:

(1) Any two objects in $\overline{\mathcal{P}}^{-1}(A)$ can be related by a sequence of tetragonal constructions.

(2) The differential $d\mathcal{P}$ is surjective over A.

In §5.4 we see that (1) holds for various examples, including generic Jacobians $\in \mathcal{J}_4$: for generic $C \in \mathcal{M}_4$, $\overline{\mathcal{P}}^{-1}(J(C))$ consists of Wirtinger

covers $\widetilde{C} \to C'$ (with normalilzation C) and of trigonals T, and the
two types are exchanged by λ. It is easier to check surjectivity of $d\mathcal{P}$
at the Wirtingers: by theorem (1.6), this amounts to showing that the
Prym-canonical curve $\Psi(X) \subset \mathbf{P}^3$ is contained in no quadrics. By [DS]
IV, Propo. 3.4.1, $\Psi(X)$ consists of the canonical curve $\Phi(C)$ together
with an (arbitrarily chosen) chord. Since $\Phi(C)$ is contained in a unique
quadric Q, which does not contain the generic chord, we are done.
[Another argument: it suffices to show that no one quadric contains
$\Psi(T)$ for all trigonal T in $\mathcal{P}^{-1}(J(C))$. By [DS], III 2.3 we have

$$\cup_T \Psi(T) \supset \Phi(C),$$

so the only possible quadric would be Q. Consider the g_4^1 on C given
by $\omega_C(\text{-}p\text{-}q)$, where $p, q \in C$ are such that the chord $\overline{\Phi(p), \Phi(q)}$ is not in
Q. Let T be the trigonal curve associated to $(C, \omega_C(\text{-}p\text{-}q))$, and choose
a plane $A \subset \mathbf{P}^3$ through $\Phi(p), \Phi(q)$, meeting Q and $\Phi(C)$ transversally,
say

$$A \cap \Phi(C) = \Phi(p + q + \sum_{i=1}^{4} x_i),$$

then by [DS],III 2.1, $\Psi(T)$ contains the point

$$\overline{\Phi(x_1), \Phi(x_2)} \cap \overline{\Phi(x_3), \Phi(x_4)}$$

which cannot be in Q.]

Q.E.D.

§5.4   Special fibers.

We want to illustrate the behavior of the Prym map over some
special loci in $\overline{\mathcal{A}}_4$. The common feature to all of these examples is that
the cubic threefold X given in Theorem (5.1) acquires a node. We thus
begin with a review of some results, mostly from [CG], on nodal
cubics.

(5.11) Nodal cubic threefolds

There in a natural correspondence between nodal cubic threefolds
$X \subset \mathbf{P}^4$ and nonhyperelliptic curves B of genus 4. Either object can
be described by a pair of homogeneous polynomials F_2, F_3, of degrees
2 and 3 respectively, in 4 variables $x_1, ..., x_4$: X has homogeneous
equation $0 = F_3 + x_0 F_2$ (in $\mathbf{P}^4$), and the canonical curve $\Phi(B)$ has
equations $F_2 = F_3 = 0$ in $\mathbf{P}^3$.

More geometrically, we express the Fano surface $F(X)$ in terms of B. Assume the two g_3^1's on B, $\mathcal{L}'$ and $\mathcal{L}''$, are distinct. They give maps

$$\tau', \ \tau'' : B \hookrightarrow S^2 B$$

sending $r \in B$ to $p+q$ if $p+q+r$ is a trigonal divisor in $|\mathcal{L}'|$, $|\mathcal{L}''|$ respectively. We then have the identification

(5.11.1) $$F(X) \approx S^2 B / (\tau'(B) \sim \tau''(B)).$$

Indeed, we have an embedding

$$\tau : B \hookrightarrow F(X),$$

identifying B with the family of lines through the node $n = (1,0,0,0,0)$. This gives a map $S^2 B \to F(X)$ sending a pair ℓ_1, ℓ_2 of lines through n to the residual intersection with X of the plane (ℓ_1, ℓ_2). this map identifies $\tau'(B)$ with $\tau''(B)$, and induces the isomorphism (5.11.1).

(5.11.2) A line $\ell \in F(X)$ determines a pair $(Q, \tilde{Q}) \in \overline{\mathcal{RQ}^-}$, which must be in $\partial^{II} \mathcal{RQ}^-$, i.e. for generic ℓ we obtain a nodal quintic Q with étale double cover $\tilde{Q}$. We can interpret (5.11.1) in terms of these nodal quintics: Start with a divisor $p+q \in S^2 B$. Then $\omega_B(-p-q)$ is a g_4^1 on B, so the trigonal construction produces a double cover $\tilde{T} \to T$, where $T \in \mathcal{M}_5$ comes with a trigonal bundle $\mathcal{L}$. The linear system $|\omega_T \otimes \mathcal{L}^{-1}|$ maps T to a plane quintic Q, with a single node given by the divisor $|\omega_T \otimes \mathcal{L}^{-2}|$ on T.

(5.11.3) In the special case that there exists $r \in B$ such that $p+q = \tau''(r)$, i.e. $p+q+r \in |\mathcal{L}''|$ is a trigonal divisor, our g_4^1 acquires a base point:

$$\omega_B(-p-q) \approx \mathcal{L}'(r).$$

As seen in (2.10.ii), the trigonal construction produces the nodal trigonal curve

$$T := B/(p' \sim q')$$

with its Wirtinger double cover $\tilde{T}$, where $p', q' \in B$ are determined by:

$$p' + q' + r \in |\mathcal{L}'|,$$

i.e. $p' + q' = \tau'(r)$. In this case, the quintic Q is the projection of $\Phi(B)$ from $\Phi(r)$, with 2 nodes $p \sim q$, $p' \sim q'$, and $\tilde{Q}$ is the reducible double cover with crossings over both nodes.

(5.12) Degenerations in $\mathcal{RC}^+$.

We fix our notation as in §5.1. Thus we have:

$$X \in \mathcal{C} \qquad (X, \delta) \in \mathcal{RC}^+$$
$$(Q, \sigma) \in \mathcal{RQ}^-, \qquad (Q, \nu), \; (Q, \nu\sigma) \in \mathcal{RQ}^+$$
$$(C, \mu), \; (C', \mu') \in \mathcal{R}_5$$
$$A \in \mathcal{A}_4$$

and these objects satisfy:

$$
\begin{aligned}
\mathcal{P}(Q, \sigma) &= J(X), & \nu, \nu\sigma &\mapsto \delta \\
\mathcal{P}(Q, \nu) &= J(C), & \sigma, \nu\sigma &\mapsto \mu \\
\mathcal{P}(Q, \nu\sigma) &= J(C'), & \nu, \; \sigma &\mapsto \mu'
\end{aligned}
$$

$$
\begin{aligned}
\mathcal{P}(C, \mu) &= \mathcal{P}(C', \mu') = A \\
\lambda(C, \mu) &= \quad (C', \mu') \\
\kappa(C, \mu) &= \quad (X, \delta).
\end{aligned}
$$

Now let X degenerate, acquiring a node, with $\bar{\varepsilon} \in J_2(X) \backslash (0)$ the vanishing cycle mod. 2. From (5.11) we see that Q also degenerates, with a vanishing cycle ε which maps (via. (1.4)) to $\bar{\varepsilon}$. Lemma (5.9) of [D4] shows that ε, hence also $\bar{\varepsilon}$, must be even.

There are 3 types of degenerations of (X, δ), distinguished as in (1.7) by the relationship of $\delta, \bar{\varepsilon}$. (A fourth type, where Q degenerates but X does not, is explained in (5.13).) The possibilities are summarized below:

(I) If $\bar{\varepsilon} = \delta$ then either $\varepsilon = \nu$ or $\varepsilon = \nu\sigma$, which gives the same picture with C, C' exchanged. In case $\varepsilon = \nu$, (Q, ν) undergoes a ∂^I degeneration, while $(Q, \nu\sigma)$ is ∂^{II}. (The notation is that of (1.7).) Thus A is a Jacobian.

The double cover $\widetilde{F(X)}$ is itself a ∂^I cover. In terms of the curve B of (5.11), we have

$$\widetilde{F(X)} = (S^2 B)_0 \amalg (S^2 B)_1 \; / \; (\tau'(B)_0 \sim \tau''(B)_1, \; \tau''(B)_0 \sim \tau'(B)_1).$$

This is clear, either from the definition of $\widetilde{F(X)}$ via the Albanese map, or by considering the restriction to $\mathcal{RP}^{-1}(X, \delta)$ of the double cover

$$\mathcal{R}^2 Q^+ \xrightarrow{\pi} \mathcal{R}^2 Q^-$$

of (5.9). One of the components parametrizes the trigonal objects (C, μ), the other parametrizes the nodals (C', μ').

(II) σ is always perpendicular to ε, ν, and the condition $\langle \bar{\varepsilon}, \delta \rangle = 0$ implies $\langle \varepsilon, \nu \rangle = 0$ by (1.4.3). Both (Q, ν) and $(Q, \nu\sigma)$ then give ∂^{II}-covers, so C, C' are nodal. Again by (1.4.3), both (C, μ) and (C', μ') are ∂^{II}, so their common Prym A is in $\partial \bar{A}_4$.

From the Albanese map we see that $\widetilde{F(X)}$ is an etale cover of $F(X)$. Indeed, δ comes from a semiperiod δ' on B, giving a double cover $\widetilde{B}$ with involution ι; the normalization of $\widetilde{F(X)}$ is then $S^2 \widetilde{B}/\iota$, and $\widetilde{F(X)}$ is obtained by glueing above $\tau(B)$.

(III) In this case both (Q, ν) and $(Q, \nu\sigma)$ are ∂^{III}, so C, C' are non-singular. The node of Q represents a quadric of rank 3 through $\Phi(C)$, so $\mathcal{L}$ is cut out by the unique ruling. By the Schottky-Jung relations [M2], the vanishing theta null on C descends to one on A.

The double cover $\widetilde{F(X)}$ is again a ∂^{III}-cover, in the sense that its normalization is ramified over $\tau'(B), \tau''(B)$, the sheets being glued. Each of the quintics in (5.11.2) gives two points of $\widetilde{F(X)}$, while the two-nodal quintics (5.11.3) land in the branch locus of π (5.9.4).

Degeneration type of (X, δ)	Degeneration type of $(Q, \sigma, \nu, \nu\sigma)$	(C, μ)	(C', μ')	A
I : $\bar{\varepsilon} = \delta$	$\varepsilon = \nu$	nonsingular trigonal	nodal, ∂^I	$\mathcal{J}_4$
II : $\bar{\varepsilon} \neq \delta$, $\langle \bar{\varepsilon}, \delta \rangle = 0$	$(\varepsilon, \sigma, \nu)$ rank 3 isotropic subgroup	nodal, ∂^{II}	nodal, ∂^{II}	$\partial \bar{A}_4$
III : $\langle \bar{\varepsilon}, \delta \rangle \neq 0$	$\langle \varepsilon, \sigma \rangle = 0$ $\langle \varepsilon, \nu \rangle \neq 0$	nonsingular, has vanishing thetanull $\mathcal{L}$, $\mathcal{L}(\mu)$ even	nonsingular, has vanishing thetanull $\mathcal{L}'$, $\mathcal{L}'(\mu')$ even	θ_{null}
IV : nonsingular	$\langle \varepsilon, \nu \rangle = 0$ $\langle \varepsilon, \sigma \rangle \neq 0$	nodal, ∂^{II}	nonsingular, has vanishing thetanull $\mathcal{L}'$, $\mathcal{L}'(\mu')$ odd	$\mathcal{A}_4$

(5.13) Degenerations in $\mathcal{R}^2 \mathcal{Q}^+$.

We have just described the universe as seen by a degenerating cubic threefold. From the point of view of a degenerating plane quintic, there are a few more possibilities though they lead to no new components. We retain the notation: $Q, \nu, \sigma, \varepsilon$ etc.

0. ε cannot equal σ, since ε is even, σ odd.

I. $\varepsilon = \nu$ reproduces case I of (5.12), as does:

I'. $\varepsilon = \nu\sigma$.

II. Excluding the above, ν, σ, ε generate a subgroup of rank 3. If it is isotropic, we are in case II above.

III. If $\langle \varepsilon, \sigma \rangle = 0$ but $\langle \varepsilon, \nu \rangle = \langle \varepsilon, \nu\sigma \rangle \neq 0$, we're in case III.

The only new cases are thus:

IV. $\langle \varepsilon, \nu \rangle = 0 \neq \langle \varepsilon, \sigma \rangle$,   or :

IV.' $\langle \varepsilon, \nu\sigma \rangle = 0 \neq \langle \varepsilon, \sigma \rangle$,

which is the same as IV after exchanging C, C'.

In case IV, we find:

• X is non-singular, in fact any X can arise. What is special is the line $\ell \in F(X)$ corresponding to Q : it is contained in a plane which is tangent to X along another line, ℓ'.

• (Q, ν) is a ∂^{II} degeneration, so C is nodal, and (C, μ) is a ∂^{II} degeneration.

• On the other hand, $(Q, \nu\sigma)$ is ∂^{III}, so C' is non-singular, and has a vanishing theta null $\mathcal{L}'$ (corresponding, as before, to the node of Q).

• This time though, $\mathcal{L}'(\mu')$ is odd, so $A \in \mathcal{A}_4$ does not inherit a vanishing theta null. In fact, any $A \in \mathcal{A}_4$ arises from a singular quintic with degeneration of type IV.

So far, we found three loci in $\bar{A}_4$ which are related to nodal cubics:

$$
\begin{array}{rcl}
\mathcal{P} \circ \kappa^{-1}(\partial^{\mathrm{I}} \mathcal{R}\mathcal{C}^+) & \subset & \mathcal{J}_4 \\
\bar{\mathcal{P}} \circ \kappa^{-1}(\partial^{\mathrm{II}} \mathcal{R}\mathcal{C}^+) & \subset & \partial \bar{A}_4 \\
\mathcal{P} \circ \kappa^{-1}(\partial^{\mathrm{III}} \mathcal{R}\mathcal{C}^+) & \subset & \theta_{\mathrm{null}}
\end{array}
$$

We are now going to study, one at a time, the fibers of $\mathcal{P}$ above generic points in these three loci. We note that related results have recently been obtained by Izadi. In a sense, her results are more precise: she knows (cf. Remark 5.4) that χ is an isomorphism on the open complement $\mathcal{U}$ of a certain 6-dimensional locus in $\mathcal{A}_4$. In [I] she shows that for $A \in \mathcal{U}$, $\chi(A)$ is singular if and only if

$$A \in \mathcal{J}_4 \cup \theta_{\text{null}}.$$

Her description of the cubic threefold corresponding to $A \in \mathcal{J}_4$ complements the one we give below. In general her techiques, based on Γ_{00}, are very different than our degeneration arguments.

(5.14) Jacobians

Theorem 5.14 Let $B \in \mathcal{M}_4$ be a general curve of genus 4, and let $(X, \delta) = \chi(J(B))$.

(1) X is the nodal cubic threefold corresponding to B (5.11).

(2) $(X, \delta) \in \partial^I$, so $\widetilde{F(X)}$ is reducible, each component is isomorphic to $S^2 B$.

(3) Let (Q, σ, ν) be the plane quintic with rank-2 isotropic subgroup corresponding to some $\ell \in F(X)$. Then Q is nodal, with trigonal normalization T, ν is the vanishing cycle, and $(Q, \sigma) = (Q, \nu\sigma) \in \partial^{II}$.

(4) $\bar{\mathcal{P}}^{-1}(J(B))$ is isomorphic to $\widetilde{F(X)}$. The component corresponding to ν (respectively $\nu\sigma$) consists of trigonal curves $T_{p,q}$ (respectively Wirtinger covers of singular curves $S_{p,q}$), $(p, q) \in S^2 B$.

(5) The tetragonal construction takes both $S_{p,q}$ and $T_{p,q}$ to $S_{r,s}$ and $T_{r,s}$ if and only if $p + q + r + s$ is a special divisor on B. The involution λ exchanges $S_{p,q}, T_{p,q}$.

(6) Any two objects in $\bar{\mathcal{P}}^{-1}(J(B))$ can be connected by a sequence of two tetragonal moves (generally, in 10 ways).

Proof

Since at least some of these results are needed for the proof of (5.8)(4), we do not use Theorem (5.3). For $(p, q) \in S^2 B$, we consider:

• $\tilde{T}_{p,q} \to T_{p,q}$, the trigonal double cover associated by the trigonal construction to B with the g^1_4 given by $\omega_B(\text{-}p\text{-}q)$.

• $\tilde{S}_{p,q} \to S_{p,q}$, the Wirtinger cover of $S_{p,q} := B/(p \sim q)$. (When $p = q$, this specializes to $B \cup_p R$, where R is a nodal rational curve in which p is a non singular point.)

These objects are clearly in $\bar{\mathcal{P}}^{-1}(J(B))$. Beauville's list ([B1], (4.10)) shows that they exhaust the fiber. This proves part (4). Now clearly κ, as defined in (5.1.6), takes any of these objects to our (X, δ); so the analysis in (5.12)(I) applies, proving (1)-(3). (Note: this already suffices to complete the proof of (5.8)(4)!)

Let $r + s + t + u$ be an $\overline{\text{arbitrary}}$ divisor in $|\omega_B(-p - q)|$. Projection of $\Phi(B)$ from the chord $\overline{\Phi(t), \Phi(u)}$ gives (the general) g_4^1 on $S_{p,q}$. The tetragonal construction takes this to the curves $T_{t,u}$ and $S_{r,s}$. (The situation is that of (2.15.2).)

On $T_{p,q}$ there are two types of g_4^1's, of the form $\mathcal{L}(x)$ and $\omega \otimes \mathcal{L}^{-1}(-x)$, where $\mathcal{L}$ is the trigonal bundle and $x \in T_{p,q}$. Now x corresponds to a $(2,2)$ partition, say $\{\{r, s\}, \{t, u\}\}$, of some divisor in $|\omega_B(-p - q)|$. The tetragonal construction, applied to $\mathcal{L}(x)$, yields the curves $S_{r,s}$ and $S_{t,u}$; while when applied to $\omega \otimes \mathcal{L}^{-1}(-x)$, it gives $T_{r,s}$ and $T_{t,u}$. Altogether, this proves (5). We conclude with:

Lemma 5.14.7 Given any $p, q, r, s \in B$, there are points $t, u \in B$ (in general, 5 such pairs) such that both $p + q + t + u, r + s + t + u$ are special.

Proof

Let α, β be the maps of degree 4 from B to $\mathbf{P}^1$ given by $|\omega_B(-p - q)|, |\omega_B(-r - s)|$. Then

$$\alpha \times \beta : B \to \mathbf{P}^1 \times \mathbf{P}^1$$

exhibits B as a curve of type $(4,4)$ on a non-singular quadric surface, hence the image has arithmetic genus $(4 - 1)^2 = 9 \rangle 4 = g(B)$, so there must be (in general, 5) singular points; these give the desired pairs (t, u).

$$\text{QED}$$

(5.15) The Boundary.

The results in this case were obtained by Clemens [C2]. A general point A of the boundary $\partial \bar{\mathcal{A}}_4$ of a toroidal compactification $\bar{\mathcal{A}}_4$ is a $\mathbf{C}^*$-extension of some $A_0 \in \mathcal{A}_3$. The extension data is given by a point a in the Kummer variety $A_0/(\pm 1)$.

Given $a \in A_0$, consider the curve

$$\tilde{B} = \tilde{B}_a := \Theta \cap \Theta_a \subset A_0$$

(where $x \in \Theta_a \Leftrightarrow x+a \in \Theta$), and its quotient $B = B_a$ by the involution $x \mapsto -a - x$. We have

$$(B, \tilde{B}) \in \mathcal{R}_4$$

and

$$\mathcal{P}(B, \tilde{B}) \approx A_0.$$

The pair $(B, \tilde{B})$ does not change (up to isomorphism) when a is replaced by $-a$.

Theorem 5.15 ([C2]) Let $A \in \partial \bar{\mathcal{A}}_4$ be the C^*-extension of $A_0 \in \mathcal{A}_3$, a generic $PPAV$, determined by $\pm a \in A_0$. Let $(X, \delta) = \chi(A)$.

(1) X is the nodal cubic threefold corresponding to $B = B_a$.

(2) $(X, \delta) \in \partial^{II}$, so $\widetilde{F(X)}$ is the etale double cover of $F(X)$ with normalization $S^2\tilde{B}/\iota$, as in (5.12.II).

(3) The corresponding quintics Q are nodal; all three of $\sigma, \nu, \nu\sigma$ are of type ∂^{II}.

(4) $\overline{\overline{\mathcal{P}}}^{-1}(A)$ is isomorphic to $\widetilde{F(X)}$, and consists of ∂^{II}-covers $(C, \tilde{C})$ whose normalizations (at one point) are of the form $(B_b, \tilde{B}_b)$ for $b = b_1 - b_2$, $b_1, b_2 \in \psi(\tilde{B})$.

Proof

Clearly $\overline{\overline{\mathcal{P}}}^{-1}(A) \subset \partial^{II}\overline{\overline{\mathcal{R}}}_5$, so consider a pair $(C, \tilde{C}) \in \partial^{II}$, say

$$C = N/(p \sim q), \quad \tilde{C} = \tilde{N}/(p' \sim q', p" \sim q")$$

with $(N, \tilde{N}) \in \bar{\mathcal{R}}_4$. Then $\overline{\overline{\mathcal{P}}}(C, \tilde{C})$ is a C^*-extension of $P(N, \tilde{N})$, with extension data

$$\pm(\psi(p') - \psi(q')) \in P(N, \tilde{N})/(\pm 1).$$

We see that $\overline{\overline{\mathcal{P}}}(C, \tilde{C}) = A$ if and only if

(5.15.5) $$(N, \tilde{N}) \in \bar{\mathcal{P}}^{-1}(A_0),$$

and:

(5.15.6) $$\psi(p') - \psi(q') = a, \quad p', q' \in \tilde{N}.$$

Now, (5.15.5) says that $(N, \tilde{N})$ is taken, by its Abel-Prym map ψ, to $(B_b, \tilde{B}_b)$ for some $b \in A_0$, and then (5.15.6) translates to:

$$a = a_1 - a_2, \qquad a_1, a_2 \in \Theta \cap \Theta_b$$

which is equivalent to

$$b = b_1 - b_2, \qquad b_1, b_2 \in \Theta \cap \Theta_a$$

(take $b_1 = a_2 + b$, $b_2 = a_2$). This proves (4), and everything else follows from what we have already seen.

$$\text{QED}$$

(5.16) Theta nulls

Let $A \in \mathcal{A}_4$ be a generic $PPAV$ with vanishing thetanull, and $(C, \tilde{C})$ a generic element of $\mathcal{P}^{-1}(A)$. By [B1], Proposition (7.3), C has a vanishing thetanull. This implies that the plane quintic Q parametrizing singular quadrics through $\Phi(C)$ has a node, corresponding to the thetanull. The corresponding cubic threefold X is thus also nodal, and we are again in the situation of (5.11.III). I do not see, however, a more direct way of describing the curve B (or the cubic X) in terms of A.

(5.17) Pentagons and wheels.

In [V], Varley exhibits a two dimensional family of double covers $(C, \tilde{C}) \in \mathcal{R}_5$ whose Prym is the unique non-hyperelliptic $PPAV$ $A \in \mathcal{A}_4$ with 10 vanishing thetanulls. The curves C involved are Humbert curves, and each of these comes with a distinguished double cover $\tilde{C}$. As an illustration of our technique, we work out the fiber of $\overline{\overline{\mathcal{P}}}$ over A and the tetragonal moves on this fiber. This is, of course, a very special case of (5.12)(III) or (5.16).

We recall the construction of Humbert curves and their double covers. Start by marking 5 points $p_1, \cdots, p_5 \in \mathbf{P}^1$. Take 5 copies L_i of $\mathbf{P}^1$, and let E_i be the double cover of L_i branched at the 4 points p_j, $j \neq i$. Let

(5.17.1) $$A := \coprod_{i=1}^5 L_i, \qquad B := \coprod_{i=1}^5 E_i.$$

The pentagonal construction applied to

(5.17.2) $$B \xrightarrow{g} A \xrightarrow{f} \mathbf{P}^1$$

(f is the forgetful map, of degree 5), yields a 32-sheeted branched cover $f_* B \to \mathbf{P}^1$ which splits, by (2.1.1), into 2 copies of the Humbert curve C, of degree 16 over $\mathbf{P}^1$.

Let β_I, $I \subset S := \{1, \cdots, 5\}$, be the involution of (5.17.2) which fixes A and acts non-trivially on E_i, $i \in I$. It induces an involution α_I on $f_* B$, hence on its quotient C. Let

$$G := \{\alpha_I \mid I \subset S\} / (\alpha_S).$$

Then C is Galois over $\mathbf{P}^1$, with group $G \approx (\mathbf{Z}/2\mathbf{Z})^4$. Let G_i, $1 \leq i \leq 5$, be the image in G of

$$\{\alpha_I | i \notin I, \ \#(I) = \text{ even}\}.$$

Then

$$C/\alpha_i \approx C/G_i \approx E_i,$$

and the quotient map

$$E_i \approx C/\alpha_i \to C/G_i \approx E_i$$

becomes multiplication by 2 on E_i. In particular, the Humbert curve C has 5 bielliptic maps $h_i : C \to E_i$. The branch locus of h_i consists of the 8 points $x \in E_i$ satisfying $g(2x) = p_i$.

For ease of notation, set $E := E_5$, $\quad p = p_5 \in \mathbf{P}^1$,

$$C \xrightarrow{h} E \xrightarrow{g} \mathbf{P}^1,$$

and

$$\{p^0, p^1\} := g^{-1}(p) \subset E.$$

Then for $j = 0, 1$, E has a natural double cover C^j, branched at the four points $\frac{1}{2}p^j$ and given by the line bundle $\mathcal{O}_E(2p^j)$. The fiber product

(5.17.3) $$\tilde{C} := C^0 \times_E C^1$$

gives a Cartesian double cover of C.

Replacing E_5 by another E_i, we get an isomorphic double cover $\tilde{C}$. Here is an invariant description of this cover:

Let $p_{i,j} := L_i \cap f^{-1}(p_j) \in A$, and consider the curve

$$Q := A/(p_{i,j} \sim p_{j,i}, \quad i \neq j).$$

Then Q can be embedded in $\mathbf{P}^2$ as a pentagon, or completely reducible plane quintic curve: embed $\mathbf{P}^1$ as a non-singular conic, and take L_i to be the tangent line of the conic at p_i. We have two natural branched double covers of Q:

(5.17.4) $$\tilde{Q}_\sigma := (\amalg_{i=1, \varepsilon=0}^{5 \quad 1} L_i^\varepsilon)/(p_{i,j}^0 \sim p_{j,i}^1, \quad i \neq j)$$

(5.17.5) $$\tilde{Q}_\nu := B/(\tilde{p}_{i,j} \sim \tilde{p}_{j,i}, \quad i \neq j),$$

where $\tilde{p}_{i,j} \in E_i$ is the unique (ramification) point above $p_{i,j} \in L_i$. We may think of $\tilde{Q}_\sigma$ as a "totally ∂^I" degeneration, and of $\tilde{Q}_\nu$ as a "totally ∂^{III}" degeneration. We then find:

(5.17.6) $(Q, \tilde{Q}_\nu) \in \overline{\mathcal{RQ}}^+$ is the quintic double cover corresponding to the Humbert curve $C \in \mathcal{M}_5$.

(5.17.7) The double cover $\tilde{Q}_\sigma$ of Q corresponds, via (1.4.2), to the double cover $\tilde{C}$ of C.

We note that $\tilde{Q}_\sigma$ is itself an odd cover, so it corresponds to some (singular) cubic threefold. A moment's reflection shows that this must be Segre's cubic threefold Y which we have already met in (4.8). Indeed, the Fano surface $F(Y)$ consists of the six rulings R_i, $0 \leq i \leq 5$, plus the 15 dual planes $\Pi^*_{i,j}$ of lines in $\Pi_{i,j}$ (notation of (4.8)). We see that:

(5.17.8) The discriminant of projection of Y from a line $\ell \in R_i$ is a plane pentagon Q, with its double cover $\tilde{Q}_\sigma$ as above.

The other covers, $\tilde{Q}_\sigma$, fit together to determine a point $(Y, \delta) \in \overline{\mathcal{RC}}^+$:

(5.17.9) $$(Y, \delta) = \kappa(C, \tilde{C}),$$

for any Humbert cover $(C, \tilde{C})$. The tetragonal construction takes any $(Q, \tilde{Q}_\sigma)$ to any other (in two steps), so we recover Varley's theorem:

(5.17.10) $A := \mathcal{P}(C, \tilde{C}) \in \mathcal{A}_4$ is independent of the Humbert cover $(C, \tilde{C})$.

But this is not the complete fiber: we have only used one of the two component types of $F(Y)$. We note:

(5.17.11) The discriminant of projection of Y from a line $\ell \subset \Pi_{ij}$ consists of a conic plus three lines meeting at a point; the double cover is split.

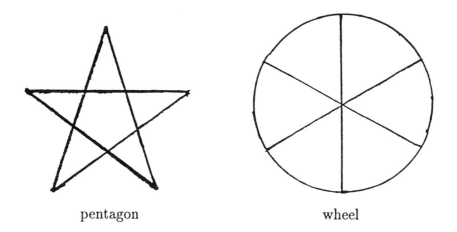

pentagon   wheel

Consider a tritangent plane, meeting Y in lines $\ell_i \in R_i$, $\ell_j \in R_j$, and $\ell_{ij} \in \Pi^*_{ij}$. It corresponds to a tetragonal construction involving two pentagons and a wheel. The other kind of tritangent plane intersects

Y in lines $\ell_{ij} \in \Pi^*_{ij}$, $\ell_{kl} \in \Pi^*_{kl}$, $\ell_{mn} \in \Pi^*_{mn}$, where $\{i, j, k, l, m, n\} = \{0, 1, 2, 3, 4, 5\}$; the tetragonal construction then relates three wheels.

Theorem 5.18 Let $A \in \mathcal{A}_4$ be the non-hyperelliptic *PPAV* with 10 vanishing thetanulls.

(1) $\chi(A)$ consists of the Segre cubic threefold Y, with its degenerate semi-period δ (5.17.9).

(2) The corresponding curve $B \in \bar{\mathcal{M}}_4$ (5.11) consists of six $\mathbf{P}^1$'s:

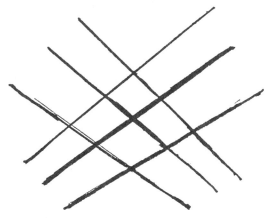

(3) The Fano surface $F(Y)$ consists of the 6 rulings R_i $(0 \le i \le 5)$ and the 15 dual planes $\Pi^*_{i,j}$. The plane quintics are pentagons, for $\ell \in R_i$, and wheels, for $\ell \in \Pi^*_{ij}$, all with split covers σ (5.17.4, 5.17.11). (The ν covers are branched over all the double points.)

(4) The fiber $\bar{\bar{\mathcal{P}}}^{-1}(A)$ is contained in the fixed locus of the involution $\lambda : \bar{\bar{\mathcal{R}}}_5 \to \bar{\bar{\mathcal{R}}}_5$ (5.1.6), so it is a quotient of $F(Y)$.

(5) $\bar{\bar{\mathcal{P}}}^{-1}(A)$ consists of two components:

- Humbert double covers $\tilde{C} \to C$ (5.17.3).
- Allowable covers $\widetilde{X_0} \cup \widetilde{X_1} \to X_0 \cup X_1$, where X_0, X_1 are elliptic, meeting at their 4 points of order 2.

All of this follows from our previous analysis, except (5). The new, allowable, covers are obtained by applying Corollary (3.7), with $n = 3$, to the Cartesian cover $\tilde{C} \to C$ in (5.17.3). It is also easy to see that the plane quintic parametrizing singular quadrics through the canonical curve $\Phi(X_0 \cup X_1)$ is a wheel, and vice versa, that the generalized Prym of any wheel (with its ∂^{III}-cover) is the generalized Jacobian $J(X_0 \cup X_1)$ of such a curve. Thus every line in $F(Y)$ is accounted for, so we have the complete fiber $\bar{\bar{\mathcal{P}}}^{-1}(A)$.

QED

§6   Other genera

For $g \leq 4$, it is relatively easy to describe the fibers of $\mathcal{P} : \bar{\mathcal{R}}_g \to \mathcal{A}_{g-1}$. Indeed, every curve in $\mathcal{M}_g$ is trigonal, and every $A \in \mathcal{A}_{g-1}$ is a Jacobian (of a possibly reducible curve), so the situation is completely controlled by Recillas' trigonal construction. Similar results can be obtained, for $g \leq 3$, by using Masiewicki's criterion [Ma].

(6.1) $\underline{g = 1}$. Here $\bar{\mathcal{P}}$ sends $\bar{\mathcal{R}}_1 \approx \mathbf{P}^1$ to $\mathcal{A}_0$ (= a point). The of $\bar{\mathcal{P}}, \mathcal{P}$ are then $\mathbf{P}^1, \mathbf{C}^*$ respectively.

(6.2) $\underline{g = 2}$. All curves of genus 2 are hyperelliptic, and all covers are Cartesian (3.2). An element of $\mathcal{R}_2$ is thus given by 6 points in $\mathbf{P}^1$, with 4 of them marked, modulo $PGL(2)$; an element E of $\mathcal{A}_1$ is given by 4 points of $\mathbf{P}^1$ modulo $PGL(2)$; and $\mathcal{P}$ forgets the 2 unmarked points. The fiber of $\mathcal{P}$ is thus rational; it can be described as S/G where

$$S := S^2 \left(\mathbf{P}^1 \setminus (4 \text{ points}) \right) \setminus (\text{diagonal})$$

and $G \approx (\mathbf{Z}/2\mathbf{Z})^2$ is the Klein group, whose action on S is induced from its action on $\mathbf{P}^1$ permuting the 4 marked points.

We note that S is $\mathbf{P}^2$ minus a conic C and four lines L_i tangent to it. To compactify it we add:

- a ∂^{I} cover for each point of $C \setminus \cup L_i$,

- a ∂^{III} cover for each point of $L_i \setminus C$, and

- an "elliptic tail" cover [DS, IV 1.3] for each point in the exceptional divisor obtained by blowing up one of the points $L_i \cap C$. (The limiting double cover obtained is

$$(E_0 \amalg E_1)/ \approx \ \longrightarrow \ E/\sim$$

where $\sim$ places a cusp at one of the four marked points p_i on E and $\approx$ places a tacnode above it. These curves are unstable, and the family of elliptic-tail covers gives their stable models, each elliptic tail being blown down to the cusp.)

The resulting $\overline{S}$ is $\mathbf{P}^2$ with 4 points in general position blown up, and the compactified fiber is $\overline{S}/G$, or $\mathbf{P}^2/G$ with one point blown up.

(6.3) $\underline{g = 3}$. Fix $A \in \mathcal{A}_2$. The Abel-Prym map sends pairs $(C, \tilde{C}) \in \mathcal{P}^{-1}(A)$ to curves $\psi(\tilde{C})$ in the linear system $|2\Theta|$ on A, uniquely defined modulo translation by the group $G = A_2 \approx (\mathbf{Z}/2\mathbf{Z})^4$. The fiber is therefore, birationally, the quotient $\mathbf{P}^3/G$. Since some curves in $|2\Theta|$

are not stable, some blowing up is required to obtain the biregular model of $\bar{\mathcal{P}}^{-1}(A)$. This is carried out in [Ve]. The quotient $\mathbf{P}^3/G$ is identified with Siegel's modular threefold, or the minimal compactification $\bar{\mathcal{A}}_2^{(2)}$ of the moduli space of $PPAV$'s with level-2 structure. To obtain $\bar{\mathcal{P}}^{-1}(A)$, Verra shows that we need to blow $\bar{\mathcal{A}}_2^{(2)}$ up at a point A', corresponding to a level-2 structure on A itself, and along a rational curve. The 2 exceptional divisors then parametrize hyperelliptic and elliptic-tail covers, respectively.

(6.4) $\underline{g = 4}$.

As we noted in (5.15), the fiber $\mathcal{P}^{-1}(A)$, $A \in \mathcal{A}_3$, consists of covers $(B_a, \tilde{B}_a)$, $a \in A/(\pm 1)$:

$$\tilde{B}_a = \Theta \cap \Theta_a, \quad B_a = \tilde{B}_a/(x \sim (-a - x)).$$

The fiber is thus (birationally) the Kummer variety $A/(\pm 1)$.

(6.5) $\underline{g \geq 7}$.

In this case, it was proved in [FS], [K], and [W], that $\mathcal{P}$ is generically injective. The results in §3 show that it is never injective: on the hyperelliptic loci there are positive-dimensional fibers, and various coincidences occur on the bielliptic loci. In [D1] we conjectured:

Conjecture 6.5.1 Any two objects in a fiber of $\mathcal{P}$ are connected by a sequence of tetragonal constructions.

We state this for $\mathcal{P}$, rather than $\overline{\mathcal{P}}$, since various other phenomena can contribute to non-trivial fibers at the boundary. For example, all fibers of $\overline{\mathcal{P}}$ on ∂^I are two-dimensional. On the other hand, from the local pictures (2.14) it is clear that the tetragonal construction can take a nonsingular curve to a singular one. In fact proposition (3.8) shows that it is possible for two objects in $\mathcal{R}_g$ to be tetragonally related through an intermediate object of $\partial\mathcal{R}_g$, so some care must be taken in clarifying which class of tetragonal covers should be allowed. The conjecture is consistent with our results for $g \leq 6$. For $g \geq 13$, Debarre [Deb2] proved it for curves which are neither hyperelliptic, trigonal, or bielliptic. Naranjo [N] extended this to generic bielliptics, $g \geq 10$. The following result was communicated to me by Radionov:

Theorem 6.5.2 [Ra] For $g \geq 7, \mathcal{R}_g^{\text{Tet}}$ is an irreducible component of the noninjectivity locus of the Prym map, and for generic $(C, \tilde{C}) \in \mathcal{R}_g^{\text{Tet}}$, $\mathcal{P}^{-1}(\mathcal{P}(C, \tilde{C}))$ consists precisely of three tetragonally related objects.

REFERENCES

[ACGH] E. Arbarello, M . Cornalba, P. Griffiths, J. Harris, *Geometry of algebraic curves*, Vol. I, Springer-Verlag, New York (1985).

[B1] A. Beauville, *Prym varieties and the Schottky problem*, Inv. Math. 41 (1977), 149-196.

[B2] A. Beauville, *Sous-variétés spéciales des variétés de Prym*, Compos. Math. 45, 357-383 (1982).

[C1] H. Clemens, *Double Solids*, Advances in Math. 47 (1983) pp. 107-230.

[C2] H. Clemens, *The fiber of the Prym map and the period map for double solids, as given by Ron Donagi*, U. of Utah, preprint.

[CG] H. Clemens, P. Griffiths, *The intermediate Jacobian of the cubic threefold*, Ann. Math. 95 (1972), 281-356.

[D1] R. Donagi, *The tetragonal construction*, AMS Bull. 4 (1981), 181-185.

[D2] R. Donagi, *The unirationality of $\mathcal{A}_5$*, Ann. Math. 119 (1984), 269-307.

[D3] R. Donagi, *Big Schottky*, Inv. Math. 89 (1987), 569-599.

[D4] R. Donagi, *Non-Jacobians in the Schottky loci*, Ann. of Math. 126 (1987), 193-217.

[D5] R. Donagi, *The Schottky problem, in: Theory of Moduli*, LNM 1337, Springer-Verlag (1988), 84-137.

[D6] R. Donagi, *On the period map for Clemens' double solids*, preprint.

[DS] R. Donagi, R. Smith, *The structure of the Prym map*, Acta Math. 146 (1981) 25-102.

[Deb1] O. Debarre, *Variétés de Prym, conjecture de la trisécante et ensembles d'Andreotti et Mayer*, Univ. Paris Sud,Thesis, Orsay (1987).

[Deb2] O. Debarre, *Sur les variétés de Prym des courbes tétragonales*, Ann. Sci. E.N.S. 21 (1988), 545-559.

[Dem] M. Demazure, *Seminaire sur les singularités des surfaces*. LNM 777 , Springer-Verlag (1980), 23-69.

[FS] R. Friedman, R. Smith, *The generic Torelli Theorem for the Prym map*, Inv. Math. 67 (1982), 473-490.

[vG]   B. van Geemen, *Siegel modular forms vanishing on the moduli space of curves*, Inv. Math. 78 (1984), 329-349.

[vGvdG]   B. van Geemen, G. van der Geer, *Kummer varieties and the moduli space of curves*, Am. J. of Math. 108 (1986), 615-642.

[vGP]   B. van Geemen, E. Previato, *Prym varieties and the Verlinde formula*, MSRI preprint, May 1991.

[I]   E. Izadi, *On the moduli space of four dimensional principally polarized abelian varieties*, Univ. of Utah Thesis, June 1991.

[K]   V. Kanev, *The global Torelli theorem for Prym varieties at a generic point*, Math. USSR Izvestija 20 (1983), 235-258.

[M1]   D. Mumford, *Theta characteristics on an algebraic curve*, Ann. Sci. E.N.S. 4 (1971), 181-192.

[M2]   D. Mumford, *Prym varieties I. Contributions to Analysis*, 325-350, New York, Acad. Press, 1974.

[Ma]   L. Masiewicki, *Universal properties of Prym varieties with an application to algebraic curves of genus five*, Trans. Amer. Math. Soc. 222 (1976), 221-240.

[N]   J. C. Naranjo, *Prym varieties of bi-elliptic curves*, Univ. de Barcelona preprint no. 65, June 1989.

[P]   S. Pantazis, *Prym varieties and the geodesic flow on $SO(n)$*, Math. Ann. 273 (1986), 297-315.

[R]   S. Recillas, *Jacobians of curves with a g_4^1 are Prym varieties of trigonal curves*, Bol. Soc. Math. Mexicana 19 (1974), 9-13.

[Ra]   D. Radionov, *letter*.

[SR]   J.G. Semple, L. Roth, *Introduction to Algebraic Geometry*, Oxford U. Press, 1949.

[SV]   R. Smith, R. Varley, *Components of the locus of singular theta divisors of genus 5*, LNM 1124, Springer-Verlag (1983), 338-416.

[T]   A. Tjurin, *Five lectures on three dimensional varieties*, Russ. Math. Surv. 27 (1972).

[V]   R. Varley, *Weddle's surfaces, Humbert's curves, and a certain 4-dimensional abelian variety*, Amer. J. Math. 108 (1986), 931-952.

[Ve]   A. Verra, *The fibre of the Prym map in genus three*, Math. Ann. 276 (1987), 433-448.

[W]   G. Welters, *Recovering the curve data from a general Prym variety*, Amer. J. of Math 109 (1987), 165-182.

Contemporary Mathematics
Volume **136**, 1992

Some Riemann Surfaces Whose Jacobians Have Strange Product Structures

CLIFFORD J. EARLE

ABSTRACT. We prove the existence of compact Riemann surfaces whose Jacobians have a factor that does not admit a principal polarization.

1. The theorem

Throughout this paper, X will denote a compact Riemann surface of genus $g \geq 1$, and $J(X)$ will denote its Jacobi variety. Many examples are known of surfaces X such that $J(X)$ is isomorphic (as a complex torus) to the product of two lower dimensional tori. (See for example [**1**], [**2**], [**7**], and [**8**].) During the 1990 Joint Summer Research Conference on Schottky problems it was asked whether these lower dimensional tori will necesarily be Jacobians themselves, or at least will admit principal polarizations. The answer is emphatically no. We shall prove this

THEOREM. *There are compact Riemann surfaces X of genus three with the following two properties:*

(1.1) $J(X)$ *admits isomorphisms onto the product of complex tori M_1 and M_2 of dimension one and two respectively, but*

(1.2) *for any such isomorphism the torus M_2 admits no principal polarization.*

Our proof relies on fundamental results of Baker [**1**], Martens [**5**], and Rauch and Lewittes [**6**]. According to Baker the Jacobian $J(X_0)$ of the Klein-Hurwitz surface of genus three with 168 automorphisms is isomorphic to a product $M_1 \times M_2$, and the work of Rauch and Lewittes leads to explicit such isomorphisms (see [**3**]). Martens observed that the canonical polarization of $J(X_0)$

1991 *Mathematics Subject Classification.* Primary 14H40, 32G20..

This work was partly supported by NSF Grant DMS-8901729.

This paper is in final form and no version of it will be submitted for publication elsewhere.

induces a nonproduct polarization on $M_1 \times M_2$ and that any nonproduct polarization imposes interesting restrictions on M_1 and M_2. It also induces polarizations on M_1 and M_2, and in the case of $J(X_0)$ the induced polarization on M_2 is not principal.

$J(X_0)$ does not satisfy condition (1.2) of the theorem, since M_2 in this case is the product of two one-dimensional tori. The idea of our proof is to perturb M_1 and M_2, retaining the nonproduct polarization on $M_1 \times M_2$, until we obtain the required Jacobians $J(X)$. Unfortunately the argument, although quite elementary, is both intricate and very computational.

For the reader's benefit we have split the proof into 6 lemmas. We also provide in Section 2 a brief summary of the facts we need to know about tori and principal polarizations. In Section 3 we state the first three lemmas and derive the theorem from them. Lemmas 1 and 2 have short proofs, presented in Sections 4 and 5. The proof of Lemma 3 requires additional lemmas. These are stated in Section 6, used in Section 7 to prove Lemma 3, and proved in the remainder of the paper. The reader who wants to avoid the most tedious details can read Sections 2 through 7 and take a look at Section 8 to see the simple idea used to prove Lemmas 4 and 5.

It is quite possible that these "strange" factorizations of $J(X)$ are generic among all splittings of Jacobians. Our ad hoc methods shed little light on the general case.

2. Principal polarizations

We begin with some standard facts and definitions (see for instance Gunning [4] and Martens [5]). Let Λ be an $n \times 2n$ complex matrix whose columns are linearly independent over $\mathbb{R}$, and let L be the lattice subgroup of $\mathbb{C}^n$ that they generate. We say Λ is a *period matrix* for the complex torus M if M is isomorphic (biholomorphically equivalent) to $\mathbb{C}^n/L$.

We shall always denote the transpose of a matrix A by ^tA. If Λ is a period matrix for M, a *principal polarization* of M is determined by a skew-symmetric matrix P in $SL(2n, \mathbb{Z})$ such that

$$(2.1) \qquad\qquad\qquad \Lambda P\,^t\Lambda = 0$$

and

$$(2.2) \qquad \text{the Hermitian matrix } \ i\,\Lambda P\,^t\overline{\Lambda} \ \text{ is positive definite.}$$

When (2.1) and (2.2) hold we shall say that P *polarizes* M or Λ.

Two complex tori M and M' with respective period matrices Λ and Λ' are isomorphic if and only if there are matrices C in $GL(n, \mathbb{C})$ and T in $GL(2n, \mathbb{Z})$ such that $C\Lambda = \Lambda'T$. Under these circumstances, if P polarizes M then $TP\,^tT$ polarizes M'.

If Λ_1 and Λ_2 are period matrices for M_1 and M_2 respectively, then the direct sum

$$(2.3) \qquad \Lambda = \begin{pmatrix} \Lambda_1 & 0 \\ 0 & \Lambda_2 \end{pmatrix}$$

is a period matrix for $M_1 \times M_2$.

3. The main lemmas

In this section we shall state our three main lemmas and derive the theorem from them.

LEMMA 1. *Let* $\mathcal{U}$ *be the family of three-dimensional complex tori* $M = M_1 \times M_2$ *such that* M_1 *and* M_2 *have the respective period matrices*

$$(3.1) \qquad \Lambda_1 = (1, \tau), \qquad \Lambda_2 = \begin{pmatrix} 1 & 0 & \alpha & \beta \\ 0 & 1/3 & \beta & \gamma \end{pmatrix}$$

satisfying

$$(3.2) \qquad \tau = \beta, \quad 3\gamma = 5\beta, \quad and \quad \mathrm{Im}(5\alpha) > \mathrm{Im}(3\beta) > 0.$$

Each M *in* $\mathcal{U}$ *is principally polarized by the matrix*

$$(3.3) \qquad P_0 = \begin{pmatrix} 0 & 2 & 0 & 0 & 0 & 1 \\ -2 & 0 & -1 & -5 & 0 & 0 \\ 0 & 1 & 0 & 0 & 1 & 0 \\ 0 & 5 & 0 & 0 & 0 & 3 \\ 0 & 0 & -1 & 0 & 0 & 0 \\ -1 & 0 & 0 & -3 & 0 & 0 \end{pmatrix}.$$

Of course we use the period matrix (2.3) for M. It is clear from (3.2) that the family $\mathcal{U}$ is a two-dimensional manifold and that α and β provide a global coordinate system.

LEMMA 2. *There is a nonempty open set* $\mathcal{U}_0 \subset \mathcal{U}$ *such that each* M *in* $\mathcal{U}_0$ *is isomorphic to* $J(X)$ *for some compact Riemann surface* X *of genus three.*

LEMMA 3. *There is a dense set* $\mathcal{U}' \subset \mathcal{U}$ *such that if* $M \in \mathcal{U}'$ *and* M *is isomorphic to a product* $M_1' \times M_2'$, *then the two-dimensional factor* M_2' *admits no principal polarization.*

These lemmas obviously imply the theorem. By Lemmas 2 and 3, the set $\mathcal{U}_0 \cap \mathcal{U}'$ is not empty. If $M \in \mathcal{U}_0 \cap \mathcal{U}'$, then $M \cong J(X)$ by definition of $\mathcal{U}_0$, M has property (1.1) by definition of $\mathcal{U}$, and M satisfies (1.2) by definition of $\mathcal{U}'$.

4. Proof of Lemma 1

Let P_0 be given by (3.3) and Λ by (2.3), with Λ_1 and Λ_2 satisfying (3.1) and (3.2). It is easy to verify that $P_0 \in SL(6, \mathbb{Z})$ and that (2.1) holds. To verify (2.2) we compute the Hermitian matrix

$$i\Lambda P_0{}^t\overline{\Lambda} = 2 \begin{pmatrix} 2\operatorname{Im}(\beta) & \operatorname{Im}(\beta) & \dfrac{5}{3}\operatorname{Im}(\beta) \\[2mm] \operatorname{Im}(\beta) & \operatorname{Im}(\alpha) & \operatorname{Im}(\beta) \\[2mm] \dfrac{5}{3}\operatorname{Im}(\beta) & \operatorname{Im}(\beta) & \dfrac{5}{3}\operatorname{Im}(\beta) \end{pmatrix}.$$

That matrix is positive definite whenever $\operatorname{Im}(5\alpha) > \operatorname{Im}(3\beta) > 0$ since the matrix

$$\begin{pmatrix} 2 & 1 & \dfrac{5}{3} \\[2mm] 1 & \dfrac{3}{5} & 1 \\[2mm] \dfrac{5}{3} & 1 & \dfrac{5}{3} \end{pmatrix}$$

is positive semi-definite.   Q.E.D.

5. Proof of Lemma 2

For $n \geq 1$, let I_n be the $n \times n$ identity matrix and let J_n be the $2n \times 2n$ skew symmetric matrix

$$(5.1) \qquad\qquad J_n = \begin{pmatrix} 0 & I_n \\ -I_n & 0 \end{pmatrix}.$$

Let X be a compact Riemann surface of genus $n \geq 1$. Any canonical homology basis on X determines an $n \times n$ Riemann period matrix Z in the standard way (see for instance [3]). The matrix $\Lambda = (I_n, Z)$ is a period matrix for $J(X)$, and the matrix J_n defines a canonical principal polarization on $J(X)$. All this is well known.

In particular let X_0 be the Klein-Hurwitz surface of genus three with 168 automorphisms. In [6] Rauch and Lewittes chose a canonical homology basis for X_0 and found that the corresponding Riemann period matrix is

$$Z_0 = \frac{1}{2} \begin{pmatrix} 3\lambda - 1 & -2\lambda & \lambda - 1 \\ -2\lambda & 4\lambda & -2\lambda \\ \lambda - 1 & -2\lambda & 3\lambda + 1 \end{pmatrix},$$

with

$$(5.2) \qquad\qquad \lambda = (1 + i\sqrt{7})\,/4 \quad \text{and} \quad 2\lambda^2 = \lambda - 1.$$

Thus $\Lambda_0 = (I_3, Z_0)$ is a period matrix for $J(X_0)$.

Now let M_1' and M_2' be the complex tori defined by the period matrices

$$\Lambda_1' = \left(1, \frac{1}{4}(2\lambda + 1)\right) \quad \text{and} \quad \Lambda_2' = \begin{pmatrix} 1 & 0 & \frac{1}{4}(2\lambda - 9) & \frac{1}{4}(2\lambda + 1) \\ 0 & \frac{1}{3} & \frac{1}{4}(2\lambda + 1) & \frac{5}{12}(2\lambda + 1) \end{pmatrix}.$$

It is clear from (3.2) and (5.2) that $M' = M_1' \times M_2'$ belongs to the family $\mathcal{U}$ defined in Lemma 1. Recall that

$$\Lambda' = \begin{pmatrix} \Lambda_1' & 0 \\ 0 & \Lambda_2' \end{pmatrix}$$

is a period matrix for M'. Put

$$C_0 = \frac{1}{12} \begin{pmatrix} -6\lambda + 9 & 0 & 6\lambda - 9 \\ 6 & 0 & 6\lambda - 3 \\ -4\lambda + 10 & 4 & 6\lambda - 5 \end{pmatrix}$$

and

$$T_0 = \begin{pmatrix} 1 & 0 & -1 & 0 & 0 & -1 \\ -1 & 0 & 1 & 1 & 0 & 0 \\ 3 & 0 & 2 & 2 & -7 & 4 \\ 3 & 1 & -2 & -1 & 0 & -3 \\ 1 & 0 & 1 & 1 & -3 & 2 \\ -1 & 0 & 0 & 0 & 2 & -1 \end{pmatrix}.$$

Direct calculation gives

$$(5.3) \qquad\qquad T_0 J_3 {}^t T_0 = P_0,$$

where $P_0 \in SL(6, \mathbb{Z})$ is defined by (3.3). This implies that $T_0 \in GL(6, \mathbb{Z})$. Another computation (using (5.2)) gives

$$(5.4) \qquad\qquad C_0 \Lambda_0 = \Lambda' T_0.$$

Equations (5.3) and (5.4) imply that M', with the principal polarization defined by P_0, is isomorphic to $J(X_0)$, with its canonical polarization. Since the canonically polarized Jacobians of genus three are an open subset of the space of all principally polarized complex tori of dimension three (see for instance [**3**]), an open neighborhood $\mathcal{U}_0$ of M' in $\mathcal{U}$ will satisfy the requirements of Lemma 2.

REMARK. This entire paper hinges on the fact that the period matrix Λ_0 of the Klein-Hurwitz surface satisfies equations (5.3) and (5.4). These equations are easy to verify directly, once they have been derived in the first place. To obtain them we began with equation (6.5) of [**3**], which describes an isomorphism between $J(X_0)$ and a product $M_1 \times M_2$. The canonical polarization of $J(X_0)$

induces a nonproduct polarization on $M_1 \times M_2$ (see [5]). The polarizing matrix P_0 has the form

$$P_0 = \begin{pmatrix} P_1 & -{}^tQ \\ Q & P_2 \end{pmatrix},$$

where P_1 and P_2 are skew-symmetric integer matrices of size 2×2 and 4×4 respectively. We are interested in the set $\mathcal{U}$ of all products $M_1 \times M_2$ that are polarized by P_0. By changing bases in $\mathbb{Z}^2$ and $\mathbb{Z}^4$ we can put P_2 into canonical form and modify Q until the set $\mathcal{U}$ has a reasonably simple description. We stopped this process upon reaching equations (5.3) and (5.4), since the defining conditions (3.2) for $\mathcal{U}$ seem about as nice as possible.

6. Three more lemmas

It remains to prove Lemma 3. Our proof will require three more lemmas. The first two impose restrictions on the principal polarizations and automorphisms of generic members of the family $\mathcal{U}$.

LEMMA 4. *Let P in $SL(6, \mathbb{Z})$ be a skew-symmetric matrix that satisfies condition (2.1) for all Λ in some nonempty open subset of $\mathcal{U}$. Then*

$$(6.1) \qquad P = \begin{pmatrix} kJ_1 & -\ell\, {}^tQ \\ \ell Q & mP' + nP'' \end{pmatrix},$$

where k, ℓ, m, and n are integers, J_1 is given by (5.1),

$$(6.2) \quad Q = \begin{pmatrix} 0 & 1 \\ 0 & 5 \\ 0 & 0 \\ -1 & 0 \end{pmatrix}, \quad P' = \begin{pmatrix} 0 & 0 & 1 & 0 \\ 0 & 0 & 0 & 3 \\ -1 & 0 & 0 & 0 \\ 0 & -3 & 0 & 0 \end{pmatrix}, \quad P'' = QJ_1\, {}^tQ.$$

and

$$(6.3) \qquad m^2 \left[k(3m + 5n) - 5\ell^2 \right]^2 = 1.$$

LEMMA 5. *Let T in $SL(6, \mathbb{Z})$ satisfy $T^2 = I_6$ but not $T = \pm I_6$. Suppose that for each Λ in some nonempty open subset of $\mathcal{U}$ there is C in $g\ell(3, \mathbb{C})$ such that $C\Lambda = \Lambda T$. Then*

$$(6.4) \qquad T = \begin{pmatrix} a & 0 & 0 & b & 0 & 0 \\ 0 & a & 0 & 0 & 3b & 5b \\ c & 0 & d & e & 0 & 0 \\ 5c & 0 & 0 & -a & 0 & 0 \\ 0 & 0 & 0 & 0 & d & 0 \\ 0 & c & 0 & 0 & 3e & -a \end{pmatrix},$$

where a, b, c, d and e are integers satisfying

$$(6.5) \qquad d^2 = a^2 + 5bc = 1 \quad \text{and} \quad a + d + 5e = 0.$$

One such matrix is

$$(6.6) \qquad T_1 = \begin{pmatrix} -1 & 0 & 0 & 0 & 0 & 0 \\ 0 & -1 & 0 & 0 & 0 & 0 \\ 0 & 0 & 1 & 0 & 0 & 0 \\ 0 & 0 & 0 & 1 & 0 & 0 \\ 0 & 0 & 0 & 0 & 1 & 0 \\ 0 & 0 & 0 & 0 & 0 & 1 \end{pmatrix} = \begin{pmatrix} -I_2 & 0 \\ 0 & I_4 \end{pmatrix}.$$

LEMMA 6. *Suppose that $S \in GL(6, \mathbb{Z})$ and that $T = ST_1S^{-1}$ satisfies the assumptions of Lemma 5. Then*

$$(6.7) \qquad S = \begin{pmatrix} xI_2 & B \\ y\,QJ_1 & D \end{pmatrix} \begin{pmatrix} A & 0 \\ 0 & I_4 \end{pmatrix},$$

where x and y are integers and $A \in GL(2, \mathbb{Z})$. In addition there is an integer z such that $B \equiv zUD \pmod 5$, where

$$(6.8) \qquad U = \begin{pmatrix} 0 & 1 & 0 & 0 \\ 0 & 0 & 3 & 5 \end{pmatrix}.$$

7. Proof of Lemma 3

We shall deduce Lemma 3 from Lemmas 4 and 6. Let $\mathcal{O}$ be a nonempty open subset of $\mathcal{U}$. Suppose that for each M in $\mathcal{O}$ there are principally polarized tori M_1' and M_2' (of respective dimensions one and two) such that M and $M_1' \times M_2'$ are isomorphic (ignoring the polarization). We shall derive a contradiction.

Let

$$\Lambda = \begin{pmatrix} \Lambda_1 & 0 \\ 0 & \Lambda_2 \end{pmatrix}$$

be the standard period matrix for M, obtained from the matrices Λ_1 and Λ_2 in Lemma 1. Since the tori M_1' and M_2' are principally polarized they have period matrices Λ_1' and Λ_2' that are polarized by J_1 and J_2 respectively. The period matrix

$$\Lambda' = \begin{pmatrix} \Lambda_1' & 0 \\ 0 & \Lambda_2' \end{pmatrix}$$

for $M_1' \times M_2'$ is then polarized by

$$(7.1) \qquad P^{\#} = \begin{pmatrix} J_1 & 0 \\ 0 & J_2 \end{pmatrix}.$$

Since M and $M_1' \times M_2'$ are isomorphic there are matrices S in $GL(6, \mathbb{Z})$ and H in $GL(3, \mathbb{C})$ such that

$$(7.2) \qquad H\Lambda' = \Lambda S$$

Therefore $P = SP^{\#}\,{}^tS$ polarizes Λ. In particular

$$(7.3) \qquad \Lambda \left(SP^{\#}\,{}^tS \right) {}^t\Lambda = 0.$$

The involution $(x_1, x_2) \mapsto (-x_1, x_2)$ of $M_1' \times M_2'$ is described in terms of Λ' by the equation

$$C_1\Lambda' = \Lambda'T_1,$$

where T_1 is the matrix (6.6) and

$$C_1 = \begin{pmatrix} -1 & 0 & 0 \\ 0 & 1 & 0 \\ 0 & 0 & 1 \end{pmatrix}.$$

In view of (7.2) we have

(7.4) $C\Lambda = \Lambda(ST_1S^{-1}),$

with $C = HC_1H^{-1}$.

The form of Λ_1 and Λ_2 shows that C in (7.4) is determined by Λ and S and depends continuously on Λ for fixed S. Therefore, for each S the set of Λ in $\mathcal{O}$ for which equations (7.3) and (7.4) hold is closed. By assumption, each Λ in $\mathcal{O}$ satisfies (7.3) and (7.4) for some S, so Baire's theorem implies that there is a fixed S in $GL(6, \mathbb{Z})$ such that both (7.3) and (7.4) hold for all Λ in a nonempty open subset of $\mathcal{O}$. Lemma 6 says that S has the form (6.7), and Lemma 4 says that $SP^{\#}\,{}^tS$ has the form (6.1). We are now ready to obtain a contradiction.

Direct computation using (6.7) and (7.1) gives

(7.5) $SP^{\#}\,{}^tS = \begin{pmatrix} x^2 (\det A) J_1 + BJ_2\,{}^tB & * \\ * & J + y^2 (\det A) P'' \end{pmatrix},$

where $J = DJ_2\,{}^tD$, $\det A = \pm 1$, and $P'' = QJ_1\,{}^tQ$, as in (6.2). Comparison of (7.5) and (6.1) gives

(7.6) $\pm x^2 J_1 + BJ_2\,{}^tB = k J_1,$

and shows also that J can be written in the form

$$J = uP' + vP'' \quad \text{for some integers } u \text{ and } v.$$

Now Lemma 6 says that $B \equiv zUD \pmod 5$, where U is given by (6.8). Therefore

$$BJ_2\,{}^tB \equiv z^2 UJ\,{}^tU \pmod 5.$$

Direct computation using (6.2) and (6.8) gives $UP'\,{}^tU \equiv 0$ and $UP''\,{}^tU \equiv 0$ (mod 5). Therefore $BJ_2\,{}^tB \equiv 0 \pmod 5$, and (7.6) gives

$$\pm x^2 J_1 \equiv k J_1 \pmod 5,$$

so $k \equiv \pm x^2 \pmod 5$. But (6.3) says that $m = \pm 1$ and $3km \equiv \pm 1 \pmod 5$. Therefore $x^2 \equiv \pm k \equiv \pm 2 \pmod 5$, which is the desired contradiction.

8. Proof of Lemma 4

Write P in the form

$$P = \begin{pmatrix} P_1 & R \\ -{}^tR & P_2 \end{pmatrix},$$

where P_1 and P_2 are skew-symmetric of size 2×2 and 4×4 respectively. Obviously P_1 is an integral multiple of J_1. Condition (2.1) yields the matrix equations (see Martens [5])

$$\Lambda_1 R\,{}^t\Lambda_2 = 0 \quad \text{and} \quad \Lambda_2 P_2\,{}^t\Lambda_2 = 0.$$

In view of (3.1) and (3.2), the first of these equations reads

$$(1,\beta) \begin{pmatrix} r_{11} & r_{12} & r_{13} & r_{14} \\ r_{21} & r_{22} & r_{23} & r_{24} \end{pmatrix} \begin{pmatrix} 1 & 0 \\ 0 & \dfrac{1}{3} \\ \alpha & \beta \\ \beta & \dfrac{5}{3}\beta \end{pmatrix} = (0,0).$$

This holds for every α and β in some nonempty open set, so it is an identity in α and β. It follows that R is a multiple of

$$\begin{pmatrix} 0 & 0 & 0 & -1 \\ 1 & 5 & 0 & 0 \end{pmatrix},$$

as claimed.

Similarly, the equation $\Lambda_2 P_2\,{}^t\Lambda_2 = 0$ produces an identity in α and β from which it follows that $P_2 = mP' + nP''$, as given in (6.1) and (6.2). Therefore P has the required form, and an easy calculation shows that

$$\det P = m^2 \left[k(3m + 5n) - 5\ell^2 \right]^2,$$

which proves (6.3).   Q.E.D.

9. Proof of Lemma 5

Fix T in $SL(6,\mathbb{Z})$ and suppose that for each Λ in a nonempty open subset of $\mathcal{U}$ the equation

$$(9.1) \qquad \Lambda T = C\Lambda = C \begin{pmatrix} 1 & \beta & 0 & 0 & 0 & 0 \\ 0 & 0 & 1 & 0 & \alpha & \beta \\ 0 & 0 & 0 & \dfrac{1}{3} & \beta & \dfrac{5}{3}\beta \end{pmatrix}$$

holds. Let V_j be the j-th column of the matrix ΛT. Equation (9.1) implies the equations

$$V_2 = \beta V_1, \quad V_5 = \alpha V_3 + 3\beta V_4, \quad V_6 = \beta(V_3 + 5V_4).$$

By hypothesis these equations are identities in α and β. They imply the existence of integers a, b, c, d, and e such that

$$T = \begin{pmatrix} a & 0 & 0 & b & 0 & 0 \\ 0 & a & 0 & 0 & 3b & 5b \\ c & 0 & d & e & 0 & 0 \\ 5c & 0 & 0 & d+5e & 0 & 0 \\ 0 & 0 & 0 & 0 & d & 0 \\ 0 & c & 0 & 0 & 3e & d+5e \end{pmatrix}.$$

The added requirements that $T^2 = I_6$ but $T \neq \pm I_6$ lead easily to the conditions (6.5).   Q.E.D.

10. Proof of Lemma 6

We are given a matrix S in $GL(6, \mathbb{Z})$ that satisfies $ST_1 = TS$, where T_1 is the matrix (6.6) and T satisfies (6.4) and (6.5). We write S and T in block form

$$S = \begin{pmatrix} A' & B \\ C & D \end{pmatrix}, \qquad T = \begin{pmatrix} aI_2 & bU \\ cV & W \end{pmatrix},$$

where

$$U = \begin{pmatrix} 0 & 1 & 0 & 0 \\ 0 & 0 & 3 & 5 \end{pmatrix}, \quad V = \begin{pmatrix} 1 & 0 \\ 5 & 0 \\ 0 & 0 \\ 0 & 1 \end{pmatrix}, \quad W = \begin{pmatrix} d & e & 0 & 0 \\ 0 & -a & 0 & 0 \\ 0 & 0 & d & 0 \\ 0 & 0 & 3e & -a \end{pmatrix},$$

and corresponding blocks of S and T have the same size. Observe that $V = -QJ_1$, where Q and J_1 are given by (6.2) and (5.1) respectively.

By (6.5), $a = -(d + 5e)$ and $d = \pm 1$. If $d = -1$, then $W - I_4$ has rank 4, so the $+1$ eigenspace of T has dimension at most two. That contradicts the given equation $T = ST_1S^{-1}$, so we have

(10.1) $d = 1$ and $1 - a = 2 + 5e.$

Now

$$\begin{pmatrix} -A' & B \\ -C & D \end{pmatrix} = ST_1 = TS = \begin{pmatrix} * & aB + bUD \\ cVA' + WC & * \end{pmatrix},$$

so $B = aB + bUD$ and $-C = cVA' + WC$. In view of (10.1) the first of these equations implies $2B \equiv bUD \pmod{5}$, so B is congruent $\pmod 5$ to a multiple of UD, as required.

Next, (10.1) and the equation $(W + I_4)C = -cVA'$ imply the equations

$2C_1 + eC_2 = -cA_1'$, $(2+5e)C_2 = -5cA_1'$, $C_3 = 0$,   and $(2+5e)C_4 = -cA_2'$, where C_j and A'_j are the j-th rows of C and A' respectively. It follows that $(1-a)C = (2+5e)C = -cVA'$. If $c = 0$, then $C = 0$. If $c \neq 0$ we obtain an equation

$$xC = -yVA' = y\,QJ_1A'$$

with x and y relatively prime integers. In either case there are integers x and y and a 2×2 integral matrix A such that $A' = xA$ and $C = y\,QJ_1A$. Thus

$$S = \begin{pmatrix} xI_2 & B \\ y\,QJ_1 & D \end{pmatrix} \begin{pmatrix} A & 0 \\ 0 & I_4 \end{pmatrix}.$$

Since

$$\det A = \det \begin{pmatrix} A & 0 \\ 0 & I_4 \end{pmatrix}$$

divides $\det S$, $\det A = \pm 1$ as required. Q.E.D.

References

1. H.F. Baker, *An introduction to the theory of multiply periodic functions*, Cambridge University Press, New York, 1907.
2. C.J. Earle, *Some Jacobian varieties which split*, Complex analysis, Joensuu 1978, Lecture Notes in Mathematics, vol. 747, Springer-Verlag, Berlin, Heidelberg, New York, 1979, pp. 101–107.
3. _____, *H.E. Rauch, function theorist*, Differential geometry and complex analysis, Springer-Verlag, Berlin, Heidelberg, New York, 1985, pp. 15–31.
4. R.C. Gunning, *Riemann surfaces and generalized theta functions*, Springer-Verlag, New York, 1976.
5. H.H. Martens, *Riemann matrices with many polarizations*, Complex analysis and its applications, vol. III, International Atomic Energy Agency, Vienna, 1976, pp. 35–48.
6. H.E. Rauch and J. Lewittes, *The Riemann surface of Klein with 168 automorphisms*, Problems in analysis, a symposium in honor of Solomon Bochner, Princeton University Press, Princeton, N.J., 1970, pp. 297–308.
7. J.F.X. Ries, *Splittable Jacobi varieties*, in this volume.
8. C.L. Tretkoff and M.D. Tretkoff, *Combinatorial group theory, Riemann surfaces, and differential equations*, Contemporary Math. **33** (1984), 467–519.

DEPARTMENT OF MATHEMATICS, CORNELL UNIVERSITY, ITHACA, NY 14853-7901

Contemporary Mathematics
Volume **136**, 1992

THE SCHOTTKY RELATION IN GENUS 4

Leon Ehrenpreis

I. Introduction.

Let W be a Riemann surface of genus g which we identify with its image in its Jacobian J under the Abel-Jacobi map. We denote by π the period matrix of C and by z the coordinate in C^g. Thus we can think of $J = C^g/(I,\pi)$ where (I,π) is the additive group generated by the columns of I and π. An arbitrary point τ in the Siegel upper half plane defines a PPAV (principally polarized abelian variety) $A = C^g/(I,\tau)$.

In addition to the Cartesian coordinate system z which is used to define PPAV there is a "coordinate system"in J associated to each W. To understand this, call W^r the r fold sum $W + \cdots + W$. Thus $W^g = J$ and W^{g-1} is, up to some fixed translation, the zero divisor of the theta function. Given any (generic) point $z \in J$ the translate $W + z$ meets W^{g-1} in g points whose sum is z, up to a constant translation. These g points on $W + z$ can be thought of as coordinates of z.

In order to have some insight into the meaning of the coordinate system defined by W, consider the case when $\pi^0 = (\pi_{ii}^0)$ is diagonal, which is a limit of actual period matrices. The corresponding W_{π^0} is (up to an unimportant translation) the union of the coordinate axes Z_j factored by (I,π^0). It is clear that $C^g = W^g$, and that the intersection of W_{π^0} with its translate $W_{\pi^0,a}$ by a is equal, generally, to g points whose sum is a. In this degenerate situation the Cartesian and W_{π^0} coordinate systems are essentially the same.

Note that the coordinate axes $\cup Z_j$ factored by (I,π^0) is the union of g elliptic curves joined at the origin. It is not easy to see from the geometry of Riemann surfaces in what sense this union is the limit of irreducible curves embedded in their Jacobians. Certainly one would be tempted, e.g. for $g = 2$, to consider a geometric deformation and expect that the limit period matrix would be diagonal.

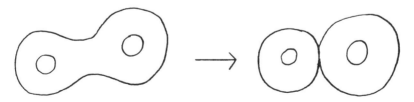

1980 *AMS (MOS) Subject Classifications* (1985 *revision*) Primary 14H40 Riemann Surfaces
Work Supported by NSF Grant No. 33-1807-231.
This paper is in final form and no version of it will be submitted for publication elsewhere.

The trouble is that it is difficult to make the calculations using the expression of π in terms of differentials of the first kind because it is difficult to renormalize the differentials in the process of deformation.

This paper is mainly concerned with an exposition of Schottky's original computation of the one relation in genus $g = 4$ which determines the Jacobians amongst all abelian varieties. We should point out that Schottky's "proof"as given in both papers [11] contains a serious defect. Perhaps for this reason he sought a different proof in [12]. That proof was also incomplete; it remained for Farkas and Rauch to complete it (see [5]). In fact, it is quite strange that Schottky referred to Poincaré's asymptotic relation for the periods (see [2]) as "a different proof of his relation"although no one has been able to derive the global (Schottky) relation from the asymptotic one. In fact, the derivation of Poincaré's relation from Schottky's is not at all trivial; it appears for the first time in Rauch [9].

The crucial Lemma 1 below is not proven by Schottky although he makes use of it. (In fact, it is assumed in both papers and is somewhat misstated in his 1903 paper.) We shall show, in fact, that the union of the coordinate axes W_{π^0} constitutes the limit of W_π as $\pi \to \pi^0$. This leads to a proof of Lemma 1 by degeneration to π^0. Our proof of this degeneration depends on a far reaching generalization of Torelli's theorem. Hopefully the degeneration argument can be combined with the ideas of [1], [2] to give results in higher genera.

We can also visualize the reverse process, that is, the deformation going from W_{π^0} to W_π. In the simplest case of genus 2 this consists of replacing the coordinate axes $W_0 = Z_1 \cup Z_2$ by the hyperbola $z_1 z_2 = \epsilon$. It is easily seen that the hyperbola also enjoys the property that its intersection with a translates consists of two points whose sum equals the amount translated.

Of course, unlike the coordinate axes, the hyperbola represents only an approximation to the actual W_π which is translated. Geometrically the passage from W_{π^0} to the hyperbola makes the two real circles into one simple closed curve while preserving the other cycles, i.e.

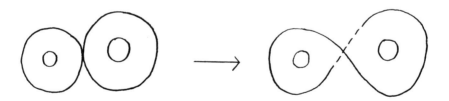

For $g > 2$ we have to detach the intersections one at a time. We hope to return to this deformation theory.

The starting point for all investigations is the theta functions; they are defined using the Cartesian coordinate system but are intimately related to the W coordinate system. We shall need theta functions with characteristics. These are

$$(1) \qquad \begin{bmatrix} m \\ m' \end{bmatrix} = \begin{bmatrix} m_1, \cdots, m_g \\ m'_1, \cdots, m'_g \end{bmatrix}$$

where m_j and m_j' take the valiues 0 or 1 and are closely related to half-periods. We set (for first order theta functions)

(2)
$$\theta \begin{bmatrix} m \\ m' \end{bmatrix} (\tau, z) =$$

$$\sum \exp i\pi \left[\sum (n_i + m_i/2)\tau_{ij}(n_j + m_j/2) + 2\sum (n_i + m_i/2)(z_i + m_i'/2) \right]$$

the sum being taken over lattice points $n = (n_1, \cdots, n_g)$.

By $\binom{\mu}{\mu'}$ we denote the half period

(3)
$$\binom{\mu}{\mu'} = \left(\mu_1'/2 + \sum \tau_{1i} u_i/2 \cdots, \mu_g'/2 + \sum \tau_{gi} u_i/2 \right).$$

We have

(4)
$$\theta \begin{bmatrix} m \\ m' \end{bmatrix} \left(\tau, z + \binom{\mu}{\mu'} \right) =$$

$$\exp i\pi \left[-\frac{1}{4} \sum \mu_i \tau_{ij} \mu_j - \frac{1}{2} \sum \mu_i(m_i' + \mu_i') - \sum \mu_i z_i \right] \theta \begin{bmatrix} m + \mu \\ m' + \mu' \end{bmatrix} (\tau, z).$$

The structure of the exponential factor is crucial for making explicit computations. The main point is that the term involving τ is independent of characteristic $\begin{bmatrix} m \\ m' \end{bmatrix}$.

Thus if we have a proposed identity among $\theta \begin{bmatrix} m \\ m' \end{bmatrix} (\tau, z)$ for varying $\begin{bmatrix} m \\ m' \end{bmatrix}$ then we can translate z by the half period $\binom{\mu}{\mu'}$ and cancel the part of the exponential containing τ. What remains from the exponential is a root of unity which we can handle. This is

<u>Computation Principle.</u> To obtain explicit identities for $\theta \begin{bmatrix} m \\ m' \end{bmatrix} (\tau, z)$ apply translation by half periods $\binom{\mu}{\mu'}$ to theoretic identities (i.e. identities obtained by computing dimensions of linear spaces).

This Computation Principle constitutes the main tool of 19th century methods of obtaining θ identities. In order to apply it one needs a systematic way of writing all $\theta \begin{bmatrix} m \\ m' \end{bmatrix}$. This is discussed in Section II.

Note that the Computation Principle is really meaningful in the Cartesian coordinate system and depends crucially on the group $(Z_2)^{2g}$.

The characteristic $\begin{bmatrix} m \\ m' \end{bmatrix}$ is called even or odd according to the value of $m \cdot m'$ mod z. The parity of $\begin{bmatrix} m \\ m' \end{bmatrix}$ determines the parity of $\theta \begin{bmatrix} m \\ m' \end{bmatrix}$ as function of z.

In addition to first order theta functions we shall deal with second order theta functions. These behave under translation like products of first order theta functions meaning

(5)
$$\theta_2 \begin{bmatrix} m \\ m' \end{bmatrix} (\tau, z + e^{(k)}) = (-1)^{m_k} \theta_2 \begin{bmatrix} m \\ m' \end{bmatrix} (\tau, z)$$

$$\theta_2 \begin{bmatrix} m \\ m' \end{bmatrix} (\tau, z + \tau^{(k)}) = (-1)^{m'_k} e^{i\pi(-2\tau_{kk} - 4z_k)} \theta_2 \begin{bmatrix} m \\ m' \end{bmatrix} (\tau, z).$$

The dimension of the space of $\theta_2 \begin{bmatrix} m \\ m' \end{bmatrix}$ for fixed $\begin{bmatrix} m \\ m' \end{bmatrix}$ is 2^g. If $\begin{bmatrix} m \\ m' \end{bmatrix} = \begin{bmatrix} 0 \\ 0 \end{bmatrix}$ they are all even while if $\begin{bmatrix} m \\ m' \end{bmatrix} \neq \begin{bmatrix} 0 \\ 0 \end{bmatrix}$ then there are 2^{g-1} linearly independent even $\theta \begin{bmatrix} m \\ m' \end{bmatrix}$ and 2^{g-1} linearly independent odd $\theta \begin{bmatrix} m \\ m' \end{bmatrix}$.

We need some more definitions:

Given two half periods $\underline{\lambda} = \begin{pmatrix} \lambda \\ \lambda' \end{pmatrix}$ and $\underline{\mu} = \begin{pmatrix} \mu \\ \mu' \end{pmatrix}$ we define

(6)
$$|\underline{\lambda}, \underline{\mu}| = (-1)^{\lambda \cdot \mu' - \mu \cdot \lambda'}.$$

$\underline{\lambda}$ and $\underline{\mu}$ are called <u>syzygetic</u> or <u>azygetic</u> acoording as $|\underline{\lambda}, \underline{\mu}| = \pm 1$.

A set of half periods is called syzygetic (azygetic) if the elements are mutually syzygetic (azygetic). A maximal syzygetic group is called a <u>Göpel group</u>.

The notion of syzygetic and azygetic has a counterpart in characteristics. A set S of characteristics is called azygetic if there is a characteristic $\begin{bmatrix} \alpha \\ \alpha' \end{bmatrix}$ in the set and an azygetic set of half periods $\begin{pmatrix} \mu \\ \mu' \end{pmatrix}$ so that S is the union of $\begin{bmatrix} \alpha \\ \alpha' \end{bmatrix}$ and the $\begin{bmatrix} \alpha + \mu \\ \alpha' + \mu' \end{bmatrix}$ where addition is taken   mod 2.

Finally, a <u>principal set</u> of g characteristics is a set of $2g + 1$ characteristics which are azygetic and are uniform, i.e. all even or all odd. A central role is played by principal sets in the compuatation of theta identities. As will become apparent, they play the role of a coordinate system in the space of characteristics. Of course, a good coordinate system is always useful for difficult computations.

The algebra of these characterstics was well known in the 19th century. However, it is very difficult to learn about it from 19th century papers. Fortunately, the book of Farkas and Rauch [5] has brought about the possibility of our understanding these beautiful ideas. I shall dicuss the necessary ingredients in Section II below.

Every approach to the Schottky relation starts with a special property of $\theta(\pi, z)$. Schottky's special property and the approach using the trisecant formula discussed below depends on the coordinate system W^g.

How does one construct special theta identities that hold on the Jacobian locus? If one thinks in terms of the W coordinate system, then, in addition to W, we are given objects like $W^r = W + \cdots + W$, and $W - W$, and $W^2 - W^2$, etc. and their translates.

Suppose we start with a general theta function identity, that is, one that holds for all τ. Restrict this identity, e.g. to W so one obtains a relation on W. The usual theta identities are homogeneous polynomials in various θ_m with coefficients depending on theta nulls. Thus we have on W a relation amongst such products.

Now, the restrictions of θ_m to W are suitably defined multivalued functions. Thus we have produced what could be termed a <u>Noether relation</u> on W, that is, a relation amongst products of functions with fixed multivalued behaviour.

Suppose we had some more intrinsic way of determining such Noether relations on W. Then the coefficients of the original generic theta relations, i.e. expressions in theta nulls, could, hopefully, be expressed in terms of the coefficients of the intrinsic relations. But these latter cofficients, themselves, satisfy relations which we may term secondary Noether relations meaning relations amongst the coefficients of the (primary) Noether relations.

Applying this to the theta nulls which are coefficients of the original generic theta relations yields a special relation that holds on the Jacobian locus.

I do not know if this computation can be carried out as it is stated. Schottky uses this idea on $W - W$ rather than on W itself. Thus he is concerned with generic theta identities which involve only products of odd thetas. He then uses an idea of Riemann: On $W - W$ we have the "splitting"

$$(7) \qquad \frac{\theta_m}{\theta_n}(x - x') = \frac{\Phi_m(x)}{\Phi_n(x)}\frac{\Phi_m(x')}{\Phi_n(x')}.$$

We have abbreviate $\theta_m = \theta\begin{bmatrix} m \\ m' \end{bmatrix}$, etc.. The functions Φ_m are Wurzelfunktionen (root functions). They transform like the square root of differentials on W.

The theta function identities involve products of thetas homogeneously. Thus we can use (7) to replace the theta functions in the identities by the corresponding products of Wurzelfunktionen. In particular, a homogeneous quadratic relation will lead to a relation amongst products of two root functions; these are Prym differentials. They become single valued on a suitable two sheeted covering of W.

For the identities we consider all Pryms are related to the same two sheeted covering. The dimension of the space of such Pryms is $g - 1$.

Following the idea sketched above, we search for the linear relations (Noether relations) amongst the products of Wurzelfunktionen, that is, the linear relations satisfied by the Pryms. It is Lemma 1 that allows us to find these relations as well as the expression of the general theta identities reduced to $W - W$ in terms of these identities. The relation amongst the Noether relations is the Schottky relation.

If we think in terms of Pryms as above, it is natural to study the corresponding two sheeted covering $\widehat{W}$ of W. A natural starting point for such investigations is the expression of the theta function of $\widehat{W}$ in terms of that functions on W and Prym thetas, i.e. thetas defined by λ which is a point of the Siegel upper half plane of degree $g - 1$. λ can be thought of as the period matrix associated to the Pryms on $\widehat{W}$. Farkas showed (see [5]) that for a suitable two sheeted covering

$$(8) \qquad \begin{aligned} &\theta\begin{bmatrix} g_1 & \varepsilon & \delta \\ h_1 & \varepsilon' & \delta' \end{bmatrix}(\hat{\pi}, \hat{z}) \\ &= \sum_{P \in Z_2^{g-2}} \theta\begin{bmatrix} g_1 & \frac{\varepsilon + \delta}{2} - P \\ h_1 & \varepsilon' + \delta' \end{bmatrix}(2\pi, u)\, \eta\begin{bmatrix} \frac{\varepsilon - \delta}{2} + P \\ \varepsilon' - \delta' \end{bmatrix}(2\tau, v). \end{aligned}$$

Here $\begin{bmatrix} g_1 \\ h_1 \end{bmatrix}$ is a 1 characteristic and $\begin{bmatrix} \varepsilon \\ \varepsilon' \end{bmatrix}$, $\begin{bmatrix} \delta \\ \delta' \end{bmatrix}$ are $g-1$ characteristics, and η is the $g-1$ theta function.

To find theta relations we want to make the left side of (8) vanish. There is, however, a technical difference between these relations and the ones used by Schottky. In (8) both θ and η are functions so we do not get a natural theta identity involving $u = 0$ and v variable which would be some analog of general genus g relations reduced to $W - W$. (Of course the second variable x' of (7) is absent here so this may correspond to restricting a theta identity to W rather than to $W - W$.)

Farkas and Rauch succeed in finding many characteristics for which the left side of (8) vanishes. These lead to convolution relations on the half period group H of the form

$$(9) \qquad \sum_{P \in H} \theta[m - P](2\pi, 0)\, \widetilde{\eta}[P](2\lambda, 0) = 0$$

where $m = \begin{bmatrix} 0 & m \\ 1 & 0 \end{bmatrix}$ and $P = \begin{bmatrix} 0 & P \\ 0 & 0 \end{bmatrix}$

$$\widetilde{\eta}[P](\lambda', z') = (-1)^{(m-P)\cdot m'}\, \eta\begin{bmatrix} P \\ 0 \end{bmatrix}(\lambda', z').$$

From (9) there follows the Schottky-Jung proportionality between θ and $\widetilde{\eta}$. Thus $\theta[m - P](2\pi, 0)$ must satisfy the same genus 3 null identities that $\widetilde{\eta}$ does and this leads to the Schottky relation.

It is significant that it is 2π and 2λ that appear in (9) rather than π and λ. Theta functions of 2π or 2λ are linear combinations of product of theta functions of π (resp λ). In particular $\eta(2\lambda, 0)$ is closely related to products of Pryms. This seems to indicate that the Farkas-Rauch approach is related to an analog of Schottky's method in which we use a group Γ of rank 2 rather than Schottky's use of a group rank 1. (This will be clarified in what follows.) Thus, instead of being related to $W - W$ this method is more naturally related to $W^2 - W^2$. We shall presently explain how the Novikov approach is also based on $W^2 - W^2$.

The Schottky-Jung method does not seem to use the coordinate system W^g.

There is a third approach to Schottky relations – one which bears the name "Novikov conjecture." This starts from the trisecant relation of Fay [6]: For x, y, u, v distinct points of W

$$(10) \qquad \begin{aligned} &\theta(z)\theta(z + (x + y) - (u + v)) \\ &= c_1\theta(z + (x - u))\theta(z + (y - v)) + c_2\theta(z + (x - v))\theta(z + (y - u)). \end{aligned}$$

In this equation π is fixed and the characteristic of θ is 0. [(10) becomes a geometric trisecant condition upon applying the Kummer map.]

For fixed z we can regard (10) as a decomposition property of θ in the spirit of the Wurzelfunktionen of (7). We can think of $\theta(z + (x + y) - (u + v))$ as a function on $z + (W^2 - W^2)$. Thus $W^1 - W^1$ which is basic for (7) is replaced by translates of $W^2 - W^2$. Instead of splitting into a single product of objects attached to W^1 as in (7), we split into a sum of two products, each related to tranlates of $W^2 - W^2$.

One of the main interests in (10) is its relation to the Kadomtsev-Petviashvilli (K-P) equation. By allowing the four points x, y, u, v to coalesce suitably, (10) becomes the K-P equation

(11) $$\frac{3}{4} u_{ss} = \frac{\partial}{\partial r} \left[u_t - \frac{1}{4}(6uu_r + u_{rrr}) \right]$$

for u of the form

(12) $$u(r, s, t) = 2 \frac{\partial^2}{\partial r^2} \log \theta(\tau, rR + sS + tT + z_0) + c$$

where R, S, T are suitable points in the Jacobian, z_0 is arbitrary, and c is a constant.

Allowing all the points to coalesce means that we are "restricting"to the "distinguished diagonal"in $W^2 - W^2$. This is in contrast to Schottky's method which factors out the part of the identity which vanishes on the diagonal. Perhaps the relation between the methods is akin to Alexander duality.

It is shown by Shiota [10] and Mulase [7] that (12) and certainly (10) imply that for any genus g, τ is in the closure of the Jacobian locus, this being a positive solution to the Novikov conjecture. Much work has been done in this area but so far this approach has not led to the explicit Schottky relation in genus 4.

It might seem puzzling that we have two totally different relations for $g = 4$ which are equivalent, namely Schottky's, which involves many characteristics and Novikov's which involves derivatives of the theta with zero characteristic. Although we do not understand fully the equivalence of K-P with Schottky, we should remark that in genus $g = 1$ there is a relation between half integer characteristics and derivatives of theta with zero characteristic. The precurser of this is

$$\frac{d}{dx} \sin \pi x = -\pi \sin \pi (x - \frac{1}{2}).$$

Thus

$$\pi \sin \pi x \sin \pi (x - \frac{1}{2}) = \frac{d}{dx} \sin^2 \pi x.$$

Hopefully a complete understanding of such relations for general g would lead to a direst verification that K-P and Schottky are the same for $g = 4$.

CONJECTURE. *Using the asymptotic method of Poincaré, as developed in [1], [2] one can prove that (10) and (11) agree asymptotically with Poincaré's asymptotic form of Schottky's identity.*

ACKNOWLEDGEMENT

This paper could not have been written without the extensive assitance I received from Hershal Farkas.

II. General Theta Identities.

By "general theta identities" we mean polynomial identities for the $\theta_m(\tau, z)$ with coefficients which are independent of z. Such identities are to hold for τ in the Siegel upper half plane in contradistinction to special identities which are derived only when $\tau = \pi$ is a Jacobian.

There are two methods for deriving general theta identities. The first is based on the calculus of half periods and characteristics. The second is based on Riemann's theta formula: Start with 4 points $\zeta, \eta, \xi, \omega \in C^g$. Then set

$$
(13) \qquad
\begin{aligned}
2\zeta' &= \zeta + \eta + \xi + \omega \\
2\eta' &= \zeta + \eta - \xi - \omega \\
2\xi' &= \zeta - \eta + \xi - \omega \\
2\omega' &= \zeta - \eta - \xi + \omega.
\end{aligned}
$$

Let $\begin{bmatrix} n \\ n' \end{bmatrix}$ and $\begin{bmatrix} m \\ m' \end{bmatrix}$ be characteristics and $\left(\begin{smallmatrix} \mu \\ \mu' \end{smallmatrix}\right), \left(\begin{smallmatrix} \nu \\ \nu' \end{smallmatrix}\right)$ half periods. Let τ be fixed. Define

$$
(14) \quad y\begin{bmatrix} n \\ n' \end{bmatrix} = (-1)^{(\mu+\nu)\cdot n'}\, \theta\begin{bmatrix} n + \mu + \nu \\ n' + \mu' + \nu' \end{bmatrix}(\zeta)\theta\begin{bmatrix} n + \mu \\ n' + \mu' \end{bmatrix}(\eta)
$$
$$
\theta\begin{bmatrix} n + \nu \\ n' + \nu' \end{bmatrix}(\xi)\,\theta\begin{bmatrix} n \\ n' \end{bmatrix}(\omega)
$$

and

$$
(15) \quad x\begin{bmatrix} m \\ m' \end{bmatrix} = (-1)^{(\mu+\nu)\cdot m'}\theta\begin{bmatrix} m + \mu + \nu \\ m' + \mu' + \nu' \end{bmatrix}(\zeta')\,\theta\begin{bmatrix} m + \mu \\ m' + \mu' \end{bmatrix}(\eta')
$$
$$
\theta\begin{bmatrix} m + \nu \\ m' + \nu' \end{bmatrix}(\xi')\theta\begin{bmatrix} m \\ m' \end{bmatrix}(\omega').
$$

Then

$$
(16) \qquad 2^g y\begin{bmatrix} n \\ n' \end{bmatrix} = \sum_m |n, m| x\begin{bmatrix} m \\ m' \end{bmatrix}.
$$

(Clearly the roles of x and y can be interchanged.) The additions $m + n$, etc. are all taken mod 2.

From a theoretical point of view the Riemann theta formula tells all. For, Mumford [8] has proven that all general theta identities follow from it. But from a practical point of view it appears very difficult to derive explicit identities from the Riemann theta formula, so we shall use the calculus of characteristics to derive the formulae we need. (One can draw an anlaogy with the formulations of Schottky's relation discussed in Section I. The Novikov conjecture gives necessary and sufficient conditions for τ to be a limit of Jacobians π. But the explicit Schottky relation has thus far eluded the Novikov or trisecant approaches.)

Any good computation starts with a good coordinate system. For us this means a systematic way of writing characteristics and half periods; this was developed by Frobenius. Let me remark first that the distinction between half periods and characteristics is very subtle in view of (4). In an abstract sense we can regard characteristics and half periods as dual objects. Thus, if we ignore the factor in (4) depending on τ, the theta functions with characteristics behave somewhat like characters of the half period group under translation. Thus they should be thought of as character-like fucntions on the half period group.

This viewpoint is significant for the computations that follow because we think of the δ functions of half periods as linear functions on the linear space spanned by theta functions and their products.

The notational device of Frobenius works for both characteristics and half periods. Yet it preserves their distinction. It is defined as follows:

Start with a polynomial set H of $2g + 1$ characteristics. Order them in some fashion. The elements of H are called first order characteristics; each is assigned a single number between 1 and $2g + 1$. For any subset C of H of odd cardinality we define the characteristic C as the sum (mod 2) of the c in C. Instead of C we write the ordered word in the numbers corresponding to the elements of C. For example, if $C = \{1, 7, 3, 5, 9\}$ then the index of the characteristic defined by C is $1\,3\,5\,7\,9$. This is called a $|C|$ order index.

The value of this indexing is two fold. In the first place the map $C \to$ characteristics is one-one from subsets of $\{1, 2, \cdots, 2g + 1\}$ of odd cardinality onto all characteristics, i.e. every characteristic has a unique index. Furthermore the characteristics of given order form a uniform set (are all even or odd in z) and changing the number of indices by 2 changes the parity. The parity can be determined by the fact that the principal set H is even if $g \equiv 0, 1 \pmod 4$ and odd otherwise. Corresponding to this notation we shall index theta functions by the index of their characteristics.

The second use of indices comes from half periods. These are defined exactly as for characteristics except that now we use an even number of indices.

We define the composition of indices by addition mod 2 of the corresponding characteristics or half periods. Since addition is mod 2 we can cancel terms that appear twice. Thus, e.g. the sum of the characteristic $1\,2\,3$ and the half period $3\,4$ is the characteristic $1\,2\,4$. This idea of composition makes sense in view of equation (4).

A final notational remark. It is sometimes convenient to consider $1\,2\,3 \cdots 2g + 1$ as the first order index $2g + 2$. This makes sense when $2g + 1$ has the same parity as other first order indices meaning that g is even. In this case every index is the same as its complement in the set $(1, 2, \cdots, 2g + 2)$. For example, for $g = 3$

$$(1\,2\,3) = (4\,5\,6\,7\,8)$$

which is clear since $(8) = (1\,2\,3\,4\,5\,6\,7)$.

To illustrate how things work, let us derive the classical theta identity in genus 1. There are 3 even characteristics $\begin{bmatrix} 0 \\ 0 \end{bmatrix}, \begin{bmatrix} 0 \\ 1 \end{bmatrix}, \begin{bmatrix} 1 \\ 0 \end{bmatrix}$ which we label 1,2,3. They clearly form a principal set. (123) is the odd characteristic $\begin{bmatrix} 1 \\ 1 \end{bmatrix}$.

There are only 2 linearly independent second order theta functions. Thus there is a relation

(17) $$c_1\theta_1^2(z) + c_2\theta_2^2(z) + c_3\theta_3^2(z) = 0.$$

We now translate equation (17) by the half period (12). (This is Computational Principal.) Then, after cancelling the common exponential factor

(18) $$\pm c_1\theta_2^2(z') \pm c_2\theta_1^2(z') \pm c_3\theta_{123}^2(z') = 0.$$

Set $z' = 0$. Since θ_{123} is odd we derive (assuming $c_2 \neq 0$)

(19) $$\pm\frac{c_1}{c_2} = \frac{\vartheta_1^2}{\vartheta_2^2}.$$

We have written ϑ for the nullverte of θ, that is, $\theta(0)$.
   Similarly

(20) $$\pm\frac{c_3}{c_2} = \frac{\vartheta_3^2}{\vartheta_2^2}.$$

This yields

(21) $$\vartheta_1^2\theta_1^2(z) \pm \vartheta_2^2\theta_2^2(z) \pm \vartheta_3^2\theta_3^2(z) = 0$$

On setting $z = 0$ we derive the classical

(22) $$\vartheta^4\begin{bmatrix}0\\0\end{bmatrix} \pm \vartheta^4\begin{bmatrix}0\\1\end{bmatrix} \pm \vartheta^4\begin{bmatrix}1\\0\end{bmatrix} = 0$$

except that we have to determine the $\pm$ signs. By checking the first two terms of the power series we find that both signs are $-$. Thus

(23) $$\vartheta^4\begin{bmatrix}0\\0\end{bmatrix} - \vartheta^4\begin{bmatrix}0\\1\end{bmatrix} - \vartheta^4\begin{bmatrix}1\\0\end{bmatrix} = 0.$$

With this "warm-up" we are now ready to tackle general genus 4 identities.
   We have remarked that the identities we are most concerned with start with second order thetas; for our purposes these are suitable products of first order thetas. More generally, for any syzygetic group Γ of half periods and any characteristic m we define the Frobenius product

(24) $$P[m] = P_m = \prod_{\gamma\in\Gamma} \theta[m + \gamma].$$

Observe that our notation is consistent. For m depends on an odd number of indices and γ on an even number, so $m + \gamma$ depends on an odd number so is thought of as a characteristic. We shall write $p[m]$ or p_m for the nullverte $P_m(0)$.

Of special interest are uniform products meaning that all factors have the same parity. Such products are called odd or even according to this parity. Of course any uniform function P is always even because $|\Gamma|$ is always even; in fact, $|\Gamma| = 2^d$ where d is the order Γ.

The number of uniform products is $2^{2(g-d)}$. Of them $2^{g-d-1}(2^{g-d}+1)$ are even and $2^{g-d-1}(2^{g-d}-1)$ are odd. They are all linearly independent (see [5]).

There is an analog of Frobenius indexing for products P as in (24). Instead of using a general azygetic set of characteristics, we deal with a Γ azygetic set. This is an azygetic set which has the additional property that the elements are syzygetic to all the elements of Γ.

We can clarify this last as follows: Let Γ' be the group of half periods syzygetic to Γ. It is clear that $|\Gamma'| = 2^{2g-d}$. The quotient $|\Gamma'/\Gamma| = 2^{2(g-d)}$ and this quotient replaces the set of all half periods. Thus a Γ principal set has $2(g-d)+1$ elements and can be used for indexing the products P of (24).

In commenting on equation (4) we mentioned that the term depending on τ will drop out from any homogeneous relation. It remains to study the remaining part which is a power of i. We denote by $(K;m)$ the power of i involved in the passage from θ_m to θ_{mk}. (m is a characteristic and K is a half period.)

We deduce easily from (4)

$$(25) \qquad (K;m) = -K' \cdot m + (mK)' \cdot [m + K - (mK)].$$

In a similar manner we can compute the power H of i in

$$\frac{\theta_{nL}\theta_{nM}}{\theta_n\theta_{nLM}} \to i^H \frac{\theta_{nKL}\theta_{nKM}}{\theta_{nK}\theta_{nKLM}}.$$

We find

$$(26) \qquad H = (K;nL) + (K;nM) - (K;n) - (K;nLM).$$

We write

$$(27) \qquad i^H = (-1)^{\frac{1}{2}H} = (K,L,M).$$

Clearly (K,L,M) is symmetric in all its indices. Moreover

$$(28) \qquad (K,L,M)(K,LN) = (K,L,MN)$$

Finally

$$(29) \qquad (L,M,LM) = \pm 1$$

depending on whether L,M is syzygetic or azygetic.

In particular, thinking in terms of the Frobenius products (24) for $d = 1$ we write $P_a = P_a P_{aK}$ when $\Gamma = \{1, K\}$. Then on adding the half period L we send

$$(30) \qquad \frac{P_a}{P_b} \to (L,M,MK)\frac{P_{aL}}{P_{bL}} = (L,M,MK)\frac{\theta_{aL}\theta_{aKL}}{\theta_{bL}\theta_{bKL}}$$

We have written $M = ab$. Since Γ or K is fixed we also use the notation

(31) $$(L, M, MK) = (L|M).$$

We find

(32) $$(L|M)(M|L) = \pm 1$$

depending on whether L and M are syzygetic or azygetic.

We are now in a position to derive a general theta identity in genus 4. We set $\Gamma = \{1, K\}$ as above. Then there are $2^6 = 64$ uniform products of which 36 are even and 28 are odd. We shall deal exclusively with the latter.

A Γ principal set has 7 elements denoted by $1, 2, \cdots, 7$. Since $g = 4$ the first order elements are odd. This means that 1 and 5 indices are odd and 3 and 7 are even. We write $\varepsilon = 1234567$.

The dimension of the space of P_a is 8. Since all odd P_a vanish at the origin, the space of odd P_a lies in the proper linear subspace defined by the linear equations $p_a = 0$ for all a. (Since not all P_a vanish at the origin this is a proper linear subspace.) Thus the dimension of the odd P_a is < 8. This means that, in particular, there is a nontrivial linear relation

(33) $$F P_{67\varepsilon} = \sum_{a=1}^{7} F_a P_a.$$

We can compute the coefficients explicitly by using the half period group.

Upon applying L to (33) we find

(34) $$F P_{67\varepsilon L} = \sum F_a (L|a67\varepsilon) P_{aL}.$$

Setting $z = 0$

(35) $$F p_{67\varepsilon L} = \sum F_a (L|a67\varepsilon) p_{aL}.$$

Now, if L has 6 indices, so $L = \beta\varepsilon$ where β has one index, then aL will have 7 indices, and so will correspond to an odd P_{aL} (5 indices) unless $a = \beta$ in which case $aL = a\beta\varepsilon = \varepsilon$. This shows that

(36) $$\begin{aligned} p_{aL} &= 0 \quad \text{for } a \neq \beta \\ p_{aL} &= p_\varepsilon \quad \text{for } a = \beta \\ p_{67\varepsilon L} &= p_{67\beta} \end{aligned}$$

Comparing (36) and (35) yields

(37) $$F p_{\beta 67} = (\beta\varepsilon|\beta 67\varepsilon) F_\beta p_\varepsilon$$

so that, up to a proportionality factor which is irrelevant for (33)

(38) $$\begin{aligned} F &= p_\varepsilon \\ F_a &= (a\varepsilon|a67\varepsilon) p_{a67}. \end{aligned}$$

(The proportionality factor cannot be 0 because (33) is a nontrivial relation.) But $p_{a67} = p_6 = 0$ for $a = 7$ and $p_{a67} = p_7 = 0$ for $a = 6$ because one index is odd. Thus we have our central formula

$$(39) \qquad p_\varepsilon P_{67\varepsilon} = \sum_{a=1}^{5} (a\varepsilon|a67\varepsilon) p_{a67} P_a .$$

So far we have used L in (34) with 6 indices. We can use other L to derive further identities. In particular, $\det L = 57$. Then $p_{67\varepsilon L} = p_{56\varepsilon} = 0$ because it has 5 indices. Also $p_{5L} = p_{7L} = 0$ because they have 1 index. Since F_6 and F_7 are zero, we can now with (35) in the form

$$(40) \qquad \sum_{a=1}^{4} F_a (57|a67\varepsilon) p_{a57} = 0 .$$

Relation (40) can be simplied somewhat. We can use (28) to show that

$$(41) \qquad (57|67)(57|a67\varepsilon) = (57|a\varepsilon) .$$

Thus, by multiplying (40) by $(57|67)$ we replace $(57|a67\varepsilon)$ by $(57|a\varepsilon)$. By the same process we have

$$(a\varepsilon|57)(a\varepsilon|a67\varepsilon) = (a\varepsilon|a56\varepsilon) .$$

Moreover, by (32) we can write $(57|a\varepsilon) = (a\varepsilon|57)$. Also everything we have done up to now is symmetric in the indices 5 and 7. Thus, using the values (38) for F_a, we derive, upon permuting 5 and 7

$$(42) \qquad \sum_{a=1}^{4} (a\varepsilon|a\varepsilon 67) p_{a56} p_{a57} = 0 .$$

III. The Schottky Relation.

Equation (39) is exactly of the form that we need to apply the Wurzelfunktionen idea. We now assume that $\tau = \pi$ is a Jacobian. Since P_a and $P_{67\varepsilon}$ are odd products, if we replace z by a point $x - x'$ on $W - W$ we can replace $P_{67\varepsilon}$ and P_a by $w_{67\varepsilon} w'_{67\varepsilon}$ and $w_a w'_a$ respectively. Here $w_m = \Phi_m(x)\Phi_{mK}(x)$ and $w'_m = \Phi_m(x')\Phi_{mK}(x')$ so the w_m, w'_m are Pryms, being the products of two Wurzelfunktionen. This means that

$$(43) \qquad p_\varepsilon w_{67\varepsilon} w'_{67\varepsilon} = \sum_{a=1}^{5} (a\varepsilon|a67\varepsilon) p_{a67} w_a w'_a .$$

This is the central point of Schottky's work. It is a linear relation amongst the 6 products $\{w_a w'_a\}$ and $w_{67\varepsilon} w'_{67\varepsilon}$. As we have mentioned in Section I, the dimension of the space of w_m is 3. Thus we have to choose a basis for the $\{w_m\}$ and compute the coefficients of the identity in terms of the linearly independent products $w_m w'_m$. The coefficients of these identities themselves satisfy identities. These latter identities involve the p_a and that is Schottky's identity.

It remains to set up the correct algebra which leads from (43) to an identity on the coefficients of the $w_a w'_a$ and $w_{67\varepsilon} w'_{67\varepsilon}$. We have already noted that there are 28 odd P_m. They give rise to 28 Pryms w_m. Of crucial importance to us is

Lemma 1. *The dimension of the space of such Pryms is 3. Any 3 of the w_m are linearly independent.*

As remarked in the Introduction, this Lemma is assumed by Schottky in his first paper of [11] and incorrectly stated (and certainly not proven) in his second paper. Our proof uses tools which were, presumably, unavailable to Schottky.

Proof. Since we are dealing with products where the sum of characteristics is constant, all the Pryms w_m belong to a fixed two sheeted covering of the given Riemann surface. By the Riemann-Roch theorem the dimension of the space of all Pryms corresponding to this covering is 3. Hence the $\{w_m\}$ span a space of dimension ≤ 3. (This was certainly known to Schottky.)

To prove the linear dependence we use a degeneration argument. We must first show that W_{π^0} is the limit of W_π as $\pi \to \pi^0$. (Recall that π^0 is a diagonal matrix and W_{π^0} is the union of the coordinate axes with identifications on each axis determined by π^0.)

To show that $W_\pi \to W_{\pi^0}$ we use a strong version of Torelli's theorem. Although not specifically stated in the form we need it, we appeal to H. Martens' proof of Torelli's theorem as found in Farkas and Kra [4]. What Martens really shows is: Suppose we are given a curve γ in an abelian variety A of dimension q. With γ we associate $\gamma^r = \gamma + \cdots + \gamma$. Suppose $\gamma^q = A$. Suppose, moreover that certain specific properties hold for suitable intersections $(\gamma^r + a) \cap (\gamma^s + b)$. Then γ is certainly determined by γ^{g-1} meaning any two such γ would differ by a translation and (possibly) a reflection.

The intersection properties are readily shown to be preserved (with a suitable interpretation) under limits, thus they hold for $W_0 = \lim_{\pi \to \pi^0} W_\pi$. On the other hand W_{π^0} satisfies all the requisite intersection properties as can be verified directly with no difficulty. Since the limit W_0^{g-1} is the union of the coordinate hyperplanes (modulo π^0) we have $W_0^{g-1} = W_{\pi^0}^{g-1}$. It follows from Torelli's theorem that, actually, $W_0 = W_{\pi^0} + a$. We normalize so that $W_0 = W_{\pi^0}$ is the union of the coordinate axes.

Actually there should be a very general combinational result which asserts that, under suitable combinational conditions, a set δ is determined by δ^r and δ^{r+1}. We shall discuss this in detail in a forthcoming article.

To complete the proof of Lemma 1 it suffices to prove the linear independence of any 3 of the w_m on W_{π^0}.

Once having established that $W_{\pi^0} = \lim W_\pi$, all questions of linear independence can be established by genus 1 calculations. Thus if linear independence is verified at π^0 it holds in a neighborhood of π^0 hence generically on Jacobians.

We assume that all the entries of π^0 are distinct. An odd genus theta is of the form

$$\theta_m(\pi^0, z) = \vartheta_{i_1}(\pi_{11}^0, z_1)\vartheta_{i_2}(\pi_{22}^0, z_2)\vartheta_{i_3}(\pi_{33}^0, z_3)\vartheta_{i_4}(\pi_{44}^0, z_4).$$

The $\vartheta_{i_j}(\pi_{jj}^0, z_j)$ are genus 1 theta functions. When we restrict to W_{π^0} this means that z lies in the union of the coordinate axes.

At least one of the $i_j = 1$, meaning it is odd. When we pass to "Pryms", we use L'Hospital's rule so this theta is replaced by a constant.

Although it is tedious to verify Lemma 1 completely in this manner, it is easy to check the four cases we need, namely $w_2, w_3, w_s; w_4, w_5, w_s$ as needed for (48);

w_1, w_4, w_s used for (50); and $w_{67\epsilon}, w_5, w_\alpha$ needed for (57).

In this way we have a complete proof of Lemma 1.

Let ξ, η, ζ be a basis for the space of Pryms, and ξ', η', ζ' the corresponding space for functions of x'. We write

$$(44) \qquad\qquad w_m = A_m\xi + B_m\eta + c_m\zeta$$

and form the determinants (non vanishing since w_m, w_n, w_r are linearly independent)

$$(45) \qquad\qquad \begin{vmatrix} A_m & B_m & C_m \\ A_n & B_n & C_n \\ A_r & B_r & C_r \end{vmatrix} = [m, n, r].$$

Now, for any x the system of equations

$$(46) \qquad \begin{aligned} w_m(x)t_1 - A_m t_2 - B_m t_3 - C_m t_4 &= 0 \\ w_n(x)t_1 - A_n t_2 - B_n t_3 - C_n t_4 &= 0 \\ w_r(x)t_1 - A_r t_2 - B_r t_3 - C_r t_4 &= 0 \\ w_p(x)t_1 - A_p t_2 - B_p t_3 - C_p t_4 &= 0 \end{aligned}$$

has the nontrivial solution $(1, \xi(x), \eta(x), \zeta(x))$. Thus the determinant vanishes, meaning

$$(47) \qquad [m, n, r]w_p = [p, n, r]w_m + [p, r, m]w_n + [p, m, n]w_r.$$

The same relations holds for w'_s replacing w_s.

The w_s are multivalued functions on W so, in accordance with our previous constructions, we should search for points on W at which some of the terms in (47) vanish. But this does <u>not</u> work. There may not exist such points. We have to pass to "ideal points." From an algebraic aspect, we start with 3 linearly independent w_a, w_b, w_c. Then we observe that the various products $w_a w'_b, \cdots$ are linearly independent in the tensor product; they are linearly independent fucntions on $W - W$. Hence we can compute the coefficients of the homogeneous quadratic (43) in this basis.

For example, use the bases w_2, w_3, w_s and $w'_4, w'_5, w'_{s'}$. We want to compute the coefficient of $w_s w'_{s'}$. If we apply (47) to $m, n, r, p = s, 2, 3, p$ then the terms on the right side involving w_2 and w_3 are irrelevant. Thus

$$(48) \qquad\qquad [s, 2, 3]w_p \sim [p, 2, 3]w_s$$

as the other terms do not matter. Hence we are at the ideal point defined by $w_2 = w_3 = 0$ (which is certainly not on W).

If we apply (48) to varying p, then we may as well drop the factor $[s, 2, 3]$ as we are dealing with homogeneous relations.

Thus, using the bases w_2, w_3, w_s and $w'_4, w'_5, w'_{s'}$ and examining the coefficient of $w_s w'_{s'}$ is equivalent to setting $w_2 = w_3 = w'_4 = w_5 = 0$. Using (48) and dropping the coefficients $[s, 2, 3]$ and $[s', 4, 5]$ in the fundamental relation (43) we find

$$(49) \qquad p_\epsilon[67\epsilon, 2, 3][67\epsilon, 4, 5] = (1\epsilon | 1\epsilon 67)p_{167}[1, 2, 3][1, 4, 5]$$

since $[2,2,3] = [3,2,3] = [4,4,5] = [5,4,5] = 0$.

We now use the bases w_2, w_3, w_s and $w'_4, w'_1, w'_{s'}$. The only non trivial term in the sum (43) occurs for $a = 5$. We find

(50) $p_\varepsilon[67\varepsilon, 2, 3][67\varepsilon, 41] = (5\varepsilon|567\varepsilon)p_{567}[5, 2, 3][1, 4, 5].$

Dividing (49) by (50) gives

(51) $$\frac{[67\varepsilon, 4, 5]}{[67\varepsilon, 4, 1]} = -\delta_1\delta_5 \frac{p_{167}[1, 2, 3]}{p_{567}[5, 2, 3]}.$$

We have written δ_j for $(j\varepsilon|j\varepsilon 67)$. Similarly

(52) $$\frac{[67\varepsilon, 4, 5]}{[67\varepsilon, 4, 2]} = -\delta_2\delta_5 \frac{p_{267}[2, 1, 3]}{p_{567}[5, 1, 3]}$$

(53) $$\frac{[67\varepsilon, 2, 4]}{[67\varepsilon, 2, 3]} = -\delta_3\delta_4 \frac{p_{367}[3, 5, 1]}{p_{467}[4, 5, 1]}.$$

On multiplying (49), (52), (53) we find

(54) $$p_\varepsilon[67\varepsilon, 4, 5]^2 = \delta_1\delta_2\delta_3\delta_4 \frac{p_{167}p_{267}p_{367}}{p_{467}p_{567}}[1, 2, 3]^2.$$

It is readily verified that $\delta_1\delta_2\delta_3\delta_4 = 1$. Thus we derive

(55) $$\frac{[\varepsilon 67, 4, 5]^2}{[1, 2, 3]^2} = \frac{p_{167}p_{267}p_{367}}{p_\varepsilon p_{467}p_{567}}.$$

By symmetry, if $\alpha, \beta, \gamma, \delta$ represent the numbers 1,2,3,4 in some order, then

(56) $$\frac{[\varepsilon 67, \alpha, 5]^2}{[\beta, \gamma, \delta]^2} = \frac{p_{\beta 67}p_{\gamma 67}p_{\delta 67}}{p_\varepsilon p_{\alpha 67}p_{567}}.$$

Next we use a new basis for Pryms. This corresponds to the ideal point $w_{67\varepsilon} = w_5 = w'_{67\varepsilon} = w'_5 = 0$. Hence, up to a constant this entails $w_\alpha = w'_\alpha = [\alpha, 5, 67\varepsilon]$. According to (43)

(57) $$\sum_{\alpha=1}^{4}(\alpha\varepsilon|\alpha 67\varepsilon)p_{\alpha 67}[\alpha, 5, 67\varepsilon]^2 = 0$$

Replace $[a, 5, 67\varepsilon]^2$ by its value given by (56). This yields our final equation

(58) $$\sum_{\alpha=1}^{4}(\alpha\varepsilon|\alpha 67\varepsilon)\frac{[\beta, \gamma, \delta]^2}{p_{\alpha 67}} = 0.$$

Up to now all our computations have been formal consequences of the transformation rules for theta functions, the relation to Wurzelfunktionen on $W - W$, and Lemma 1 concerning the linear dependence and independence of the Wurzelfunktionen. To complete the computation we need one further independence property. To formulate this we set

(59)
$$\Phi_{123} = \frac{[1,2,3]^2}{p_{456}p_{457}p_{467}p_{567}}.$$

Similar definitions are made for $\Phi_{\beta\gamma\delta}$.

From our basic equation (58) we obtain 3 equations by permuting 5 with 6 and also 5 with 7:

(60)
$$\sum(\alpha\varepsilon|\alpha\varepsilon 67)p_{\alpha 56}p_{\alpha 57}\Phi_{\beta\gamma\delta} = 0$$
$$\sum(\alpha\varepsilon|\alpha\varepsilon 75)p_{\alpha 67}p_{\alpha 65}\Phi_{\beta\gamma\delta} = 0$$
$$\sum(\alpha\varepsilon|\alpha\varepsilon 56)p_{\alpha 75}p_{\alpha 76}\Phi_{\beta\gamma\delta} = 0.$$

Regard these as equations for the 4 unknowns $\Phi_{123}, \Phi_{124}\Phi_{134}, \Phi_{234}$.

It is important to observe that the general theta null identity (42) means that the sum of the coefficients in each equation in (60) vanishes. Thus one solution of (60) is

(61)
$$\Phi_{\beta\gamma\delta} \text{ is independent of } \alpha.$$

Lemma 2. *(61) is the only solution of (60) for generic Jacobian* π.

Lemma 2 asserts that the rank of the coefficient matrix in (60) is 3. We shall prove this below but for the present we assume it and show how it implies Schottky's relation.

The numbers $[\beta, \gamma, \delta]$ are not independent. In fact we have

Lemma 3. $[2,3,5][1,4,5] + [3,1,5][2,4,5] + [1,2,5][3,4,5] = 0.$

Proof of Lemma 3. It is clear that for any u_j, v_j

$$\begin{vmatrix} u_2 & v_2 \\ u_3 & v_3 \end{vmatrix}\begin{vmatrix} u_1 & v_1 \\ u_4 & v_4 \end{vmatrix} + \begin{vmatrix} u_3 & v_3 \\ u_1 & v_1 \end{vmatrix}\begin{vmatrix} u_2 & v_2 \\ u_4 & v_4 \end{vmatrix} + \begin{vmatrix} u_1 & v_1 \\ u_2 & v_2 \end{vmatrix}\begin{vmatrix} u_3 & v_3 \\ u_4 & v_4 \end{vmatrix} = 0.$$

Expand the determinants in Lemma 3 by the last row and apply this identity to obtain the result.

We want to apply Lemma 3 to (59). In order to do this we have to apply a permutation of 1,2,3,4,5. This presents no difficulty since everything is clearly invariant under such a permutation. We could, in fact allow permutations involving 6,7 also with a slight change in notation; but these permutations are not needed. For $\alpha, \beta, \gamma, \delta$ some permutation of 1,2,3,4 we clearly have

(62)
$$[\alpha, \beta, 5]^2 = \Phi_{\alpha,\beta,5}p_{\alpha 67}p_{\delta 67}p_{\gamma\delta 6}p_{\gamma\delta 7}.$$

Now, replace $(1,2,3,4)$ of (58) by $(1,2,3,5)$. Then Lemma 3 shows that $\Phi_{125} = \Phi_{123}$. In fact, all Φ_{ik} related to Lemma 3 are equal. Thus the identity of Lemma 3 becomes, upon cancelling a factor $\Phi\sqrt{p_{167}p_{267}p_{367}p_{467}}$, Schottky's relation

Theorem 1.

(63) $$\sum \pm\sqrt{p_{\alpha\delta6}p_{\alpha\delta7}p_{\beta\gamma6}p_{\beta\gamma7}} = 0.$$

This sum contains 3 terms corresponding to $(\alpha\beta\gamma\delta) = (12\ 34)$ or $(13\ 24)$ or $(14\ 23)$. Note that each p is the product of 2 theta nulls. Therefore each term in the sum in (63) is the square root of the product of 8 theta nulls.

One can put (63) is a slightly more pleasant form. Let 1,2,3,4,5,6,7,8,9 be a principal set. Set $K = 89$ and $\chi = (123456789)$. Then it is easily seen that, for $\Gamma = \{0, (89)\}$ a Γ principal set is $\{(\alpha8\chi)\}_{\alpha=1,\cdots,7}$. Hence $p_\alpha = \vartheta_{\alpha8\chi}\vartheta_{\alpha9\chi}$. (Note that in this notation the single index thetas are even.) Formula (63) then becomes

(64) $$\sum \pm\sqrt{\vartheta_{\alpha\delta68\chi}\vartheta_{\alpha\delta69\chi}\vartheta_{\alpha\delta78\chi}\vartheta_{\alpha\delta79\chi}\vartheta_{\beta\gamma68\chi}\vartheta_{\beta\gamma69\chi}\vartheta_{\beta\gamma78\chi}\vartheta_{\beta\gamma79\chi}} = 0.$$

An important observation is that if we call $a = \alpha\delta68\chi$ then all the terms are obtained from a by applying the Göpel group of order 3

$$G = \{0, 89, 67, 6789, 1234, 123489, 123467, 12346789\}.$$

We have already remarked that a rank 3 Göpel group in genus 4 has only 3 even uniform products, say r_1, r_2, r_3. Thus (64) is of the form

(65) $$\pm\sqrt{r_1} \pm \sqrt{r_2} \pm \sqrt{r_3} = 0$$

Schottky proves that this relation is independent of the Göpel group G. More precisely, set

(66) $$J = r_1^2 + r_2^2 + r_3^2 - 2r_1r_2 - 2r_1r_3 - 2r_2r_3.$$

Theorem 2. *J is not identically zero on the Siegel upper half plane. It is independent of the Göpel group.*

Proof. If $J \equiv 0$ then, by (65), the vanishing of two even theta constants coming from different rank 3 products would imply the vanishing of the third. This is not true (see [3]). The fact that $J \not\equiv 0$ can also be proven by an asymptotic analysis of J. We separate the variables $\tau = \tau_{44}$ and $u = (\tau_{14}, \tau_{24}, \tau_{34})$ from the rest. Using (2) we write θ in the form

(67) $$\theta\begin{bmatrix} m \\ m' \end{bmatrix}(\tau, z) = \sum_{n_4} e^{i\pi[\tau(n_4+m_4/2)^2+(z_4+m_4'/2)(n_4+m_4/2)]}$$

$$\theta\begin{bmatrix} m \\ m_0' \end{bmatrix}(\tau_0, z_0 + (n_4 + m_4/2)u - m_4'/2).$$

An expansion in terms of τ then shows that $J \not\equiv 0$.

It is shown by E. Rauch [9] that the vanishing of J implies the (Poincaré) asymptotic form of Schottky's relation. The asymptotic form is certainly $\not\equiv 0$. This shows again that $J \not\equiv 0$. That J is independent of the Göpel group, will be dealt with in section IV.

Proof of Lemma 2. We claim that all three minors of the coefficient matrix in (60) have non vanishing determinant. Thus, call a_j, b_j, c_j the coefficients of Φ with subscript j missing in (60). We claim that

$$(68) \qquad D = \begin{vmatrix} a_1 & a_2 & a_3 \\ b_1 & b_2 & b_3 \\ c_1 & c_2 & c_3 \end{vmatrix}$$

does not vanish identically on the Jacobian locus.

We expand D by minors of the first column. Since b_1 and c_1 contain the factor p_{167} we can write

$$(69) \qquad D = a_1 \begin{vmatrix} b_2 & b_3 \\ c_2 & c_3 \end{vmatrix} + p_{167}G.$$

b_2 and c_2 contain the common factor p_{267} while p_3 and c_3 contain the common factor p_{367}. Since $a_1 = \pm p_{156}p_{157}$ we can write, using the computations at the beginning of Section II,

$$(70) \qquad D = \pm p_{156}p_{157}p_{267}p_{367}(p_{256}p_{357} - (23|67)p_{356}p_{257}) + p_{167}G.$$

We can use an interesting device to rewrite the term, say Δ, in parenthesis in (70). The point is that the null identity (42) does not depend on the specific Γ fundamental set we use. In particular it is readily verified that

$$(71) \qquad 3, \; 2, \; 46\varepsilon, \; 16\varepsilon, \; 56\varepsilon, \; 23\varepsilon, \; 76\varepsilon$$

in another Γ fundamental set. The new ε which we call ε' is the composition of all these indices, that is,

$$(72) \qquad \varepsilon' = 1457\varepsilon = 236.$$

We write (42) in these indices. There results

$$(73) \quad (26|37)p_{256}p_{357} + (36|27)p_{356}p_{257} + (1567|1523)p_{167}p_{123}$$
$$+ (4567|4523)p_{467}p_{423} = 0$$

Again using the identities for the $\pm$ signs $(u|v)$ we find easily that

$$(74) \qquad \pm\Delta = \pm p_{167}p_{123} \pm p_{467}p_{423}.$$

Going back to (70) this shows that

$$(75) \qquad D = p_{167}\widetilde{G} \pm p_{156}p_{157}p_{234}p_{267}p_{367}p_{467}.$$

If $D(\pi) \equiv 0$ then $p_{167}(\pi)$ would divide the product of the 6 terms in the product in (75). In particular, when one theta factor of $p_{167}(\pi)$ vanishes, some other theta must vanish. Hence, by a result of Weber [3], [13], π is hyperelliptic. But the theta factors vanish on a set of π of dimension 8 while the hyperelliptics form a set of dimension 7. This is a contradiction so Lemma 2 is proven.

With this proof we have completed the proof Schottky's relation (Theorem 1).

IV. The Uniqueness Of The Schottky Relation.

We want to show that J defined by (66) is independent of the rank 3 group Γ. We start with a rank 2 group Γ_0 say generated by (K, L). Since Γ_0 has corank 2 the indexing notation behaves like that of genus 2. In particular, a Γ_0 principal set has 5 uniform products Q_1, Q_2, Q_3, Q_4, Q_5 each of which is an odd product pf 4 thetas. The odd products are these 5 and Q_{12345} while the even products have 3 indices.

Lemma 4. We can choose a $\{K\}$ principal set so that $L = (67)$.

Lemma 5. We can choose a Γ_0 principal set so that in terms of the notation defined by the $\{K\}$ principal set, amongst the 10 even null Γ_0 uniform products we have the 5 relations

$$(76) \qquad (1\varepsilon|2345)q_{234} + (2\varepsilon|3415)q_{341} + (3\varepsilon|4125)q_{412} + (4\varepsilon|1235)q_{123} = 0$$

[and the permutations of $(15), (25), (35), (45)$ of (76)]. The subscripts on q_{abc} are depending by Γ_0 and the coefficients by $\{K\}$.

Proof of Lemma 4. Start with a $\{K\}$ principal set, $P_1, \cdots, P_7$. Thus L has an even number of indices. If L has 2 indices we can clearly renumber so that $L = (67)$.

If L has 4 indices, say $L = (1234)$, then we can form the following $\{K\}$ principal set

$$P_{17\varepsilon}, P_{27\varepsilon}, P_{37\varepsilon}, P_{47\varepsilon}, P_4, P_5, P_{67\varepsilon}.$$

In the new indexing L becomes (67) since $67\varepsilon = (1234)5$.

Finally, if L has 6 indices, say $L = (123456)$ then we can choose a new $\{K\}$ principal set

$$P_{16\varepsilon} \quad P_{26\varepsilon} \quad P_{36\varepsilon} \quad P_{56\varepsilon} \quad P_7 \quad P_4 \quad P_{47\varepsilon}$$

in which L becomes (67) since $(123456)4 = (12356) = 47\varepsilon$.

Proof of Lemma 5. In terms of the given $\{K\}$ notation the theta null relation (42) becomes

$$(77) \qquad \qquad \sum (a\varepsilon|a\varepsilon 67)q_{a57} = 0$$

if we set $q_{a57} = p_{a56}p_{a57} = p_{a56}p_{a56L}$.

From the $\{K\}$ principal set $\{P_a\}$ we build the $\{K, L\}$ principal set

$$(78) \qquad \qquad Q_{16\varepsilon} \quad Q_{26\varepsilon} \quad Q_{36\varepsilon} \quad Q_{46\varepsilon} \quad Q_{56\varepsilon}$$

which we now denote by Q_1, Q_2, Q_3, Q_4, Q_5. Then (77) becomes (76).

Proof of Theorem 2 concluded. We now start with the rank 2 group $\{K, L\}$ and we examine the effect of M. Now, M has 2 or 4 indices. For example, if M has 2 indices, say $M = 45$ then the only possible choice for $\{r_j\}$ is (up to permutation)

$$(79) \qquad \begin{aligned} r_1 &= q_{234}q_{235} \\ r_2 &= q_{314}q_{315} \\ r_3 &= q_{124}q_{125} \end{aligned}$$

since each r_j must be a product of two q_{abc} with 3 indices.

On the other hand, M might have 4 indices, say $M = 1245$. Then the three uniform products for $\{K, L, M\}$ must be

$$
(80) \qquad
\begin{aligned}
s_1 &= q_{234}q_{135} \\
s_2 &= q_{314}q_{235} \\
s_3 &= q_{312}q_{345}.
\end{aligned}
$$

Our first observation is

$$(81) \qquad\qquad r_1 r_2 = s_1 s_2.$$

We set $\alpha = (4\varepsilon|1235)(5\varepsilon|1234)$ and $\beta = (1\varepsilon|2345)(2\varepsilon|1345)(4\varepsilon|1235)(5\varepsilon|1234)$. Using Lemma 5 we find easily

$$(82) \qquad\qquad \alpha(r_1 + r_2 - r_3) + \beta(s_1 + s_2 - s_3) = 0.$$

Thus

$$(83) \qquad\qquad (r_1 + r_2 - r_3)^2 = (s_1 + s_2 - s_3)^2.$$

Combining this with (81) shows that J is the same for $M = 45$ as it is for $M = 1245$.

We have thus shown that $J(K, L, 45) = J(K, L, 1245)$. By permutation we find that $J(K, L, M) = J(K, L)$, in fact

$$(84) \qquad\qquad J(K, L) = J(K, M) = J(K, LM)$$

for any M which is syzygetic to both K and L.

Now, choose any N which is syzygetic to K. Then clearly N is syzygetic to at least one of L, M, LM. By (84) this means that $J(K, L) = J(K, N)$. In other words, $J(K, L) = J(K) = J(L) = J(KL)$ so long as K and L are syzygetic. Again, any N is syzygetic to one of K, L, KL so that $J(K) = J(N)$ hence is totally independent of everything.

This completes the proof of Theorem 2.

V. Higher Genus.

Naturally the first question that arizes is : What happens for $g > 4$?

The computations of Section II certainly depend only on the fact that the corank of $\{K\}$ is 3. This means that as far as the indexing goes, the $\{K\}$ principal set has 7 elements and all questions of odd and even are as though we were in genus 3 with trivial syzygetic group.

All this means is that if we start with a syzygetic group of rank $g - 3$ then all computations and notations proceed as in Section II. In particular (39) and (42) are true but with a different meaning for P and p.

The fact that $g = 4$ is used in several places in Section III. Most crucial is the fact that the dimension of Pryms is $g - 1$ which is 3 for $g = 4$.

The ratios of odd thetas reduce to ratios of products of 2^{g-3} square roots of differentials, which we may term higher order Pryms in contradistinction to the (linear) Pryms or Wurzelfunktionen that appeared in genus 4. For $g = 5$ the dimension of this space 9. Equation (43) gives relations amongst these Pryms so no analog of Lemma 1 can hold.

Any extension of the method of Section III to $g > 4$ requires extensive new ideas. It may be possible to start with theta relations which are more complicated than (39) and (42) and then apply the linear Prym theory. Such relations would start from a group of rank 1 rather than rank $g - 3$.

However, the proof of the independence of J from the particular syzygetic group of rank $g - 1$ used to define it works for any genus. The uniform products r_1, r_2, r_3 now contain 2^{g-1} terms.

BIBLIOGRAPHY

1. L. Ehrenpreis, H. Farkas, *Some Refinements of the Poincaré Period Relation, Discontinuous Groups and Riemann Surfaces*, Annals of Math Studies 79, Princeton Univ. Press, 1974.
2. L. Ehrenpreis, H. Farkas, H. Martens, H.E. Rauch, *On the Poincaré Relation*, Contributions to Analysis, Academic Press, 1974.
3. H. Farkas, *Special divisors and analytic subloci of Teichmueller space*, Amer. Jour. of Math. **88** (1966), 881–901.
4. H. Farkas, I. Kra, *Riemann Surfaces*, Springer-Verlag, 2nd edition, 1991.
5. H. Farkas, H.E. Rauch, *Theta Functions with Applications to Riemann Surfaces*, Williams and Wikins Co., Baltimore, 1974.
6. J. Fay, *Theta Functions on Riemann Surfaces*, Lecture Notes in Math. 352, Springer-Verlag, 1973.
7. M. Mulase, *Cohomological structure in solution equations and Jacobian varieties*, J. Diff. Geom. **19** (1984), 403–430.
8. D. Mumford, *On the equations defining abelian variety*, Invent. Math. 1 (1966).
9. H.E. Rauch, *Schottky Implies Poincaré*, Advances in the Series of Riemann Surfaces, Princeton Univ. Press, 1971.
10. T. Shioto, *Characterization of Jacobian varieties in terms of Soliton equation*, Invent. Math. **83** (1986), 333–382.
11a. F. Schottky, *Zur Theorie der Abelschen Funktionen von vier Variablen*, J. F. Math **102** (1888), 304–352.
11b. _____, *Über die Moduln der Thetafunktionen*, Acta Math. **27** (1903), 235–288.
12. F. Schottky, H. Jung, *Neue Sätze über Symmetralfunkctionen und die Abelschen Funktionen der Riemann'schen Theorie*, S-B Press Akad. Wiss. (Berlin), Phys. Math. Kl. 1 (1909), 282–297.
13. Weber, *Ueber Gewisse in der Theorie der Abelschen Funktionen, Authetende Ausnahmfalle*, Math. Ann. **31** (1878), 35–48.

Department of Mathematics, Temple University, Philadelphia, PA 19122

Contemporary Mathematics
Volume **136**, 1992

The Trisecant Formula and Hyperelliptic Surfaces

Hershel M. Farkas *

Oct. 1, 1990

1   Introduction

The cross ratio of four points on the sphere has been generalized by Gunning
[5] to a compact Riemann surface of arbitrary genus $g > 0$. The definition
made use of the properties of the normalized differentials of third kind. We
recall that the normalized differential of third kind τ_{PQ} on a compact surface
is a meromorphic differential with simple poles at P and Q with residue 1
at P and residue -1 at Q. Furthermore $\int_{\gamma_l} \tau_{PQ} = 0$ for the cycles γ_l, where
$\gamma_1, ... \gamma_g, \delta_1, ..., \delta_g$ is a cannonical homology basis for the surface. The reader
can check using the bilinear relations as found in [1], that the expression

$$exp\left(\int_{P_3}^{P_1} \tau_{P_2 P_4}\right)$$

depends on the path of integration one takes from P_3 to P_1. So if we think
of the three points P_2, P_3, P_4 as fixed and the point P_1 as variable, the above
expression picks up well defined exponential multipliers depending on the
path one chooses from P_3 to P_1. The theory of the theta function as given
in [1] shows that the following is true as well.

---

*Research partially supported by a U.S. Israel BSF grant and the Landau Center for
research in Analysis.
1991 *Mathematics Subject Classification.* Primary 30F10, 14H42; Secondary 30F30, 14H40,
and 14H55.
The final version of this paper will be submitted for publication elsewhere.

Let α be a nonsingular element of the theta divisor. For any such α, and any four points on the Riemann surface consider the following expression:

$$\frac{\theta(\alpha + \int_{P_2}^{P_1})\theta(\alpha + \int_{P_4}^{P_3})}{\theta(\alpha + \int_{P_4}^{P_1})\theta(\alpha + \int_{P_2}^{P_3})} = \lambda(P_1, P_2, P_3, P_4).$$

In the above expression we can think of the last three points as fixed and the point P_1 as a variable point on the surface. In this case the above expression is a multivalued meromorphic function on the surface. When the point P_1 is continued over cycles the expression picks up exponential multipliers. As a multivalued function however it has a zero at $P_1 = P_2$ and a pole at $P_1 = P_4$. If we restrict P_1 to lie in a fundamental polygon for the surface (or in a simply connected region whose closure is the surface) then the value of the function (now single-valued) at $P_1 = P_3$ is one.

It is now an easy task to show that the exponential multipliers which the two multivalued functions defined above pick up are the same so that their quotient is a holomorphic function on the surface and thus constant and equal to 1. The above properties show that $\lambda(P_1, P_2, P_3, P_4)$ is in fact Gunnings generalized cross ratio. We quickly remark that it is possible for the expression not to be well defined for a given α and a given four points. However, if we are given the α there are at most $g - 1$ points on the surface where this can occur. It is therefore always possible to define this quantity. This expression would not be of much use if it were to change with changes of the point α. It is however clear from our previous remarks that $\lambda(P_1, P_2, P_3, P_4)$ is independent of the point α used to define it.

If we now denote the expression $\lambda(P_i, P_j, P_k, P_l)$ by λ_{ijkl}, it follows immediately from the definitions that

$$\lambda_{ijkl} = \lambda_{jilk} = \lambda_{klij} = \lambda_{lkji}.$$

In addition, with a little more work one can also show that if we denote λ_{1234} by λ and λ_{1342} by μ and finally λ_{1423} by ν then $\lambda_{1432} = 1/\lambda$, $\lambda_{1243} = 1/\mu$ and $\lambda_{1324} = 1/\nu$ and $\lambda\mu\nu = -1$.

It in fact follows from the remarks we have already made that

$$\lambda_{ijkl} = \frac{\theta\begin{bmatrix} \mu \\ \mu' \end{bmatrix}(\int_{P_j}^{P_i})\theta\begin{bmatrix} \mu \\ \mu' \end{bmatrix}(\int_{P_l}^{P_k})}{\theta\begin{bmatrix} \mu \\ \mu' \end{bmatrix}(\int_{P_l}^{P_i})\theta\begin{bmatrix} \mu \\ \mu' \end{bmatrix}(\int_{P_j}^{P_k})}$$

for any non singular odd theta characteristic $\begin{bmatrix} \mu \\ \mu' \end{bmatrix}$.

In the classical case of the sphere, it turns out that $\mu = \frac{\lambda-1}{\lambda}$ and $\nu = \frac{1}{1-\lambda}$, so that we have the relation $\lambda + \frac{1}{\nu} = 1$. This relation is false in the general case. The relation which replaces it however as shown by C. Poor [6] is

$$\lambda\frac{\theta(z + \int_{P_4}^{P_1})\theta(z + \int_{P_2}^{P_3})}{\theta(z + \int_{P_2}^{P_1})\theta(z + \int_{P_4}^{P_3})} + \frac{1}{\nu}\frac{\theta(z)\theta(z + \int_{P_2}^{P_1} + \int_{P_4}^{P_3})}{\theta(z + \int_{P_2}^{P_1})\theta(z + \int_{P_4}^{P_3})} = 1.$$

Since the above is true for every $z \in C^g$ it follows that we can replace the theta function with zero characteristic in the above formula with the theta function with any non singular even characteristic. This yields the relation

$$\lambda\frac{\theta\begin{bmatrix} \epsilon \\ \epsilon' \end{bmatrix}(z + \int_{P_4}^{P_1})\theta\begin{bmatrix} \epsilon \\ \epsilon' \end{bmatrix}(z + \int_{P_2}^{P_3})}{\theta\begin{bmatrix} \epsilon \\ \epsilon' \end{bmatrix}(z + \int_{P_2}^{P_1})\theta\begin{bmatrix} \epsilon \\ \epsilon' \end{bmatrix}(z + \int_{P_4}^{P_3})} + \frac{1}{\nu}\frac{\theta\begin{bmatrix} \epsilon \\ \epsilon' \end{bmatrix}(z)\theta\begin{bmatrix} \epsilon \\ \epsilon' \end{bmatrix}(z + \int_{P_2}^{P_1} + \int_{P_4}^{P_3})}{\theta\begin{bmatrix} \epsilon \\ \epsilon' \end{bmatrix}(z + \int_{P_2}^{P_1})\theta\begin{bmatrix} \epsilon \\ \epsilon' \end{bmatrix}(z + \int_{P_4}^{P_3})} = 1.$$

The above formula is known as Fay's trisecant formula. It is the form derived in [3]. Poor derives this formula using the cross ratio ideas and a construction on the Jacobian of the surface while the derivation in [3] uses a construction on the surface itself.

2   The trisecant formula

The trisecant formula has been one of the key tools and ideas in the solution of the Schottky problem; however, when it comes to Schottky's original relation involving theta constants the trisecant formula till now has not shed much light. Can the trisecant formula be used to obtain Schottky type relations?

The Scottky type relations are derived usually as an application of the Schottky-Jung proportionalities which involve the theory of Prym varieties. For a discussion of this topic see [4] and the literature there cited. In this section we shall show that at least for the case of hyperelliptic surfaces Schottky type relations are deriveable from the trisecant formula. That this is true is perhaps not so surprising in view of the fact that Schottky type relations for hyperelliptic surfaces have been derived previously for hyperelliptic surfaces without using the Schottky-Jung proportionalities [2], however, there is some hope that the general case may yet be handled in this way as well.

The drawback of the trisecant formula for the solution of the Schottky Problem is the appearance in it of points on the curve rather than pure theta constants. In the case of a hyperelliptic surface the points can be chosen so that this is not the case and it is therefore of interest to see what the results are in this case.

We shall be considering a compact Riemann surface of genus $g > 0$ which has a representaion as a branched two sheted cover of the sphere with $2g + 2$ branch points. The branching will be assumed to be over the $2g + 2$ points $\lambda_1, ..., \lambda_{2g+2}$ on the sphere and we shall denote by $P_k = z^{-1}(\lambda_k)$ where z is the function on the surface mapping the surface onto the sphere branched over the λ_k. A canonical homology basis is chosen as indicated in Figure 1.

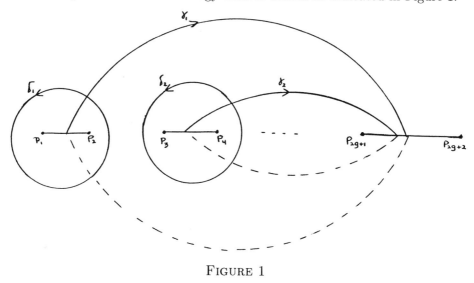

FIGURE 1

Let $\phi_1, ..., \phi_g$ be a basis for the holomorphic differentials dual to the given canonical homology basis and denote the associated period matrix by $(\mathbf{I}\ \Pi)$. A period in the Jacobi variety is then an integral linear combination of the columns of the period matrix and a half period is a linear combination with half integers. The half period, written in matrix notation as $1/2(\Pi\epsilon + I\epsilon')$ will be denoted by the symbol $\begin{pmatrix} \epsilon \\ \epsilon' \end{pmatrix}$. Here ϵ and ϵ' are vectors of integers.

The theta function of n variables and charactertistic $\begin{bmatrix} \epsilon \\ \epsilon' \end{bmatrix}$ is defined as

$$\theta \begin{bmatrix} \epsilon \\ \epsilon' \end{bmatrix} (z, \pi) = \sum_{N \in Z^n} exp2\pi i[1/2(N + \epsilon/2)\Pi(N + \epsilon/2) + (N + \epsilon/2)(z + \epsilon'/2)].$$

This function has the following property:

$$\theta \begin{bmatrix} \epsilon \\ \epsilon' \end{bmatrix} \left(z + \begin{pmatrix} \mu \\ \mu' \end{pmatrix} , \pi \right)$$

$$= exp\pi i[-1/4\mu\Pi\mu - 1/2\mu(\epsilon' + \mu') - \mu z]\theta \begin{bmatrix} \epsilon + \mu \\ \epsilon' + \mu' \end{bmatrix} (z, \pi).$$

We can clearly choose a torus and four points on it so that $\int_{P_1}^{P_2} = 1/2$, $\int_{P_2}^{P_3} = \frac{\tau}{2}$, and such that $\int_{P_1}^{P_4} = \frac{\tau}{2}$. Here and in the sequel the meaning of $\int_P^Q$ is $\int_P^Q \vec{\Phi}$ where $\vec{\Phi}$ denotes a vector whose entries are a basis for the holomorphic differentials on the surface. This is the usual choice of the four points of order two in the period parallelogram. If we now substitute this into the trisecant formula and use the property of the theta function listed above together with the following property

$$\theta \begin{bmatrix} \epsilon \\ \epsilon' \end{bmatrix} (z, \pi) = exp\pi i\epsilon v' \theta \begin{bmatrix} \hat{\epsilon} \\ \hat{\epsilon}' \end{bmatrix} (z, \pi)$$

where $\hat{\epsilon} = \epsilon + 2v$ and $\hat{\epsilon}' = \epsilon' + 2v'$, we find that on the torus $\lambda = -\dfrac{\theta^2 \begin{bmatrix} 1 \\ 0 \end{bmatrix} (0,\tau)}{\theta^2 \begin{bmatrix} 0 \\ 1 \end{bmatrix} (0,\tau)}$

and $\dfrac{1}{\nu} = \dfrac{\theta^2 \begin{bmatrix} 0 \\ 0 \end{bmatrix} (0,\tau)}{\theta^2 \begin{bmatrix} 0 \\ 1 \end{bmatrix} (0,\tau)}$. Hence substituting into Fay's trisecant formula in this case gives rise to the well known elliptic theta identity:

$$\theta^2 \begin{bmatrix} 0 \\ 0 \end{bmatrix} (0)\theta^2 \begin{bmatrix} 0 \\ 0 \end{bmatrix} (z) = \theta^2 \begin{bmatrix} 1 \\ 0 \end{bmatrix} (0)\theta^2 \begin{bmatrix} 1 \\ 0 \end{bmatrix} (z) + \theta^2 \begin{bmatrix} 0 \\ 1 \end{bmatrix} (0)\theta^2 \begin{bmatrix} 0 \\ 1 \end{bmatrix} (z).$$

This by setting $z = 0$ now becomes a theta constant identity.

We now turn to the general hyperelliptic surface of genus $g \geq 2$ with canonical homology basis as given in Figure 1. The theory of hyperelliptic surfaces as indicated in the beginning of chapter VII in [1] shows that for $k = 1, ..., g - 1$ it is the case that

$$\int_{P_1}^{P_2} = 1/2\pi^1$$

$$\int_{P_1}^{P_{2k+1}} = 1/2(\pi^1 + ... + \pi^k + e^1 + e^{k+1})$$

$$\int_{P_1}^{P_{2k+2}} = 1/2(\pi^1 + ... + \pi^{k+1} + e^1 + e^{k+1})$$

and that the vector of Riemann constants with base point at P_1,

$$K_{P_1} = 1/2(ge^1 + e^2 + ... + e^g + g\pi^1 + (g-1)\pi^2 + ... + \pi^g).$$

In the above the equalities are meant to be equalities in the Jacobian. Thus the path of integration is irrelevant. From the above one can write down all the vanishing and non-vanishing properties of hyperelliptic theta functions. Since this is not the intention of this work we, for the sake of simplicity consider the case of genus four and write down all the details. The simplification is that we write down all the vanishing even thetas and note that there are no singular odd thetas. In this way we are sure that we have well defined objects. Furthermore since we will be computing λ and ν we will be a bit more careful with our integrations.

The ten singular even points of order two are easy to compute and are found to be:

$$\begin{pmatrix} 0 & 1 & 0 & 1 \\ 0 & 1 & 1 & 1 \end{pmatrix} = K_{P_1}, \begin{pmatrix} 1 & 1 & 0 & 1 \\ 0 & 1 & 1 & 1 \end{pmatrix}, \begin{pmatrix} 1 & 0 & 0 & 1 \\ 1 & 0 & 1 & 1 \end{pmatrix}, \begin{pmatrix} 1 & 1 & 0 & 1 \\ 1 & 0 & 1 & 1 \end{pmatrix}$$

$$\begin{pmatrix} 1 & 0 & 0 & 1 \\ 1 & 1 & 0 & 1 \end{pmatrix}, \begin{pmatrix} 1 & 0 & 1 & 1 \\ 1 & 1 & 0 & 1 \end{pmatrix}, \begin{pmatrix} 1 & 0 & 1 & 1 \\ 1 & 1 & 1 & 0 \end{pmatrix}, \begin{pmatrix} 1 & 0 & 1 & 0 \\ 1 & 1 & 1 & 0 \end{pmatrix}$$

$$\begin{pmatrix} 1 & 0 & 1 & 0 \\ 1 & 1 & 1 & 1 \end{pmatrix}, \begin{pmatrix} 0 & 1 & 0 & 1 \\ 1 & 1 & 1 & 1 \end{pmatrix}.$$

In order to compute λ and $\frac{1}{\nu}$ we need the values of the following integrals:

$$\int_{P_2}^{P_1} = \begin{pmatrix} 1 & 0 & 0 & 0 \\ 0 & 0 & 0 & 0 \end{pmatrix}$$

$$\int_{P_3}^{P_4} = \begin{pmatrix} 0 & -1 & 0 & 0 \\ 0 & 0 & 0 & 0 \end{pmatrix}$$

$$\int_{P_1}^{P_4} = \begin{pmatrix} -1 & -1 & 0 & 0 \\ 1 & -1 & 0 & 0 \end{pmatrix}$$

$$\int_{P_2}^{P_3} = \begin{pmatrix} 0 & 0 & 0 & 0 \\ 1 & -1 & 0 & 0 \end{pmatrix}.$$

Consider the following 10 odd charactersistics where we take

$$\begin{bmatrix} \mu \\ \mu' \end{bmatrix} \text{ as } \begin{bmatrix} 1 & 0 & \epsilon_1 & \epsilon_2 \\ 1 & 1 & \epsilon_1' & \epsilon_2' \end{bmatrix}$$

where $\begin{bmatrix} \epsilon_1 & \epsilon_2 \\ \epsilon_1' & \epsilon_2' \end{bmatrix}$ is an arbitrary even two characteristic. These ten different odd characteristics give rise to ten different expressions for $\lambda(P_1, P_2, P_3, P_4)$. In particular we obtain

$$\frac{\theta \begin{bmatrix} 1 & 0 & \epsilon_1 & \epsilon_2 \\ 1 & 1 & \epsilon_1' & \epsilon_2' \end{bmatrix} (\int_{P_2}^{P_1}) \theta \begin{bmatrix} 1 & 0 & \epsilon_1 & \epsilon_2 \\ 1 & 1 & \epsilon_1' & \epsilon_2' \end{bmatrix} (\int_{P_4}^{P_3})}{\theta \begin{bmatrix} 1 & 0 & \epsilon_1 & \epsilon_2 \\ 1 & 1 & \epsilon_1' & \epsilon_2' \end{bmatrix} (\int_{P_4}^{P_1}) \theta \begin{bmatrix} 1 & 0 & \epsilon_1 & \epsilon_2 \\ 1 & 1 & \epsilon_1' & \epsilon_2' \end{bmatrix} (\int_{P_2}^{P_3})}$$

is independent of the two characteristic. If we now substitute the values of the integrals into the above we find

$$\frac{\theta \begin{bmatrix} 1 & 0 & \epsilon_1 & \epsilon_2 \\ 1 & 1 & \epsilon_1' & \epsilon_2' \end{bmatrix} (\begin{pmatrix} 1 & 0 & 0 & 0 \\ 0 & 0 & 0 & 0 \end{pmatrix}) \theta \begin{bmatrix} 1 & 0 & \epsilon_1 & \epsilon_2 \\ 1 & 1 & \epsilon_1' & \epsilon_2' \end{bmatrix} (\begin{pmatrix} 0 & 1 & 0 & 0 \\ 0 & 0 & 0 & 0 \end{pmatrix})}{\theta \begin{bmatrix} 1 & 0 & \epsilon_1 & \epsilon_2 \\ 1 & 1 & \epsilon_1' & \epsilon_2' \end{bmatrix} (\begin{pmatrix} 1 & 1 & 0 & 0 \\ -1 & 1 & 0 & 0 \end{pmatrix}) \theta \begin{bmatrix} 1 & 0 & \epsilon_1 & \epsilon_2 \\ 1 & 1 & \epsilon_1' & \epsilon_2' \end{bmatrix} (\begin{pmatrix} 0 & 0 & 0 & 0 \\ 1 & -1 & 0 & 0 \end{pmatrix})}$$

is independent of the two charactertistic and using the property mentioned in the introduction with $z = 0$, we obtain the preceding equal to

$$exp(\frac{\pi i}{2} \pi_{12}) \frac{\theta \begin{bmatrix} 2 & 0 & \epsilon_1 & \epsilon_2 \\ 1 & 1 & \epsilon_1' & \epsilon_2' \end{bmatrix} \theta \begin{bmatrix} 1 & 1 & \epsilon_1 & \epsilon_2 \\ 1 & 1 & \epsilon_1' & \epsilon_2' \end{bmatrix}}{\theta \begin{bmatrix} 2 & 1 & \epsilon_1 & \epsilon_2 \\ 0 & 2 & \epsilon_1' & \epsilon_2' \end{bmatrix} \theta \begin{bmatrix} 1 & 0 & \epsilon_1 & \epsilon_2 \\ 2 & 0 & \epsilon_1' & \epsilon_2' \end{bmatrix}}.$$

The above together with the property mentioned previously gives us the following pretty symmetric form of the above computation.

Theorem.

$$\lambda(P_1, P_2, P_3, P_4) = exp(\frac{\pi i}{2}\pi_{12})\frac{\theta\begin{bmatrix} 0 & 0 & \epsilon_1 & \epsilon_2 \\ 1 & 1 & \epsilon_1' & \epsilon_2' \end{bmatrix}\theta\begin{bmatrix} 1 & 1 & \epsilon_1 & \epsilon_2 \\ 1 & 1 & \epsilon_1' & \epsilon_2' \end{bmatrix}}{\theta\begin{bmatrix} 0 & 1 & \epsilon_1 & \epsilon_2 \\ 0 & 0 & \epsilon_1' & \epsilon_2' \end{bmatrix}\theta\begin{bmatrix} 1 & 0 & \epsilon_1 & \epsilon_2 \\ 0 & 0 & \epsilon_1' & \epsilon_2' \end{bmatrix}}$$

and

$$\lambda(P_1, P_3, P_2, P_4) = exp(\frac{\pi i}{2}\pi_{12})\frac{\theta\begin{bmatrix} 0 & 0 & \epsilon_1 & \epsilon_2 \\ 0 & 0 & \epsilon_1' & \epsilon_2' \end{bmatrix}\theta\begin{bmatrix} 1 & 1 & \epsilon_1 & \epsilon_2 \\ 0 & 0 & \epsilon_1' & \epsilon_2' \end{bmatrix}}{\theta\begin{bmatrix} 0 & 1 & \epsilon_1 & \epsilon_2 \\ 0 & 0 & \epsilon_1' & \epsilon_2' \end{bmatrix}\theta\begin{bmatrix} 1 & 0 & \epsilon_1 & \epsilon_2 \\ 0 & 0 & \epsilon_1' & \epsilon_2' \end{bmatrix}}$$

are independent of the even two characteristic $\begin{bmatrix} \epsilon_1 & \epsilon_2 \\ \epsilon_1' & \epsilon_2' \end{bmatrix}$.

Proof.   The proof of the first statement is the computation made above, and the proof of the second statement is a similar computation.

Let us now choose for our nonsingular even characteristic the four characteristic $\begin{bmatrix} \delta \\ \delta' \end{bmatrix}$ as $\begin{bmatrix} 0 & 0 & \delta_1 & \delta_2 \\ 0 & 0 & \delta_1' & \delta_2' \end{bmatrix}$ with $\begin{bmatrix} \delta_1 & \delta_2 \\ \delta_1' & \delta_2' \end{bmatrix}$ an arbitrary even two characteristic. In this case the trisecant formula reads:

$$\theta\begin{bmatrix} 0 & 1 & \delta_1 & \delta_2 \\ 0 & 0 & \delta_1' & \delta_2' \end{bmatrix}\theta\begin{bmatrix} 1 & 0 & \delta_1 & \delta_2 \\ 0 & 0 & \delta_1' & \delta_2' \end{bmatrix}\theta\begin{bmatrix} 0 & 1 & \epsilon_1 & \epsilon_2 \\ 0 & 0 & \epsilon_1' & \epsilon_2' \end{bmatrix}\theta\begin{bmatrix} 1 & 0 & \epsilon_1 & \epsilon_2 \\ 0 & 0 & \epsilon_1' & \epsilon_2' \end{bmatrix} +$$

$$\theta\begin{bmatrix} 1 & 1 & \delta_1 & \delta_2 \\ 1 & 1 & \delta_1' & \delta_2' \end{bmatrix}\theta\begin{bmatrix} 0 & 0 & \delta_1 & \delta_2 \\ 1 & 1 & \delta_1' & \delta_2' \end{bmatrix}\theta\begin{bmatrix} 0 & 0 & \epsilon_1 & \epsilon_2 \\ 1 & 1 & \epsilon_1' & \epsilon_2' \end{bmatrix}\theta\begin{bmatrix} 1 & 1 & \epsilon_1 & \epsilon_2 \\ 1 & 1 & \epsilon_1' & \epsilon_2' \end{bmatrix} =$$

$$\theta\begin{bmatrix} 1 & 1 & \delta_1 & \delta_2 \\ 0 & 0 & \delta_1' & \delta_2' \end{bmatrix}\theta\begin{bmatrix} 0 & 0 & \delta_1 & \delta_2 \\ 0 & 0 & \delta_1' & \delta_2' \end{bmatrix}\theta\begin{bmatrix} 1 & 1 & \epsilon_1 & \epsilon_2 \\ 0 & 0 & \epsilon_1' & \epsilon_2' \end{bmatrix}\theta\begin{bmatrix} 0 & 0 & \epsilon_1 & \epsilon_2 \\ 0 & 0 & \epsilon_1' & \epsilon_2' \end{bmatrix}$$

for all even two characteristics $\begin{bmatrix} \epsilon \\ \epsilon' \end{bmatrix}$ and $\begin{bmatrix} \delta \\ \delta' \end{bmatrix}$.

If we now divide through by the first term of the previous equation we obtain an equation of the form:

$$1 + A(\delta, \delta')A(\epsilon, \epsilon') = B(\delta, \delta')B(\epsilon, \epsilon').$$

We note however that $A(\delta, \delta')$ and $B(\delta, \delta')$ are independent of the two characteristics. The reason for this is of course the form of $A(\delta, \delta')$. We note that $A(\delta, \delta')$ and $B(\delta, \delta')$ are precisely the expressions given in our theorem aside from the exponential multiplier. Putting all of the above together gives us:

$$1 + \sqrt{A(\delta, \delta')A(\tilde{\delta}, \tilde{\delta'})A(\epsilon, \epsilon')A(\tilde{\epsilon}, \tilde{\epsilon'})} = \sqrt{B(\delta, \delta')B(\tilde{\delta}, \tilde{\delta'})B(\epsilon, \epsilon')B(\tilde{\epsilon}, \tilde{\epsilon'})}$$

which is a Schottky type relation.

References

[1] H. M. Farkas and I. Kra. *Riemann Surfaces*, volume 71 of *Graduate Texts in Mathematics*. Springer-Verlag, 1980.

[2] H.M. Farkas. Period relations for hyperelliptic riemann surfaces. *Israel Journal of Mathematics*, 10:289–301, 1971.

[3] H.M. Farkas. On Fay's trisecant formula. *Journal D'analyse Mathematique*, 44:205–217, 1984/85.

[4] H.M. Farkas. H.E. Rauch, theta function practitioner. *Differential Geometry and Complex Analysis; H.E. Rauch Memorial Volume, Springer-Verlag*, pages 33–47, 1985.

[5] R.C. Gunning. *Lecture Notes on Theta Functions. Parts A and B. (to appear)*. Princeton University Press.

[6] C. Poor. Another proof of Fay's trisecant formula. *(to appear)*.

INSTITUTE OF MATHEMATICS

THE HEBREW UNIVERSITY OF JERUSALEM

JERUSALEM ISRAEL

Contemporary Mathematics
Volume **136**, 1992

The Non-Abelian Szegö Kernel and Theta-Divisor

JOHN FAY [*]

Introduction. Let C be a compact Riemann surface of genus $g > 1$ and R be the moduli space of rank r, degree 0 stable bundles on C equipped with holomorphic connections. R is a complex manifold of dimension $2m$, $m = r^2(g-1) + 1$, and can be identified (via the monodromy representation) with a dense open subset of $\mathrm{Hom}(\pi_1(C), \mathrm{GL}(r, \mathbb{C}))/\mathrm{PGL}(r, \mathbb{C})$. For a fixed square root Δ of the canonical bundle K_C, define the theta-divisor $(\theta) \subset R$ to be (set theoretically) those $\chi \in R$ for which $h^0(\chi \otimes \Delta) > 0$ [5]. One approach to constructing an analytic section $\theta(\chi)$ vanishing on (θ) uses the Quillen factorization of Ray-Singer analytic torsion for $\chi \otimes \Delta$ equipped with some hermitian metric. In this paper, we define directly from the Szegö kernel an intrinsic one-form ω on R holomorphic off (θ) and with a simple pole of residue 1 along (θ). As a local equation for (θ), we use a regularized determinant following that given by Malgrange [4] in his analysis of the Sato-Miwa-Jimbo tau-function for isomonodromic deformations on $\mathbb{P}_1$. This determinant, as well as the general holomorphic section $\theta(\chi)$ constructed from ω, will vanish to order $h^0(\chi \otimes \Delta)$ at any $\chi \in (\theta)$. We describe the tangent cone at singular points of (θ) and also show (¶4) that for fixed $\chi \in R$, $\theta(\chi \otimes \lambda)$ is an abelian function of $\lambda \in J_0(C)$ which can be expressed as a determinant of Szegö kernels.

1. Construction of ω. For notational convenience, we write sections of $\chi \otimes \Delta$ as vector functions $f(z)$ on the universal cover H of C satisfying

$$(1) \qquad f(\gamma z) = (cz + d)\chi_\gamma f(z), \qquad \gamma \in \Gamma = \pi_1(C)$$

if $(cz+d)$ are the automorphy factors for Δ trivialized on H. If $h^0(\chi \otimes \Delta) = 0$, then by Riemann-Roch $h^0(\chi \otimes \Delta \otimes \{z\}) = r$ for all $z \in C$, and one can form an $r \times r$ matrix $S(z', z; \chi)$ whose columns span the meromorphic

1991 *Mathematics Subject Classification.* Primary 32G13, 14K25.
[*] Research supported by the Institute for Advanced Study.
This paper is in final form and no version of it will be submitted for publication elsewhere.

sections (in z') of $\chi \otimes \Delta$ with a simple pole at $z' = z$. Furthermore, lifting $S(z', z; \chi)$ to H we may assume S is locally meromorphic in z and $\operatorname{Res}_{z'=z} S(z', z; \chi) = -I$ for z restricted to sufficiently small neighborhoods in H. Analytic continuation in z implies that

$$S(z', \gamma z; \chi) = S(z', z; \chi)\phi_\gamma(z)$$

for some invertible matrix $\phi_\gamma(z)$; and the residue condition gives

$$-(cz + d)^2 I = \operatorname*{Res}_{z'=z} S(\gamma z', \gamma z; \chi)$$

$$= \operatorname*{Res}_{z'=z}(cz' + d)\chi_\gamma S(z', z; \chi)\phi_\gamma(z) = -(cz + d)\chi_\gamma \phi_\gamma(z).$$

Thus $\phi_\gamma(z) = (cz + d)\chi_\gamma^{-1}$ and $S(z', z; \chi)$ is a section in z of $\chi^* \otimes \Delta$ with a simple pole at $z = z'$ of residue I:

(2) $S(z', z; \chi) = -{}^t S(z, z'; \chi^*),$ $\chi^* = {}^t\chi^{-1}.$

The Laurent expansion at $z' = z \in H$ has the form

(3) $S(z', z; \chi) = \dfrac{1}{z - z'} + a_0(z; \chi) + O(z' - z)$

where $a_0(z; \chi) = -{}^t a_0(z; \chi^*) \in H^0(K \otimes \operatorname{Ad}\chi)$; and from Cauchy's Theorem,

(4) $f(z') = -\sum\limits_{z \neq z'} \operatorname{Res}[S(z', z; \chi)f(z)]$

for any meromorphic section f of $\chi \otimes \Delta$ holomorphic at z'.

Now suppose $s = (s_1, \ldots, s_{2m})$ are local holomorphic coordinates for some open set of $\chi(s) \in R$. The tangent space to R at $\chi(s)$ can be identified with the cohomology group $H^1(\Gamma, \operatorname{Ad}\chi)$ which in turn is the quotient space {meromorphic $\operatorname{Ad}\chi$-differentials of the second kind/exact differentials} [3; p. 173]. So we can write:

(5) $-\chi_\gamma^{-1}(s)\Omega_i(\gamma z, s)\chi_\gamma(s) + \Omega_i(z, s) = \chi_\gamma^{-1}\partial_{s_i}\chi_\gamma(s)$ $\gamma \in \Gamma, z \in H$

for some meromorphic $\operatorname{Ad}\chi$-integrals Ω_i, $1 \leq i \leq 2m$; the local one-form on R

(6) $\Omega(z, s) = \sum\limits_{i=1}^{2m} \Omega_i(z, s)ds_i$

will thus be uniquely determined up to addition by a meromorphic section of the bundle $\operatorname{Ad}\chi$. We can and will assume that for $\chi(s)$ in sufficiently small open sets U in R, $\Omega(z, s)$ has only finitely many (possibly movable) singularities in z all inside some disc $D \subset C$ (depending on U), while $\Omega(z, s)$ is biholomorphic in z and s for z outside D. This can be achieved, for instance, by taking

(7) $\Omega(z, s) = -dN(z, s)N(z, s)^{-1}$

where $N(z, s)$ is any local family of invertible matrix cross sections of the bundle $\mathrm{Hom}(\chi(s_1), \chi(s))$ for some fixed $\chi(s_1)$, with N and N^{-1} having only finitely many singularities all inside D. Then for any local family of $A(z, s) \in H^0(K \otimes \mathrm{Ad}\,\chi)$ holomorphic in s, $tr \sum \mathrm{Res}_z\, A(z, s)\Omega(z, s)$ will define, as a Cauchy integral over ∂D, a local holomorphic one-form on R independent of the choice of Ω.

THEOREM 1. *For* $\chi(s)$ *with* $h^0(\chi(s) \otimes \Delta) = 0$, *define the one-form*

$$(8) \qquad \omega(s) = tr \sum_z \mathrm{Res}[a_0(z; \chi(s))\Omega(z, s)]$$

independent of Ω *and holomorphic on* $R - (\theta)$. *Then*

$$(9) \qquad d\omega(s) = -\frac{1}{2} tr \sum_z \mathrm{Res}(\Omega \wedge \partial_z \Omega)(z, s).$$

PROOF. The Szegö kernel is holomorphic off (θ) (see Lemma 1) and so a_0 and ω are holomorphic there also by our hypothesis on Ω. Now (5) implies that

$$f(z') := dS(z', z; \chi(s)) + \Omega(z', s)S(z', z; \chi(s)) - S(z', z; \chi(s)\Omega(z, s)$$

is a well-defined meromorphic section (in z') of $\chi(s) \otimes \Delta$ holomorphic at $z' = z$; and so by (4):

$$dS(z', z; \chi(s)) = -\Omega(z', s)S(z', z; \chi(s)) + S(z', z; \chi(s)\Omega(z, s)$$
$$- \sum_{\zeta \neq z', z} \mathrm{Res}\, S(z', \zeta; \chi(s))\Omega(\zeta, s)S(\zeta, z; \chi(s)).$$

Taking the Laurent expansion at $z' = z$, we find

$$da_0(z; \chi(s)) = -[\Omega(z, s), a_0(z; \chi(s))] + \partial_z\Omega(z, s)$$
$$- \sum_{\zeta \neq z} \mathrm{Res}\, S(z, \zeta; \chi(s))\Omega(\zeta, s)S(\zeta, z; \chi(s)),$$

and hence

$$-d\omega = -tr \sum_z \mathrm{Res}[a_0(z; \chi(s))(d\Omega + \Omega \wedge \Omega)(z, s)]$$
$$+ tr \sum_z \mathrm{Res}(\Omega a_0 \Omega + \Omega \partial_z \Omega)(z, s)$$
$$+ tr \sum_z \mathrm{Res} \left\{ \sum_{\zeta \neq z} \mathrm{Res}[S(z, \zeta; \chi(s))\Omega(\zeta, s)S(\zeta, z; \chi(s))\Omega(z, s)] \right\}.$$

From (5), $d\Omega + \Omega \wedge \Omega$ is a meromorphic section of $\mathrm{Ad}\,\chi(s)$ and so the first term above vanishes. Interchanging z and ζ and then the order of summation, the third term, say $\sum$, can be written:

$$\sum = -\sum -tr \sum_z \mathrm{Res} \left\{ \mathrm{Res}_{\zeta=z}\, S(z, \zeta; \chi(s))\Omega(\zeta, s)S(\zeta, z; \chi(s))\Omega(z, s) \right\}$$
$$= -\sum -2tr \sum_z \mathrm{Res} \left(\Omega a_0 \Omega + \frac{1}{2}\Omega \partial_z \Omega \right)(z, s)$$

and (9) then follows.

An alternate definition of ω is:

(10) $\omega(s) = \partial \log T(\chi(s) \otimes \Delta) + \dfrac{1}{\pi} tr \iint_C (h^{-1}\partial_z h)\partial_{\bar{z}}(h^{-1}\partial h)dx\,dy.$

where $T(\chi(s) \otimes \Delta)$ is Ray-Singer analytic torsion, the determinant of the Laplacian $4\rho^2 h^{-1}\partial_z(h\partial_{\bar{z}})$ for any metric $\rho^{-1}|dz|$ on C and family of smooth metrics $h(z, \chi(s)) \otimes \rho(z)$ on $\chi(s) \otimes \Delta$. The proof of (10) and the fact that $\bar{\partial}\omega = 0$ requires perturbation formulas for analytic torsion and the curvature for the Quillen metric—see [2]. The holomorphic two-form

(9)′ $d\omega(s) = \dfrac{1}{\pi} tr \iint_C \partial_z(h^{-1}\partial h)\partial_{\bar{z}}(h^{-1}\partial h)dx\,dy$

will then be independent of the metric h (note that $\Omega - h^{-1}\partial h$ is a well-defined section of $\mathrm{Ad}\,\chi$) and cohomologous to the Chern class $-2\pi i c_1(\theta) = \bar{\partial}\partial \log T(\chi(s) \otimes \Delta)$ for the line bundle determined by (θ).

As motivation for (8), we describe the situation in the abelian (rank one) case. Here, we choose coordinates $(a, b) \in \mathbb{C}^{2g}$ covering $R = \mathrm{Hom}(\Gamma, \mathbb{C}^*)$ so that the bundle $\chi(s)$ has multipliers:

(11) $\chi_{A_j} = e^{-2\pi i a_j}, \quad \chi_{B_j} = e^{2\pi i b_j} \qquad 1 \le j \le g$

for an (A_j, B_j) marking of C. Let τ be the period matrix of the abelian differentials $v = (v_1, \ldots, v_g)$ normalized for the given marking and set

$$\theta \begin{bmatrix} a \\ b \end{bmatrix}(u) = \sum_{n \in \mathbb{Z}^g} \exp\{\pi i^t(n + a)\tau(n + a) + 2\pi i^t(n + a)(u + b)\}$$

and $\theta(u) = \theta\begin{bmatrix} 0 \\ 0 \end{bmatrix}(u)$ for $u \in \mathbb{C}^g$. If Δ is chosen to be the Riemann divisor class for θ:

(12)
$$S(z', z; \chi(s)) = \dfrac{\theta\begin{bmatrix} a \\ b \end{bmatrix}(\int_{z'}^{z} v)}{\theta\begin{bmatrix} a \\ b \end{bmatrix}(0)E(z', z)}$$

$$= e^{-2\pi i \int_z^{z'} a \cdot v} \dfrac{\theta(\int_z^{z'} v - t)}{\theta(t)E(z', z)}, \qquad t = \tau a + b$$

where E is the Schottky-Klein prime form [1; p. 16]. The differential (3) is thus

(13)
$$a_0(z; \chi(s)) = \sum_1^g \partial_{u_j} \log \theta \begin{bmatrix} a \\ b \end{bmatrix}(u)\bigg|_{u=0} v_j(z)$$

$$= \sum_1^g (\partial_{t_j} \log \theta(t) + 2\pi i a_j)v_j(z).$$

Now let (M_{ij}) be the $g \times g$ matrix inverse to $(v_i(p_j))$ for distinct generic points $p_1, \ldots, \mathrm{p}_g \in C$. If $\omega_2(z, \mathrm{p})$ is the differential with double pole at

$z = p$ only and periods $0, 2\pi i v_j(p)$ along A_j, B_j, we can take for (6):

$$(14) \qquad \Omega(z, a, b) = 2\pi i \left(\int_{z_0}^z v \right) \cdot da$$

$$- {}^t \left(\int_{z_0}^z \omega_2(z, p_1), \ldots, \int_{z_0}^z \omega_2(z, p_g) \right) M(\tau da + db)$$

and then

$$(15) \qquad \sum_z \operatorname{Res} v_j(z) \Omega(z, a, b) = (\tau da + db)_j, \qquad 1 \le j \le g.$$

From (13) and (15):

$$\omega(a, b) = d \log \theta(\tau a + b) + 2\pi i a \cdot (\tau da + db)$$

and

$$d\omega(a, b) = 2\pi i \sum_1^g da_j \wedge db_j$$

is cohomologous to the pullback to R of the Riemann form $-2\pi i c_1(\theta)$ on the Jacobi variety $J_0(C)$:

$$d\omega - \pi (\operatorname{Im} \tau)^{-1} dt \wedge \overline{dt} = 2\pi i d[(a - (\operatorname{Im} \tau)^{-1} \operatorname{Im} t) \cdot dt]$$

The relation (10) is equivalent to the factorization (see [2]):

$$T(\chi(s) \otimes \Delta) = c_p \exp[-2\pi \operatorname{Im} t \cdot (\operatorname{Im} \tau)^{-1} \operatorname{Im} t] |\theta(t)|^2$$

for some anomaly c_p depending only on the metric ρ, provided $\chi(s)$ is given the harmonic metric

$$h(z) = \exp -2\pi \operatorname{Re} \int_{z_0}^z (\operatorname{Im} \tau)^{-1} ((\overline{\tau} a + b) v - (\tau a + b) \overline{v}).$$

2. $S(z', z; \chi)$ near the theta-divisor. To analyze $\omega(s)$ near (θ) we need:

LEMMA 1. $S(z', z; \chi(s))$ *is locally meromorphic on* R.

PROOF. Choose a divisor $D = \sum_1^d p_k$ with $p_k \in C$ distinct and d so large that $h^1(\chi \otimes D) = 0$ for $\chi \in R$. Let $\{\sigma_j(z, \chi), 1 \le j \le \mu = h^0(\chi \otimes D) = (d + 1 - g)r\}$ be any holomorphic local system in $\chi \in R$ forming a basis for sections (in z) of χ with poles at D. Fix distinct $z_0, \ldots, z_{d-g} \in C$ with $\sum_0^{d-g} z_i \equiv D - \Delta$ in $J_{d+1-g}(C)$ and let $M(\chi)$ be the $\mu \times \mu$ matrix $(\sigma_j(z_i))$, $1 \le j \le \mu$ and $0 \le i \le d - g$; then $\det M(\chi) = 0$ if and only if $h^0(\chi \otimes \Delta) > 0$. For each $k = 1, 2, \ldots, r$, let $g_k(z, \chi)$ be the formal expansion of the determinant of the $\mu \times \mu$ "matrix" obtained by replacing the kth row of $M(\chi)$ by the sections $(\sigma_1, \ldots, \sigma_\mu)$. Then

$$f_k(z, \chi) = g_k(z, \chi) \frac{\prod_1^d E(z, p_j) \prod_1^{d-g} E(z_0, z_i)}{\prod_1^{d-g} E(z, z_i) \prod_1^d E(z_0, p_j)} \frac{U(z)}{E(z_0, z)}$$

will be a section of $\chi \otimes \Delta$ for a suitably chosen holomorphic section $U(z) \neq 0$ of the trivial bundle $\sum_0^{d-g} z_i + \Delta - D$ (here $E(z, w)$ is any prime-form vanishing only when $z = w$). The f_k by construction are holomorphic in χ and in z except for a simple pole at $z = z_0$ with residue $U(z_0) \det M^t(0, 0, \ldots, \overset{(k)}{1}, \ldots, 0)$. Thus

$$-(f_1(z, \chi), \ldots, f_r(z, \chi)) = U(z_0) \det M(\chi) S(z, z_0; \chi)$$

and S is locally meromorphic on R.

Suppose $\chi(t) = \chi(s(t))$ is any smooth curve passing through $\chi(0) = \chi$; then we can rewrite (5) as:

$$(5)' \qquad -\chi_\gamma^{-1} \eta(\gamma z, \dot{\chi}) \chi_\gamma + \eta(z, \dot{\chi}) = \chi_\gamma^{-1} \dot{\chi}_\gamma(0) \in H^1(\Gamma, \operatorname{Ad} \chi)$$

if $\dot{\chi} = \partial_t \chi(t)|_{t=0}$ and $\eta(z, \dot{\chi}) dt = \Omega(z, s(t))_{t=0}$. Up to addition by a section of $\operatorname{Ad}\chi$, $\eta(z, \dot{\chi})$ will be holomorphic in z (that is, $\chi^{-1}\dot{\chi} = 0$ in $H^1(\operatorname{Ad}\chi)$) if and only if $\chi^{-1}\dot{\chi} \in \delta H^0(K \otimes \operatorname{Ad}\chi)$, the subspace of $H^1(\Gamma, \operatorname{Ad}\chi)$ consistency of period classes of holomorphic $\operatorname{Ad}\chi$-differentials; this subspace is the tangent space at χ to the m-dimensional manifold of representations near χ defining the same holomorphic bundle [3; p. 224]. The differential $\omega(s)$ (8) thus vanishes along the subspace $\delta H^0(K \otimes \operatorname{Ad}\chi)$ at each χ, reflecting the fact that $\omega(s)$ is the pullback to R of a local one-form on the m-dimensional moduli space of holomorphic, stable degree 0 bundles (ignoring the flat structure).

THEOREM 2. *For $\chi \in (\theta)$ and $\eta(z, \dot{\chi})$ as in $(5)'$, the pairing*

$$(16) \qquad \begin{aligned} (e, f) &\in H^0(\chi \otimes \Delta) \times H^0(\chi^* \otimes \Delta) \to \langle e, f \rangle_{\dot{\chi}} \in \mathbb{C} \\ \langle e, f \rangle_{\dot{\chi}} &= \sum \operatorname{Res}_z{}^t f(z) \eta(z, \dot{\chi}) e(z) \end{aligned}$$

is non-degenerate for generic $\chi^{-1}\dot{\chi} \in H^1(\Gamma, \operatorname{Ad}\chi)$. Along such directions, $S(z', z; \chi(t))$ has a simple pole at $t = 0$ with residue $\sum_1^n e_i(z')^t f_i(z)$ if $\{e_i\}$, $\{f_i\}$ are any bases of $H^0(\chi \otimes \Delta)$, $H^0(\chi^* \otimes \Delta)$ with $\langle e_i, f_j \rangle_{\dot{\chi}} = \delta_i^j$. The form ω restricted to the generic curve $\chi(t)$ thus has a simple pole of residue $n = h^0(\chi \otimes \Delta)$ at $t = 0$:

$$(17) \qquad tr \sum \operatorname{Res}_z \left[\operatorname{Res}_{t=0} a_0(z; \chi(t)) \Omega(z, s(t)) \right] = h^0(\chi \otimes \Delta)$$

PROOF. The tangent vector to $\chi(t)$ defines a linear map:

$$(18) \qquad e \in H^0(\chi \otimes \Delta) \to \chi^{-1}\dot{\chi} e \in H^1(\chi \otimes \Delta) \simeq H^0(\chi^* \otimes \Delta)^*$$

where by definition $(5)'$, the pairing on the right-side of (18) is:

$$f \in H^0(\chi^* \otimes \Delta) \to tr \sum \operatorname{Res}_z \eta(z, \dot{\chi}) e(z)^t f(z) = \langle e, f \rangle_{\dot{\chi}}.$$

Hence (18) is an isomorphism except on the subvariety $(\theta)_\chi$ of $\chi^{-1}\dot{\chi} \in H^1(\Gamma, \operatorname{Ad}\chi)$ where

$$(19) \qquad \det_{1 \le i, j \le n} \left(tr \sum_z \operatorname{Res} \eta(z, \dot{\chi}) e_i(z, \chi)^t e_j(z, \chi^*) \right) = 0$$

for any bases $\{e_i(z, \chi^{(*)})\}$ of $H^0(\chi^{(*)} \otimes \Delta)$. Note that $\langle\ ,\ \rangle_{\dot{\chi}}$ is identically 0 if $\chi^{-1}\dot{\chi} = 0$ in $H^1(\operatorname{Ad}\chi)$ and so $(\theta)_\chi$ is ruled by the m-dimensional subspaces $\delta H^0(K \otimes \operatorname{Ad}\chi)$. We claim $(\theta)_\chi$ is a proper subvariety of $H^1(\Gamma, \operatorname{Ad}\chi)$: this is equivalent to showing there is some direction $\chi^{-1}\dot{\chi}$ for which there is no $e \in H^0(\chi^* \otimes \Delta)$ with

$$(20) \qquad tr \sum \operatorname{Res} \eta(z, \dot{\chi}) e_i(z, \chi)^t e(z, \chi^*) = 0$$

for all $i = 1, 2, \ldots, n$. But for any $e \in H^0(\chi^* \otimes \Delta)$, $\operatorname{span}\{e_i{}^t e, 1 \le i \le n\}$ is an n-dimensional subspace of $H^0(K \otimes \operatorname{Ad}\chi) \simeq H^1(\operatorname{Ad}\chi^*)^*$ contained in the kernel (20) of a $2m - n$ dimensional family of functionals η defined from $\chi^{-1}\dot{\chi}$. As e varies, the space of all such $\chi^{-1}\dot{\chi}$ has dimension at most $(2m - n) + n - 1 = 2m - 1$ and thus (20) cannot hold for all $\chi^{-1}\dot{\chi} \in H^1(\Gamma, \operatorname{Ad}\chi)$.

To prove (17), assume $\chi(t)$ is a generic curve—that is, one for which the tangent vector $\chi^{-1}\dot{\chi}$ is off $(\theta)_\chi$. For any basis $\{e_i\}$ of $H^0(\chi \otimes \Delta)$, let $E_i(z) = E_i(z, \dot{\chi})$ be any meromorphic vectors (cochains) such that

$$-\chi_\gamma^{-1} E_i(\gamma z)(cz + d)^{-1} + E_i(z) = \chi_\gamma^{-1}\dot{\chi}_\gamma e_i(z, \chi) \in H^1(\chi \otimes \Delta).$$

Then from (18), there is a basis $\{f_j\}$ of $H^0(\chi^* \otimes \Delta)$ such that

$$\langle e_i, f_j \rangle_{\dot{\chi}} = \sum_z \operatorname{Res} {}^t E_i(z) f_j(z) = \delta_i^j.$$

Along $\chi(t)$, the Szegö-kernel has by Lemma 1 a Laurent expansion

$$S(z', z; \chi(t)) = t^{-N}(\psi(z', z) + t\psi_1(z', z) + \cdots + t^k \psi_k(z', z) + \cdots)$$

where all terms are holomorphic in z', z except for ψ_N which has a simple pole of residue $-I$ at $z' = z$. Also $N \ge 1$ since otherwise (4) would be contradicted for $f \in H^0(\chi \otimes \Delta)$. From the automorphy of S:

$$\chi_\gamma^{-1}\psi_1(\gamma z', z)(cz' + d)^{-1} - \psi_1(z', z) = \chi_\gamma^{-1}\dot{\chi}_\gamma \psi(z', z)$$

and hence, if we write $\psi(z', z) = \sum_{i=1}^n e_i(z', \chi)^t h_i(z, \chi^*)$ for sections $h_i \in H^0(\chi^* \otimes \Delta)$ not all zero,

$${}^t h_j(z, \chi^*) = \sum_{i=1}^n \sum_{z'} \operatorname{Res} {}^t f_j(z', \chi^*) E_i(z')^t h_i(z, \chi^*)$$

$$= -\sum_{z'} \operatorname{Res} {}^t f_j(z', \chi^*)\psi_1(z', z) = \begin{cases} {}^t f_j(z, \chi^*) & N = 1 \\ 0 & N > 1 \end{cases}$$

for all $j = 1, \ldots, n$. Thus $N = 1$, $h_j = f_j$ for all j and

$$tr \sum_z \text{Res} \, \psi(z, z) \eta(z, \dot{\chi}) = \sum_{i=1}^{n} \sum_z \text{Res} \, {}^t E_i(z) f_i(z) = n.$$

REMARK. In the abelian case, the tangent cone condition (19) reduces by (14) to the classical identity

$$\sum_1^g \frac{\partial^n \theta \begin{bmatrix} a \\ b \end{bmatrix}}{\partial u_{i_1} \cdots \partial u_{i_n}} (0) w_{i_1} \cdots w_{i_n} = c \, \det[{}^t(e_i(p_k, \chi^*))(e_j(p_k, \chi))]$$

for $w = \sum_1^g v(p_k)$ and $c \neq 0$ a constant depending on the choice of bases but independent of the generic $p_1, \ldots, p_g \in C$.

3. Local equation for (θ). Let D be some fixed open disc about a point $p \in C$, and identify D with the unit disc $|u| < 1$ by means of a local coordinate u at p. Let S^1 be the unit circle $|u| = 1$ with measure $\frac{du}{2\pi i u}$, and H_+^r (resp. H_-^r) the closed subspace of $(L_2(S^1))^r = H_+^r \oplus H_-^r$ spanned by $u^{-k} e_l$, $k \geq 1$ (resp. $u^k e_l$, $k \geq 0$) for $e_l = {}^t(0, 0, \overset{l}{1}, \ldots, 0)$, $1 \leq l \leq r$; the projection onto H_+^r is then

$$f_+(z) := -\frac{1}{2\pi i} \int_{|u|=1} f(u) \frac{du}{u - z} = \sum_1^\infty a_n z^{-n} \quad \text{for } |z| > 1$$

if $f(u) = \sum_{-\infty}^\infty a_n u^{-n}$ is holomorphic near S^1. Choose a square root Δ of K_C with $h^0(\Delta) = 0$ and fix a trivialization of the bundles $\chi(s) \otimes \Delta$ near p, so that sections become column vectors, as in (1). Let $V \simeq H_+^r$ be the closure of the subspace of L_2^r spanned by those functions holomorphic near S^1 which extend to holomorphic sections of $I_r \otimes \Delta$ on $C - D$ (I_r being the trivial rank r bundle). For any $r \times r$ matrix N holomorphic near S^1, define $T_N \colon V \to H_+^r$ by

$$(21) \qquad (T_N g)(z) = (Ng)_+(z) = -\frac{1}{2\pi i} \int_{|u|=1} N(u) g(u) \frac{du}{u - z}$$

for $g \in V$ and $|z| > 1$. We will be particularly interested in the case that $N(u) = N(u, s)$ is a local family of matrix cross sections of $\text{Hom}(I_r, \chi(s))$ such that N and N^{-1}, as in (7), have only finitely many singularities in $u \in D$ and are biholomorphic in u and s for all u outside D. Then $\text{Ker} \, T_N \simeq H^0(\chi(s) \otimes \Delta)$ under left multiplication by $N^{-1}(z)$, and $\text{Coker} \, T_N \simeq H^0(\chi(s)^* \otimes \Delta)^*$ by

$$f \in H^0(\chi(s)^* \otimes \Delta) \to \frac{1}{2\pi i} \int_{|z|=1} {}^t h(z) f(z) \, dz, \qquad h \in H_+^r.$$

If $h^0(\chi(s) \otimes \Delta) = 0$, T_N is invertible with

$$(22) \qquad (T_N^{-1} h)(z) = -\frac{1}{2\pi i} N^{-1}(z) \int_{|u|=1} S(z, u; \chi(s)) h(u) \, du$$

for $h \in H_+^r$ and $|z| > 1$; this follows from (4) since for small $\varepsilon > 0$:

$$\frac{1}{(2\pi i)^2} N^{-1}(z) \int_{|u|=1+\varepsilon} S(z, u; \chi(s)) \left(\int_{|w|=1} N(w)g(w) \frac{dw}{w-u} \right) du$$

$$= -N^{-1}(z) \sum_{w \neq z} \operatorname{Res} S(z, w; \chi(s)) N(w)g(w) = g(z) \in V$$

Define $T_I^{-1}: H_+^r \xrightarrow{\sim} V$ to be the isomorphism (22) for $N \equiv I$ and $\chi = I_r$. Then for $g \in V$ and any matrix A holomorphic on S^1:

$$(23) \quad (T_I^{-1} T_A g)(z) = -\frac{1}{2\pi i} \int_{|w|=1} \left[\frac{A(w)}{w-z} + (w-z)R(z, w) \right] g(w) dw$$

with R biholomorphic near S^1 since $S(z, w; I_r) = (w-z)^{-1}I + O(w-z)$ from the antisymmetry (2). Likewise for $h \in H_+^r$ and $|z| > 1$,

$$(24) \quad (T_A T_I^{-1} h)(z) = -\frac{1}{2\pi i} \int_{|w|=1} \left[\frac{A(w)}{w-z} + B(z, w) \right] h(w) dw$$

where for $\varepsilon > 0$ sufficiently small:

$$B(z, w) = -\frac{1}{2\pi i} \int_{|u|=1-\varepsilon} \frac{A(u)}{u-z} S(u, w; I_r) du$$

is holomorphic for $|z| \geq 1$, w near S^1; in particular, if $N(u) = N(u, s)$ is holomorphic in a neighborhood of $|u| = 1$:

$$(25) \quad (T_N T_I^{-1} T_{N^{-1}dN} T_I^{-1} h)(z) = -\frac{1}{2\pi i} \int_{|w|=1} \left[\frac{dN(w)}{w-z} + K(z, w) \right] h(w) dw$$

with K holomorphic for all $|z| \geq 1$ and w near S^1. For any such kernel $K(z, w) = \sum_{\alpha=1}^{\infty} \sum_{\beta=-\infty}^{\infty} (K_{\alpha\beta} z^{-\alpha} w^{\beta})$ biholomorphic on S^1, $\sum_{\alpha, \beta} |K_{\alpha\beta}| < \infty$ and the compact integral operator on H_+^r defined by K is of trace class (see [7; p. 247: (1.7), 3]) with trace

$$Tr(K) = \sum_{l=1}^{r} \sum_{n=1}^{\infty} \int_{|z|=1} \left(\int_{|w|=1+\varepsilon} {}^t e_l K(z, w) e_l w^{-n} z^n \, dw \right) \frac{dz}{2\pi i z}$$

$$= \int_{|w|=1} trK(w, w) dw.$$

Now suppose $N = N(u, s)$ is a family of sections of $\operatorname{Hom}(I_r, \chi(s))$ holomorphic in s for u outside D and $\chi(s)$ near $\chi(s_0) = \chi \in (\theta)$. Following [4, p. 415], we can define an operator $K: H_+^r \to H_+^r$ of finite rank $n = h^0(\chi(s_0) \otimes \Delta)$ such that $L_s := T_{N(\cdot, s)} T_I^{-1} + K$ is invertible near $s = s_0$: if $T_0 = T_{N(\cdot, s_0)} T_I^{-1}$, let $K | \operatorname{Ker} T_0$ be an isomorphism of $\operatorname{Ker} T_0$ onto some subspace complementary to $\operatorname{Im} T_0$ in H_+^r and set $K \equiv 0$ on some subspace complementary to $\operatorname{Ker} T_0$ in H_+^r.

THEOREM 3. *For* $\chi \in (\theta)$ *and* K *as above, the determinant*

$$(26) \qquad \theta_0(s) = \det(T_{N(\cdot,s)}T_I^{-1}(T_{N(\cdot,s)}T_I^{-1} + K)^{-1})$$

is a holomorphic function of s *near* s_0, *vanishing if and only if* $\chi(s) \in (\theta)$. *The local differential form* $\omega(s) - d\log\theta_0(s)$ *is holomorphic at* $s = s_0$.

PROOF. If $\chi(s) \notin (\theta)$ and $g \in V$,

$$(T_N^{-1}dT_N g)(z) = -\frac{1}{2\pi i}N^{-1}(z)\int_{|w|=1} S(z,w;\chi(s))dN(w)g(w)dw$$

for $|z| > 1$, and on the other hand:

$$(T_{N^{-1}dN}g)(z) = -\frac{1}{2\pi i}\int_{|w|=1} N^{-1}(w)dN(w)g(w)\frac{dw}{w-z}$$

for $|z| > 1$. From (23), (24) one then finds that $T_I(T_N^{-1}dT_N - T_I^{-1}T_{N^{-1}dN})T_I^{-1}$ is a trace class operator on H_+^r with

$$(27) \qquad \begin{aligned} \omega(s) = &-\sum_z \operatorname{Res}(N^{-1}\partial_z NN^{-1}dN)(z,s) \\ &+ Tr(T_N^{-1}dT_N - T_I^{-1}T_{N^{-1}dN}) \end{aligned}$$

by using (7) in the definition (8). Now

$$T_N T_I^{-1}L_s^{-1} = I - KL_s^{-1}: H_+^r \to H_+^r$$

where KL_s^{-1} is of trace class, holomorphic for s near s_0; so the determinant (26) exists for s near s_0 [7, p. 254] and vanishes if and only if $\operatorname{Ker} T_N$ is non-empty—i.e. $h^0(\chi(s) \otimes \Delta) > 0$. To show

$$\omega - d\log\theta_0 = \omega - Tr(T_I T_N^{-1}dT_N T_I^{-1} - L_s^{-1}dL_s)$$

is holomorphic at $s = s_0$, it suffices from (27) to show that

$$\begin{aligned} T_{N^{-1}dN}&T_I^{-1} - L_s^{-1}dL_s \\ &= L_s^{-1}K(T_{N^{-1}dN}T_I^{-1}) + L_s^{-1}(T_N T_I^{-1}T_{N^{-1}dN}T_I^{-1} - T_{dN}T_I^{-1}) \end{aligned}$$

is a trace class operator near s_0. However, $T_{N^{-1}dN}T_I^{-1}$, L_s and L_s^{-1} are bounded operators near s_0, and K and the last factor above (see (24), (25)) are trace class operators. Since left or right multiplication by bounded operators preserves the trace class property, the above operator must also be of trace class for s near s_0.

4. Normalized θ **and** $\det S$. Combining results from the previous sections, we have our main conclusion:

THEOREM 4. *Let* $\sigma = \sum_1^{2m}\sigma_i\,ds_i$ *be any local holomorphic one-form on* R *with* $d\sigma = d\omega$, *the holomorphic two-form* (9). *Then there is a holomorphic section* $\theta(\chi(s))$ *on* R *with*

$$(28) \qquad\qquad d\log\theta = \omega - \sigma.$$

At points $\chi = \chi(s_0)$ *for which* $h^0(\chi \otimes \Delta) = n$, $\theta(\chi(s))$ *vanishes to order* n *at* $s = s_0$ *with tangent cone to* (θ) *at* χ *given by* (19); *if* θ_0 *is the determinant* (26), $\theta(\chi(s))/\theta_0(s)$ *is a holomorphic unit near* $s = s_0$.

REMARK. Though ω is intrinsic, there is no preferred choice of σ determining θ. In the abelian case, $\sigma = 2\pi i a \cdot (\tau da + db)$ gives the standard θ-function, up to a universal constant. On the other hand, one can define (redundant) flow coordinates (t_n) so that $\tau a + b = \sum_1^\infty t_n v^{(n-1)}(p)/(n-1)!$ for some basepoint $p \in C$; setting

$$\Omega(z, t) = 2\pi i \int_p^z v \cdot da - \sum_1^\infty dt_n \partial_p^n \log E(z, p)/(n-1)!$$

then leads to the K-P tau-function for suitable (ad hoc) choice of σ in (28). Any explicit relationship between (8), (28) and the loop-group approach of [6], however, will require a more detailed analysis of the flows N (7) in higher-rank.

To study the dependence of θ on its "abelian part", we normalize the local holomorphic coordinates (s) as follows: let $s_k = b_k$ and $s_{k+g} = a_k$, $1 \le k \le g$, define the line bundle $\det \chi(s) \in \mathrm{Hom}(\Gamma, \mathbb{C}^*)$ with multipliers $e^{-2\pi i r a_k}$, $e^{2\pi i r b_k}$ along A_k, B_k respectively. Choose local coordinates $\hat{s} = (s_{2g+1}, \ldots, s_{2m})$ for some open subset U of representations $\chi_0(\hat{s}) \in R_0 = \mathrm{Hom}(\Gamma, \mathrm{SL}(r, \mathbb{C}))/\mathrm{PSL}(r, \mathbb{C})$. Then $s = (s_1, \ldots, s_{2m})$ are coordinates for $\mathbb{C}^{2g} \times U$ covering the open set of bundles $\chi(s) = \lambda(s_1, \ldots, s_{2g}) \otimes \chi_0(\hat{s}) \in R$ with λ the line bundle (11). The form (6) splits as $\Omega(z, s) = \Omega(z, a, b)I + \Omega_0(z, \hat{s})$ where Ω is given by (14) and $tr\Omega_0(z, \hat{s})$ is a well-defined function in z. The two-form (9) also splits and we can choose a *normalized* σ as in Theorem 4:

$$(29) \qquad \sigma = 2\pi i r \sum_{j=1}^g a_j(\tau da + db)_j + \sigma_0(\hat{s})$$

for some local holomorphic form σ_0 on R_0. The section $\theta(s)$ defined by (28) will then be a function of $\tau a + b = \frac{1}{r}\det\chi(s)$ for fixed $\hat{s}$ since, from (29) and (15), $\sigma = 0$ and

$$\omega(s) = \sum_z \mathrm{Res}(tr\sigma_0(z; \chi(s)))\Omega(z, a, b) = 0$$

when $\tau a + b$ and $\hat{s}$ are constant.

This normalized $\theta(s)$ is (for fixed $\hat{s}$) a section on the Jacobi variety which can be expressed algebraically as a determinant of Szegö kernels; for this purpose, we make use of an "insertion" residue-formula for the Szegö kernel:

LEMMA 2. *For any points* $x_j, y_j \in H$, $1 \le j \le n$, *let* $\sum_1^n(y_j - x_j) \in \mathrm{Hom}(\Gamma, \mathbb{C}^*)$ *be the line bundle* (11) *with multipliers* 1, $\exp 2\pi i \sum_1^n \int_{x_j}^{y_j} v_k$

along A_k, B_k *respectively. Then for* $h^0(\chi \otimes \Delta) = h^0(\chi \otimes \sum_1^n (y_j - x_j) \otimes \Delta) = 0$,

(30)
$$a_0 \left(z \, ; \chi \otimes \sum_1^n (y_j - x_j) \right) - a_0(z \, ; \chi) = \partial_z \log \prod_1^n \frac{E(z, x_j)}{E(z, y_j)}$$
$$- (S(z, y_k \, ; \chi))(S(x_i, y_j \, ; \chi))^{-1}(S(x_l, z \, ; \chi))$$

(with indices $i, j, k, l = 1, \ldots, n$ *).*

PROOF. For any $z', z \in H$,

$$S(z', \zeta \, ; \chi) S \left(\zeta, z \, ; \chi \otimes \sum_1^n (y_j - x_j) \right) \prod_1^n \frac{E(\zeta, x_j)}{E(\zeta, y_j)}$$

is a (matrix) abelian differential in ζ with a zero-sum of residues:

$$S \left(z', z \, ; \chi \otimes \sum_1^n (y_j - x_j) \right) \prod_1^n \frac{E(z', x_j)}{E(z', y_j)} - S(z', z \, ; \chi) \prod_1^n \frac{E(z, x_j)}{E(z, y_j)}$$
$$= \sum_{k=1}^n S(z', y_k \, ; \chi) S \left(y_k, z \, ; \chi \otimes \sum_1^n (y_j - x_j) \right) \frac{\prod_{j=1}^n E(y_k, x_j)}{\prod_{j \neq k} E(y_k, y_j)}.$$

Setting $z' = x_1, \ldots, x_n$ to eliminate the terms $S(\cdot, z \, ; \chi \otimes \sum_1^n (y_j - x_j))$, this becomes:

$$S \left(z', z \, ; \chi \otimes \sum_1^n (y_j - x_j) \right) \prod_1^n \frac{E(z', x_j)E(z, y_j)}{E(z', y_j)E(z, x_j)} - S(z', z \, ; \chi)$$
$$= -(S(z', y_k \, ; \chi))(S(x_i, y_j \, ; \chi))^{-1}(S(x_l, z \, ; \chi))$$

(a product of matrices of order $r \times rn$, $rn \times rn$, $rn \times r$ respectively), which gives (30).

THEOREM 5. *Assume that* $\theta(s)$ *is defined as above from a normalized* σ (29) *with* $\frac{1}{r} \det \chi(s)$ *the bundle* (11). *Then* $\theta(s)$ *is a theta-function of order* r *in* $\tau a + b$ *(for* $\hat{s}$ *fixed), and for all* $x_1, y_1, \ldots, x_n, y_n \in C$:

(31)
$$\frac{\theta(s + \sum_1^n (y_j - x_j))}{\theta(s) \prod_{i,j=1}^n E(x_i, y_j)^r} \prod_{i<j}^n E(x_i, x_j)^r E(y_j, y_i)^r$$
$$= \exp \left[-2\pi i r \sum_1^n \int_{x_j}^{y_j} a \cdot v \right] \det[S(x_i, y_j \, ; \chi(s))]$$

if $\sum_1^n (y_j - x_j) = (\sum_1^n \int_{x_j}^{y_j} v, 0) \in \mathbb{C}^{2m}$.

PROOF. Let Ω_j (5) be the coefficient of db_j in (14); then for $z \in C$,

$$W(\zeta, z) = \sum_1^g \Omega_j(\zeta) v_j(z) = -2\pi i \, {}^t \left(\int_{p_0}^\zeta \omega_2(\zeta, p_k) \right) M v(z)$$

is an abelian integral in ζ with $\sum \operatorname{Res}_\zeta v(\zeta) W(\zeta, z) = v(z)$ for any holomorphic abelian differential v (15). With a choice of coordinates $s_j = b_j$, $1 \leq j \leq g$, as above:

$$\partial_{x_k} \left[\log \theta \left(s + \sum_1^n (y_j - x_j) \right) + 2\pi i r \sum_1^n \int_{x_j}^{y_j} a \cdot v \right]$$

$$= -\sum_{j=1}^g \left(\partial_{s_j} \log \theta \left(s + \sum_1^n (y_j - x_j) \right) + 2\pi i r a_j \right) v_j(x_k)$$

$$= -\sum_\zeta \operatorname{Res} \left\{ tra_0 \left(\zeta; \chi(s) \otimes \sum_1^n (y_j - x_j) \right) W(\zeta, x_k) \right\}$$

$$= -tra_0 \left(x_k; \chi(s) \otimes \sum_1^n (y_j - x_j) \right)$$

$$= \partial_{x_k} \log \left\{ \prod_{j=1}^n E(x_k, y_j)^r \prod_{j \neq k}^n E(x_k, x_j)^{-r} \det[S(x_i, y_j; \chi(s))] \right\}$$

using (28), (29) and the Laurent expansion of (30) at $z = x_k$. Thus both sides of (31) have constant ratio in any x_k (or similarly y_k) and so must be equal since they have the same singularity at $x_j = y_j$, $1 \leq j \leq n$. Now ω depends only on the point $\chi \in R$; so for any $m \in \mathbb{Z}^{2g}$, integration of (28), (29) gives:

$$\log[\theta(s_1 + m_1, \ldots, s_{2g} + m_{2g}, \hat{s})/\theta(s)] = -2\pi i r \sum_1^g m_{k+g}(\tau a + b)_k + c(m)$$

for some constant $c(m)$ depending only on m. Comparing automorphy (in any x_k) of both sides of (31), we find $c(m) = -\pi i r \sum_1^g m_{k+g} \tau_{kl} m_{l+g}$ and thus $\theta(s)/\theta(a\tau + b)^r$ is a periodic function of $(a, b) \in \mathbb{C}^{2g}/\mathbb{Z}^{2g}$, as asserted.

REFERENCES

1. J. Fay: Theta Functions on Riemann Surfaces, Springer Lecture Notes 352 (1973)
2. J. Fay: Kernel Functions, Analytic Torsion and Moduli Spaces, A.M.S. Memoirs (to appear).
3. R. Gunning: Lectures on Vector Bundles over Riemann Surfaces, Princeton University Press (1967)
4. B. Malgrange: Sur les Déformations Isomonodromiques, in "Mathématique et Physique (E.N.S. Séminaire 1979–1982)", Birkhäuser (1983), pp. 401–426
5. M. S. Narasimhan, S. Ramanan: Moduli of Vector Bundles on a Compact Riemann Surface, Ann. Math. 89(1969), pp. 19–51
6. G. B. Segal, G. Wilson: Loop Groups and Equations of KdV Type, Publ. Math. I.H.E.S. 61(1985), pp. 5–65
7. B. Simon: Notes on Infinite Determinants of Hilbert Space Operators, Advances in Math. 24 (1977), pp. 244–273

DEPARTMENT OF MATHEMATICS, ST. JOSEPH'S COLLEGE, NO. WINDHAM, MAINE 04062

Contemporary Mathematics
Volume **136**, 1992

Theta functions on the Boundary of Moduli Space

GABINO GONZÁLEZ-DÍEZ

ABSTRACT. By approaching singular stable curves (Riemann surfaces with nodes) through nonsingular ones, we work out some aspects of the theory of abelian integrals and theta functions.

The topics dealt with are period matrices, embedding a curve into its jacobian, Riemann's theta function Lefschetz theorem and solutions of P.D.E's of K-dV type.

We work within the framework of Bers' theory of deformation spaces.

0. Introduction

Let $D = D(S)$ be Bers' deformation space of a Riemann surface with nodes (stable curve) S. If S has no nodes at all D is simply T_g, the Teichmüller space; in this case it is well known that there are fibre spaces $\pi: V_g \to T_g$ such that $\pi^{-1}(t) = S_t$ is the Riemann surface represented by $t \in T_g$, and $\pi_1: J(V_g) \to T_g$ such that $\pi_1^{-1}(t)$ is the jacobian of S_t, i.e. $\pi_1^{-1}(t) = J(S_t) = \mathbf{C}^g / \mathbf{Z}^g + \mathbf{Z}^g \cdot \Omega_t$ with Ω the period matrix.

One each $J(S_t)$ $t \in T_g$; one has the classical theta function $\theta(z, \Omega_t)$. This article is an attempt to understand what happens to this function when we allow t to vary in D (not only in T_g).

We start by looking at a holomorphic map of fibre spaces over T_g defined by Earle in [E],

$$\Phi: \quad \begin{array}{ccc} V_g & \to & J(V_g) \\ \pi \searrow & & \swarrow \pi_1 \\ & T_g & \end{array} ; \qquad \Phi(t, x) = (t, \frac{1}{1-g} k(t, x)),$$

$$k(t, x) = \text{Riemann's constant},$$

which embeds each nonsingular curve into its jacobian. We observe that in order to extend this map to singular stable curves one has to translate the

---

1991 *Mathematics Subject Classification.* Primary 32G20, 14K25, 14H40.

Research supported by a grant of the CICYT.M.E.C.SPAIN. The author is also grateful to the A.M.S. for support to attend this conference.

The final version of this paper will be submitted for publication elsewhere.

Riemann's constant by $(1/2) \cdot \delta(t)$, where $\delta(t)$ is the vector of diagonal entries of Ω_t.

This indicates that—as it has also been suggested by Mumford [M1, M2]—in the boundary of moduli space the appropriate function to look at is not $\theta(z, \Omega)$ but rather $\theta_\delta(z, \Omega) = \theta(z - (\delta/2), \Omega)$.

We describe θ_δ explicitly in §6. Having done it we are in position to do (degenerate) Lefschetz theory and in fact to prove Lefschetz's Theorem for generalized jacobians.

We end our work by sketching the degenerate theory of theta functions and P.D.E.'s developed in the nonsingular case by Kriechever, Novikov and others.

1. Bers' theory of deformation spaces

For any Riemann surface with nodes S, Bers [B1, B2], see also [Wo] has defined a *deformation space* $D(S)$ parametrizing all Riemann surfaces with nodes which can be deformed into S. $D(S)$ is a cell that can be realized as a bounded domain of holomorphy in $\mathbf{C}^{3g-3}$. Of course for S nonsingular $D(S) = T_g$.

Here by a *deformation* we mean a continuous surjection $f \colon S' \to S$ such that the image of a node is a node, the inverse image of a node is either a node or a Jordan curve avoiding all nodes, and the restriction of f to the complement of the inverse image of the nodes is a sense preserving homeomorphism of surfaces (see Figure 1).

Two deformations $f \colon S' \to S$ and $h \colon S'' \to S$ are regarded as *equivalent* if there are homeomorphisms $\alpha \colon S' \to S''$ and $\beta \colon S \to S$, homotopic to an isomorphism and to the identity respectively, such that $h \cdot \alpha = \beta \cdot f$.

REMARK 1.1. This roughly says that two deformations in the same equivalence class may differ by products of Dehn twists about Jordan curves mapped into nodes.

Bers defines $D = D(S)$ to be the set of equivalence classes of such deformations $[S', f]$, and realises it as a bounded domain of $\mathbf{C}^{3g-3}$. He does it as follows [Be1, §3 pp. 46–47].

(i) Let S have r components $S_1, \ldots, S_r$ and n nodes. Choose Fuchsian groups $G_1, \ldots, G_r$ acting on discs $U_1, \ldots, U_r$ with disjoint closures in $\mathbf{P}^1$ such that

(a) G_j has n_j nonconjugate maximal elliptic subgroups, each of the same fixed order > 3 with $n_j = \#$ punctures of S_j-{nodes}.

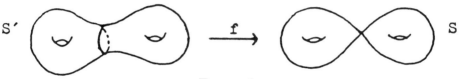

FIGURE 1

(b) $U_j^*/G_j = U_j/G_j$-{image of elliptic vertices} is isomorphic to $S_j^* = S_j$-{nodes}.

(ii) $G_1, \ldots, G_r$ generate a Kleinian group G, which is their free product, such that G has precisely one invariant component U.

(iii) For each node P_i, we can assign two nonconjugate maximal elliptic subgroups Γ_i', Γ_i'' corresponding to the two punctures x_i, y_i determined by P_i. If $x_i \in S_j$ and $y_i \in S_l$ it is assumed that $\Gamma_i' \subset G_j$ and $\Gamma_i'' \subset G_l$.

(iv) Call two elliptic fixed points not in U *related* if they are fixed under elliptic subgroups conjugate to Γ_i' and Γ_i'' for some i. Then the union of the U_j/G_j, with the images of any pair of related elliptic fixed points identified, is isomorphic to S.

(v) Let g_{i,s_i} be the unique loxodromic transformation which conjugates Γ_i' into Γ_i'', has multiplier s_i, $|s_i| > 0$ and small, and has fixed points in U_j and U_l (j and l as before).

(vi) Now, the group generated by G and the transformations g_{i,s_i} $i = 1, \ldots, n$ with $s_i \neq 0$ is a Kleinian group. Let us call it $G_{0,\mathbf{s}}$ where $\mathbf{s} = (s_1, \ldots, s_n)$

(vii) Let $\mathbf{s}$ be as before, and let w be a quasiconformal automorphism of $\mathbf{C}$ such that: w leaves $0, 1, \infty$ fixed, $w_{|U}$ is conformal and $w \cdot G_{0,\mathbf{s}} \cdot w^{-1}$ is a Kleinian group. Then $w_{|U_j}$, $j = 1, \ldots, r$ defines an element τ_j of the Teichmüller space $\mathbf{T}(G_j)$; set $\tau = (\tau_1, \ldots, \tau_r)$. If $s_i \neq 0$, set $\varepsilon_i = a_i - \hat{a}_i$, where

$$a_i = \text{repelling fixed point of } w \cdot g_{i,s} \cdot w^{-1}$$

$$\hat{a}_i = \text{fixed point of } w \cdot \Gamma_i' \cdot w^{-1} \text{ in } U_j.$$

If $s_i = 0$, set $\varepsilon_i = 0$. Also set $\boldsymbol{\varepsilon} = (\varepsilon_1, \ldots, \varepsilon_n)$. The point $(\boldsymbol{\tau}, \boldsymbol{\varepsilon}) \in \mathbf{C}^{3\mathbf{g}-3}$ determines the group $w \cdot G_{0,\mathbf{s}} \cdot w^{-1}$ completely; we denote this group by $G_{\tau, \varepsilon}$. The set of all such points is precisely $D = D(S)$.

As in the nonsingular case there is a fibre space $F(D) \to D$ which Bers constructs as follows [**Be1, pp. 47–48**].

(viii) Let $t \in D$ and denoted by $U(t)$ the part of the region of discontinuity of G_t which does not map onto U/G, and by $U^0(t)$ the complement in $U(t)$ of the set of elliptic vertices. The images of the pair Γ_i', Γ_i'' under the conjugation $G_{0,\mathbf{s}} \to w \cdot G_{0,\mathbf{s}} \cdot w^{-1}$ will be denoted by $\Gamma_i'(t), \Gamma_i''(t)$. Two elliptic fixed points in $U(t)$ are called related if they are fixed under elliptic subgroups conjugate to $\Gamma_i'(t)$ and $\Gamma_i''(t)$ for some i. The quotient $U(t)/G(t)$, with images of related elliptic fixed points identified, is a Riemann surface with nodes S_t. The genus of S_t is g and S_t has as many nodes as there are zeros in $(\varepsilon_1, \ldots, \varepsilon_n)$. Furthermore, each S_t is equipped with a deformation $f_t: S_t \to S$, determined up to equivalence.

(ix) The fibre space $F(D)$ over D is the set of points $(t; z)$ with $t \in D$, $z \in U^0(t)$. $F(D)$ is a domain in $\mathbf{C}^{3\mathbf{g}-2}$.

(x) In particular there are natural fibre spaces $V^0(D) \to D$ with fibre $S_t^0 = U^0(t)/G_t$, and $V(D) \to D$ with fibre $S_t = \overline{S}_t^0$.

2. Preparatory material

A. Canonical Cycles.

In the setting just described Bers [Be1] defines g Poincaré series on $F(D)$,

$$F_j(t, z), \qquad j = 1, \ldots, g$$

whose restriction to each fibre $U^0(t)$ provide a basis $\{w_1(t), \ldots, w_g(t)\}$ for the *regular* 1-*forms* on S, normalized by the condition

(2.1)
$$\int_{A_j(t)} w_i(t) = \delta_{ij}$$

where $A_1(t), \ldots, A_g(t)$ are simple loops which either define nonvanishing homology classes or are homotopic to nodes. He also remarks that "one may use this result to study the behaviour of a period matrix under degeneration."

In order to do that, we need to explain how to choose the loops $A_1(t), \ldots, A_g(t)$; $B_1(t), \ldots, B_g(t)$.

With the notation of the preceding §, let [S, id] be the origin $(0; 0)$ of $D = D(S)$ and let $\Delta \subset D$ be a 1-dimensional disc through it. We assume it to be transverse to the divisors of equation $\varepsilon_i = 0$, e.g. $\Delta = \{(0; \mu, \ldots, \mu)/\mu$ small) so that $V(D)_{|\Delta}$ has nonsingular curves S_μ as fibres, except for $\mu = 0$ where the fibre is isomorphic to S

It has been shown by Jambois [J], see also [B-P-V and N] that S is a strong deformation retract of V. We shall denote this deformation by $f: V \to S$ and its restriction to the fibre S_μ by f_μ.

Now if $p: \widetilde{S} \to S$ is the normalization of S and $p(x_j) = p(y_j) = P_j$ $j = 1, \ldots, n$ are the nodes, we can consider disjoint bands $V_i(\mu)$ around $\delta_i(\mu) = f_\mu^{-1}(P_i)$, P_i a node of S, sufficiently small such that we have home-omorphisms

$$f_\mu: S_\mu - V(\mu) \approx S - V, \quad \text{and} \quad p: \widetilde{S} - \widetilde{V} \approx S - V$$

where $V(\mu) = U_i V_i(\mu)$, $V_i = f_\mu(V_i(\mu))$, $V = U_i V_i$, $\widetilde{V}_i = p^{-1}(V_i)$ and $\widetilde{V} = U_i \widetilde{V}_i$, [J, 1.4]. Thus, f_μ is a deformation in the sense of Bers. We also observe that in terms of Bers' theory $\delta_i(\mu)$ can be thought as a circle with centre a_i and radius $\mu = |a_i - \hat{a}_i|$.

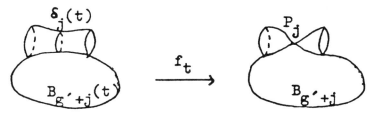

FIGURE 2

If $\tilde{g}$ is the genus of $\tilde{S}$, one may take loops $A_1, \dots, A_{\tilde{g}}; B_1, \dots, B_{\tilde{g}}$ supported on $S - V$ which give—via p^{-1}—a symplectic basis for $H_1(\tilde{S}, \mathbf{Z})$ [J, 1.5].

This enables one to define loops

$$A_i(\mu) = f_\mu^{-1}(A_i), \qquad i = 1, \dots, \tilde{g},$$
$$B_i(\mu) = f_\mu^{-1}(B_i), \qquad i = 1, \dots, \tilde{g}.$$

Next, if say $P_1, \dots, P_k$ is *maximal collection of nonseparating nodes*—i.e. maximal with the property that $S - \{P_1, \dots, P_k\}$ is connected—then we put

$$A_{\tilde{g}+j}(\mu) = \delta_j(\mu), \qquad j = 1, \dots, k.$$

It is important to observe that $k = g - \tilde{g}$ [J, 1.7].

Finally for each $j = 1, \dots, k$, $B_{\tilde{g}+j}(\mu)$ is going to be a loop in S_μ supported on $S_\mu - \{A_i(\mu), B_i(\mu), i = 1, \dots, \tilde{g}\}$ and intersecting $V(\mu)$ precisely in a generating line of the cylinder $V_j(\mu)$ around $\delta_j(\mu)$; we also assume that this line passes through the image of $\hat{a}_i$ in S_μ. Away from $V(\mu)$ it is defined as the f_μ^{-1}-image of the following path $B_{\tilde{g}+j}$ in S.

(I) If x_i, y_i lie both in the same irreducible component of S, then our $B_{\tilde{g}+j}$ will be a path joining x_j, to y_j within this component.

(II) If not, i.e. if P_j is a pole joining two components L and R, the fact that $S \backslash \{P_1, \dots, P_k\}$, is still connected, and hence that $x_j \in L$ and $y_j \in R$ can be joined by a path within $S \backslash \{P_1, \dots, P_k\}$, implies that there is a cyclic collection of nodes and components $l_i, r_i \in T_i, i = 1, \dots, m$ such that $T_1 = L, T_m = R, l_1 = x_j, r_m = y_j$ and $p(l_{i+1}) = p(r_i)$ is a node not in $\{P_1, \dots, P_k\}$.

In this case our $B_{\tilde{g}+j}$ will be a collection of paths joining l_i to r_i within T_i.

These k paths $B_{\tilde{g}+j}$ can be chosen in such a way that $A_1(\mu), \dots, B_g(\mu)$ afford a locally continuous choice of loops which form a canonical homology basis for S_μ [J, Corollary 2].

Having done this, we can associate to each S_t, $t \in D$ a canonical system of loops $\{A_1(t), \dots, A_g(t); B_1(t), \dots, B_g(t)\}$ as follows

Away from the analytic subset E of equation $\varepsilon_1 \cdots \varepsilon_n = 0$, which parametrises singular stable curves (see §1 viii), $V(D)$ is (topologically) a locally trivial fibration. Thus, if $t \notin E$ we can join t to a point $\mu \in \Delta$ by a path within $D \backslash E$ and transport the canonical basis on S_μ to S_t. If $t \in E$ we draw a disc through t transverse to E and define the loops by retraction as above.

REMARK 2.1. The $B_{\tilde{g}+i}(t)$'s are determined by the marking on S—namely $A_1, \ldots, A_g; B_1, \ldots, B_{\tilde{g}}; P_1, \ldots, P_{g-\tilde{g}}$ with prescribed orientation—only up to a **Z**-linear combination of $A_{\tilde{g}+1}(t), \ldots, A_g(t)$, *the vanishing cycles* [**J 1.8 and N**], see also Remark 1.1. At any rate, A-cycles are entirely determined; this will enable us to speak later on of a dual basis for the regular 1-forms.

Our discussion will be greatly simplified if we make the following assumption.

If the loop $A_{\tilde{g}+j}(t)$ is not homotopic to a node (i.e. if $\varepsilon_j \neq 0$) then $B_{\tilde{g}+j}(t)$ does not encounter any node (i.e. none of the $\varepsilon_{d(j)}$'s corresponding to the nodes appearing in the discussion of case II above, vanish). It is clear that if the number of such loops is $r \leq k$, then $g + r$ is precisely the genus of $\tilde{S}_t$.

In what follows we shall denote by D^* the subset of D in which this assumption holds (see Figure 3). In terms of Bers' coordinates we have

DEFINITION 2.2. D^* is the set of points $t = (\boldsymbol{\tau}, \boldsymbol{\varepsilon}) \in D$ such that if $\varepsilon_j \neq 0$ then each $\varepsilon_{d(j)} \neq 0$, for $j = 1, \ldots, k$.

REMARK 2.3. (i) D^* contains the origin of D.

(ii) D^* contains the open set of points of D for which $\varepsilon_{d(j)} \neq 0$.

(iii) In particular D^* contains the subset of point $t \in D$ corresponding to nonsingular curves.

(iv) By a suitable choice of the marking on S, any given point of D can be made to lie in D^*.

(v) If S has just one node, then $D = D^*$.

Summary.

With the ambiguity mentioned in Remark 2.1 above, we have assigned to each $t \in D^*$ a collection of $2g$ loops in S_t

$$A_1(t), \ldots, A_g(t); \qquad B_1(t), \ldots, B_g(t).$$

If S_t is nonsingular these loops provide a canonical homology basis.

If not let $\tilde{g} + r$ be the genus of its normalization $p: \tilde{S}_t \to S_t$, then—up to reordination—the first $\tilde{g} + r$ A-loops together with the first $\tilde{g} + r$ B-loops provide via p^{-1} a canonical homology basis on $\tilde{S}_t$.

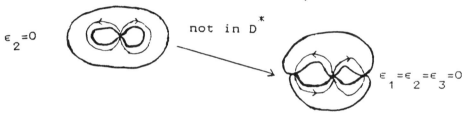

FIGURE 3

For the remainder $g - (\tilde{g} + r)$ indexes j, the parameter ε_j vanishes, which corresponds to the fact that S_t has a node $P_j(t)$. The loop $B_{\tilde{g}+j}$ encounters this node and $A_{\tilde{g}+j}$ is homotopic to it.

B. Regular 1-forms.

Next we observe that a *dual basis*—i.e. satisfying condition (2.1)—for the regular 1-forms on S_t exists and is unique for each $t \in D^*$.

UNIQUENESS. In fact if $w_1(t), \ldots, w_g(t)$ is such a basis any other 1-form w on S_t can be uniquely written as $w = \lambda_1 w_1(t) + \cdots + \lambda_g w_g(t)$ with $\lambda_i = \int_{A_i(t)} w$.

When $A_i(t)$ is homotopic to the node $P_i(t)$ this integral is, of course, the residue.

EXISTENCE. Let $g' = \tilde{g} + r$ be the genus of $\widetilde{S}_t$ and let $w_1(t), \ldots, w_{g'}(t)$ be a basis for the holomorphic 1-forms on $\widetilde{S}_t$ dual to $A_1(t), \ldots, A_{g'}(t)$. It only remains to show that for each $j = 1, \ldots, g - g'$ there exists a 1-form $w_{g'+j}(t)$ possessing the two following properties

(α) $\int_{A_i} w_{g'+j} = 0$, $j = 1, \ldots, g - g'$; $i = 1, \ldots, g'$,

(β) $w_{g'+j}$ has a pole at $P_j(t)$ with residue $1/2\pi i$ and is bounded at the other nodes $P_s(t)$ $s = 1, \ldots, g - g'$, $s \neq j$. In other words we require

$$2\pi i (\mathrm{res}_{x_s} w_{g'+j} = -\mathrm{res}_{y_s} w_{g'+j}) = \int_{A_{g+s}} w_{g'+j} = \delta_{js}.$$

Suppose that $B_{g'+j}(t)$ is as in case I above, then we apply the Riemann-Roch theorem to the component of S_t containing $B_{g'+j}(t)$ to obtain a 1-form v_j having precisely one pole at x_j and y_j with opposite residues ($= \pm 1/2\pi i$).

The desired $w_{g'+j}$ is obtained then by normalizing v_j by addition of a suitable linear combination of the holomorphic forms $w_1, \ldots, w_{g'}$.

There is a slight modification to be made in case II since $B_{g'+j}$ may involve several nodes and components. In this case by a similar argument we can find v_j with poles only at x_j, y_j; l_{i+1}, r_i all i with opposite residues $\pm 1/2\pi i$.

OBSERVATION 2.4. The dual forms have only residues equal to $\pm 1/2\pi i$.

We also observe that each of the nodes $p(l_{i+1}) = p(r_i)$ is nonseparating.

C. Period Matrix.

Now let $t^0 \in D^*$ represent a stable curve and $t \in D^*$ represent a non-singular curve near it; then with the notation as above we have for $1 \leq i$, $j \leq g' = \mathrm{genus}\,(S_{t^0})$,

$$\int_{B_j(t)} w_i(t) \to \int_{B_j(t^0)} w_i(t^0) \quad = \int_{p_*^{-1} B_j(t^0)} p^* w_i(t^0) = \Omega_{ij}(t^0).$$

The matrix $\widetilde{\Omega}$ built up with these entries is, of course, a period matrix for S_{t^0}.

More generally, if $B_j(t^0)$ does not pass through the poles of $w_i(t^0)$, $\int_{B_j(t)} w_i(t)$ has finite limit $\Omega_{ij}(t^0)$.

It remains to understand the limit of e.g., $\int_{B_g(t)} w_g(t)$ where for simplicity we assume that $w_g(t)$ has a unique pole $P_k(t)$. In order to do that we again quote Bers. With the notation of §1 (vi), one has

$$w_g(t) = \alpha(t)(ds/(s - a_k(t)) - ds/(s - b_k(t))) + \cdots$$

which is obtained as a linear combination of the basis given in [Be1, §4]. The coefficients of this linear combination, e.g. $\alpha(t)$, are holomorphic functions of t (see the proof of Proposition 7.4) and, by Observation 2.4, we must have $\alpha(t^0) = 1/2\pi i$.

Recall that $a_k = a_k(t)$, $b_k = b_k(t)$ are holomorphic functions such that $\varepsilon_k = \hat{a}_k - a_k$ and $b_k = g_k(\hat{a}_k)$.

Thus, since B_g passes through $\hat{a}_k$ we see that our integral takes the value $2(1/2\pi i)\log (\hat{a}_k - a_k) + a$ bounded function of $t \approx (1/\pi i)\log \varepsilon_k$.

In general one has

$$\Omega_{g'+ig'+j}(t) \approx (1\backslash \pi i)(\log \varepsilon_1 + \cdots + \log \varepsilon_{n_{ij}})$$

where n_{ij} is the number of poles of $w_{g'+j}$ encountered by $B_{g'+i}$ (compare [J and F]).

According to this we shall write

$$(2.2) \qquad \Omega_{t^0} = (\Omega_{ij}(t^0)) = \begin{pmatrix} t\widetilde{\Omega} & \widehat{\Omega} \\ \widehat{\Omega} & \Omega \end{pmatrix}.$$

The entries of Ω may or may not be bounded. Their size is controlled by the matrix (see Remark 6.12),

$$(2.3) \qquad N = N(t^0) = (n_{ij}) \qquad 1 \le i,\ j \le g - g'.$$

REMARK 2.5. N is the matrix which appears in Picard-Lefschetz's theory (see [J]). The well-known fact that N is positive definite can be seen by taking limit in the classical Riemann's Bilinear Relation. Namely, since

$$N \cdot \infty \approx \lim_{t \to t^0} \mathrm{Im}\overline{\Omega}_t$$

the relation $r \cdot N \cdot r = 0$ for some $r = (r_1, \dots) \in \mathbf{R}^{g-g'}$ would imply

$$\lim_{t \to t^0} \int_{S_t} (r_1 w_{g'+1} + \cdots + r_{g-g'} w_g) \wedge \overline{(r_1 w_{g'+1} + \cdots + r_{g-g'} w_g)}$$

$$= \lim_{t \to t^0} r \cdot \mathrm{Im}\overline{\Omega}_t \cdot r$$

is bounded. Contradiction.

EXAMPLE 2.6. Let S be the stable curve of genus g obtained by identifying g pairs of points in $\mathbf{P}^1$. In other words we have

$$p: S = \mathbf{P}^1 \to S; \qquad p(x_j) = p(y_j) = P_j.$$

Then $g' = 0$, $\{P_1, \ldots, P_g\}$ is a maximal collection of nonseparating nodes,

$$w_j = (1/2\pi i)d \, \log((x - x_j)/(x - y_j)),$$
$$N = (\text{identity})_{g \times g}, \quad \text{and for } i \neq j,$$

$$\Omega_{ij} = \int_{y_i}^{x_i} w_j$$
$$(1/2\pi i)\log(y_j - y_i)(x_j - x_i)/(y_j - x_i)(x_j - y_i)$$

(compare [N]).

3. The fibre space $J(D^*) \to D^*$

In this, as well as in the next section, some of the statements are not formally correct, because D^* is not in general an open subset of D. But the reader will have no difficulty to formulate the precise statement in each situation.

Denote by L_t the lattice $\mathbf{Z}^g + \mathbf{Z}^g \cdot \Omega_t$ if S_t is nonsingular and $\mathbf{Z}^g + \mathbf{Z}^{g'} \cdot (\tilde{\Omega}_t, \hat{\Omega}_t)$ otherwise. Observe that despite the multivaluedness of Ω_t (2.2) above, L_t is entirely determined (see Remark 2.1).

DEFINITION 3.1. $J(D^*) = D^* \times \mathbf{C}^g$, where we identify (t, z) to (t', z') when $t = t'$ and $z - z' \in L_t$.

THEOREM 3.2. $J(D^*)$ is a complex manifold with a complex structure such that the projection

$$J(D^*) \to D^*$$
$$(t, z) \to z$$

is analytic.

PROOF. It is not hard and can be found in [J].

REMARK 3.3. The fibre over $t \in D^*$ is $J(S_t) = \mathbf{C}^g / L_t$, the usual jacobian in the nonsingular case, and the so-called *generalized jacobian* otherwise. Note that the latter is not necessarily compact. When $D = T_g$, $J(T_g)$ is, of course, our $J(V_g)$ of §0.

4. The extension of Φ

Recall that the Riemann constant at a point P comes given by

$$k_p = (\ldots, (1/2) - (\Omega_{ii}/2) + \sum_{k \neq i} \int_{A_k} \left(\int_P^s w_i \right) w_k ds, \ldots)$$

so the extension of Earle's map (see §0) presents the problem that Ω_{ii} may not be bounded as t approaches a singular curve. We therefore set for each

$(t, P) \in V^0(D)$

$$(4.1) \quad \Phi(t, P) = \left(t, (1/(1-g)) \left(\dots, \sum_{k \neq i} \int_{A_k(t)} \left(\int_P^s w_i(t)\right) w_k(t) ds, \dots\right)\right).$$

THEOREM 4.1. Φ *provides a holomorphic map* $\Phi: V^0(D^*) \to J(D^*)$ *of fibre spaces over* D^*.

PROOF. We only have to check welldefinedness. Let $g_i(t, s) = \int_P^s w_i(t)$. Different choices of $g_i(t, s)$ correspond to different choices of paths within S_t^0. Two such paths differ by a homology cycle σ of the form

$$\sigma \approx d_1 A_1(t) + \dots + d_{g'} A_{g'}(t) + c_1 B_1(t) + \dots + c_{g'} B_{g'}(t) + \delta$$

where δ is some **Z**-linear combination of cycles around the punctures of S_t^0.

Thus, any two choices of $g_i(s, t)$ differ by

$$\int_\sigma w_i(t) = d_i + \sum c_k \Omega_{ki}(t) + \{\text{a finite sum of } 2\pi i \cdot \text{residues} = \pm 1\}$$

$$= r_i + \sum c_k \Omega_{ki}(t) \quad \text{with} \quad r_i \in Z.$$

Hence $\Phi(t, P)$ changes by

$$(1/(1-g)) \left(\dots, \sum_{k \neq i} \int_{A_k(t)} (r_i + \sum c_k \Omega_{ki}(t)) w_k(t), \dots\right)$$

$$= (1/(1-g)) \left(\dots, \left(r_i + \sum c_k \Omega_{ki}(t)\right) \sum_{k \neq i} \int_{A_k(t)} w_k(t), \dots\right)$$

$$= -\left(\dots, \left(r_i + \sum c_k \Omega_{ki}(t)\right), \dots\right) \in L_t \quad \text{as required.}$$

5. The embedding theorem

The goal of this section is to prove the following:

THEOREM 5.1. Φ *embeds* S_t^0 *in* $J(S_t)$ *if and only if* S_t *has no nonsingular rational components whose nodes are all separating.*

We begin by proving the following:

LEMMA 5.2.

$$d\Phi(t, z)/dz_{|z=P} = (w_1(P), \dots, w_g(P))$$

where w_i *stands for* $w_i(t)$.

PROOF.

$$d\Phi(t,z)/dz_{|z=P}$$

$$= (1/1-g)\left(\dots, \sum_{k\neq i}\int_{A_k(t)} d/dz\left(\int_z^s w_i\right)_{|z=P} \cdot w_k ds, \dots\right)$$

$$= (1/1-g)\left(\dots, \sum_{k\neq i}-w_i(P)\int_{A_k(t)} w_k, \dots\right)$$

$$= (1/1-g)(\dots, -w_i(P)\cdot(g-1), \dots)$$

$$= (\dots, w_i(P), \dots) \quad \text{as required.}$$

COROLLARY 5.3. *If* S_t *has no rational components, then* $\Phi_{|S_t^0}$ *is an embedding.*

PROOF. Let C be a component and let p be its genus. We saw in §2 that any basis for the regular 1-forms on S_t contains a basis for the regular 1-forms on C; hence Ω_t has a period matrix of C as a submatrix.

We have therefore a natural projection

$$\pi: J(S_t) \to J(C),$$
$$(z_1, \cdot, z_g) \to (z_1, \cdot, z_p).$$

What Lemma 5.2 says is that $\pi \cdot \Phi$ restricted to the component C has the same gradient, and hence it is a translate of the classical Abel map. This ends the proof.

LEMMA 5.4. *Let* C *be a rational component of* S_t *having at least one nonseparating node; then* Φ *restricted to* C *is still an embedding.*

PROOF. Let $D = D(S)$ and choose in S a maximal collection of nonseparating nodes containing the retraction of a node P of $C \equiv \mathbf{P}^1$ (see §§1 and 2).

Then (see Example 2.6) the dual basis for the regular 1-forms contains one, say w_g, of the form

$$w_{g|_{\mathbf{P}^1}} = (1/2\pi i)d\log((x-x_1)/(x-y_1)) \quad \text{with } p(x_1)=p(y_1)=P\in C.$$

Now as in the proof of Corollary 4.3 we compose Φ with the projection

$$\pi: J(S_t) \to \mathbf{C}/\mathbf{Z},$$
$$(z_1, \dots, z_g) \to z_g,$$

to obtain (up to translation) an isomorphism

$$\pi \cdot \Phi: \mathbf{P}^1\backslash\{x_1, y_1\} \to \mathbf{C}/\mathbf{Z},$$
$$x \to (1/2\pi i)\log((x-x_1)/(x-y_1)).$$

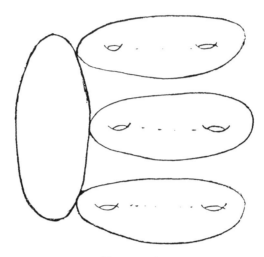

FIGURE 4

When we choose another maximal collection of nonseparating nodes (which may not contain the retraction of a node of C) we obtain, of course, a different map Φ'. However Lemma 5.2 implies that

$$d\Phi'(t, x)/dx_{|x=P} = Q \cdot d\Phi(t, x)/dx_{|x=P}$$

where Q is the nonsingular matrix relating the two dual basis of regular 1-forms.

This shows that Φ' and $Q \cdot \Phi$ differ only by a translation, hence Φ' also embeds C.

PROOF OF THEOREM 5.1. Corollary 5.3 plus Lemma 5.4 prove the "if" part.

Conversely if $C \equiv \mathbf{P}^1$ and has only separating nodes (see Figure 4) then the description of the regular forms given in §2 along with Observation 2.4 imply that the forms in the dual basis do not have poles at these nodes. Thus they are holomorphic, hence identically zero, over C.

6. Limit of thetas

What happens to the theta function $\theta(z, \Omega)$ as we approach a singular stable curve?

Recall that

$$\theta(z, \Omega) = \sum_{m \in \mathbf{Z}^g} e^{\pi i\{m \cdot \Omega \cdot m + 2m \cdot z\}}$$

$$= \sum_m \prod_{k=1}^{g} e^{\pi i m_k^2 \Omega_{kk}} \prod_{l<j} e^{2\pi i m_l m_j \Omega_{lj}} e^{2\pi i m \cdot z}.$$

We see that if $\Omega_{kk} \approx i\infty$ as in Example 2.6 then the theta function becomes just $\theta(z, \Omega) \equiv 1$ because in the above series all the terms $m \neq (0, \dots, 0)$ vanish.

This is rather disappointing since it indicates that in the limit, the theta function does not keep track of the theta divisor.

Instead, for each $t \in D^*$ with S_t nonsingular, we shall consider the function

(6.1) $$\theta_\delta(z, \Omega_t) = \theta(z - \delta\Omega_t/2, \Omega_t)$$

where δ of a matrix means its diagonal.

LEMMA 6.1. *Despite the multivaluedness of Ω_t, θ_δ is single valued.*

PROOF. Different choices of loops $B_i(t)$ give period matrices which differ by addition of a (necessarily symmetric) matrix M of A-periods (Remark 2.1), hence with integer entries. This affects θ_δ as follows:

$$\theta_\delta(z, \Omega_t + M)$$
$$= \sum_{m \in \mathbf{Z}^g} e^{\pi i\{m\cdot\Omega_t\cdot m + 2m\cdot(z - \delta\Omega_t/2) + m\cdot M\cdot m - 2m\cdot\delta M/2\}}.$$

Now

$$e^{\pi i\{m\cdot M\cdot m - 2m\cdot\delta M/2\}} = \prod_{i<j} e^{2\pi i m_i m_{ij} m_j} \prod_i e^{\pi i(m_i^2 - m_i)m_{ii}} = 1.$$

Thus, $\theta_\delta(z, \Omega_t + M) = \theta_\delta(z, \Omega_t)$ as stated.

OBSERVATION 6.2. Lemma 6.1 allows us to write $\theta_\delta(z, \Omega_t)$ more simply as $\theta_\delta(z, t)$.

Our next task is to extend $\theta_\delta(z, t)$ to t's representing singular curves. We start by looking at two particular cases.

EXAMPLE 6.3. Let $t^0 \in D^*$ correspond to a stable curve of genus g obtained by gluing Riemann surfaces S_1 of genus g_1 and S_2 of genus $g_2 = g - g_1$ along points $P_1 \in S_1$ and $P_2 \in S_2$.

Then it is clear that $\widetilde{S}_{t^0} = S_1 \cup S_2$,

$$\Omega_{t^0} = \begin{pmatrix} \Omega_1 & \\ & \Omega_2 \end{pmatrix}$$

with Ω_i a period matrix for S_i, $i = 1, 2$ and $\lim_{t \to t^0} \theta_\delta(z, t) = \theta_\delta(z_1, \Omega_1) \cdot \theta_\delta(z_2, \Omega_2)$ where $z = (z_1, z_2)$.

EXAMPLE 6.4. Let now $t^0 \in D^*$ correspond to a stable curve of genus g obtained by identifying two points of a Riemann surface S of genus $g - 1$. Then $\widetilde{S}_{t^0} = S$, and

$$\Omega_t = \begin{pmatrix} \widetilde{\Omega}(t) & \widehat{\Omega}(t) \\ \widehat{\Omega}(t) & \Omega_{gg}(t) \end{pmatrix} \to \begin{pmatrix} \widetilde{\Omega} & \widehat{\Omega} \\ \widehat{\Omega} & i\infty \end{pmatrix} \quad \text{as } t \to t^0.$$

Now writing $z = (z_1, \dots, z_{g-1}; z_g) = (\tilde{z}, z_g)$, $m = (\tilde{m}; m_g)$ we have

$$\theta_\delta(z, \Omega_t) = \sum_m \prod_{k=1}^g e^{\pi i(m_k^2 - m_k)\Omega_{kk}} \prod_{l<j} e^{2\pi i m_l m_j \Omega_{lj}} e^{2\pi i m\cdot z}$$

$$= \sum_{m_g} e^{\pi i(m_g^2 - m_g)\Omega_{gg}} e^{2\pi i m_g z_g} \left(\sum_{\tilde{m}} e^{\pi i\{\tilde{m}\cdot\widetilde{\Omega}\cdot\tilde{m} + 2\tilde{m}\cdot(\tilde{z} - \tilde{\delta}/2) + 2\tilde{m}\cdot\widehat{\Omega}\cdot m_g\}} \right).$$

Now near t^0, $\Omega_{gg} \approx (1/2\pi i)\log \varepsilon$ so we are only left with the m_g's satisfying $m_g^2 - m_g = 0$; namely $m_g = 0, 1$.

As for the series in brackets we see by rewriting

$$\pi i\{\tilde{m} \cdot \tilde{\Omega} \cdot \tilde{m} + 2m \cdot (\tilde{z} - \tilde{\delta}/2) + 2\tilde{m} \cdot \hat{\Omega} \cdot m_g\}$$
$$= \pi i\{\tilde{m} \cdot \tilde{\Omega} \cdot \tilde{m} + 2\tilde{m}(\tilde{z} + \hat{\Omega} \cdot m_g - \tilde{\delta}/2)\}$$

that its value is $\theta_\delta(\tilde{z} + \hat{\Omega} \cdot m_g, \tilde{\Omega})$. Thus, definitively $\lim_{t \to t^0} \theta_\delta(z, t) = \theta_\delta(\tilde{z}, \tilde{\Omega}) + e^{2\pi i z_g}\theta_\delta(\tilde{z} + \hat{\Omega}, \tilde{\Omega})$ [M2, F].

Before we state our next result we must again recall (see comments at the beginning of §3) that in some of the next statements, D^* should really mean, any open set in D^*.

THEOREM 6.5. θ_δ extends to a holomorphic function on $D^* \times \mathbf{C}^\mathbf{g}$.

PROOF. The preceding examples show that θ_δ extends continuously, hence analytically, to the analytic set consisting of points of D^* corresponding to curves with just one node. The remaining points form an analytic set of codimension 2 [Be1], so the result follows (see e.g. [G-R, p. 132].

REMARK 6.6. For later use we point out that in the process of proving this theorem we have in fact shown that for each m, the function

$$t \to e^{\pi i\{m \cdot \Omega_t \cdot m + 2m \cdot (z - \delta\Omega_t/2)\}}$$

is holomorphic and single valued all over D^*.

We now describe $\theta_\delta(z, t)$ for any $t \in D^*$. Let us write $\tilde{g} = $ genus of $\tilde{S}_t$ and $z = (z_1, \dots, z_{\tilde{g}}; z_{\tilde{g}+1}, \dots, z_g) = (\tilde{z}; \hat{z})$.

Then we have

THEOREM 6.7. For any $(t, z) \in D^* \times \mathbf{C}^\mathbf{g}$,

$$\theta_\delta(z, t) = \sum c_{\hat{m}} e^{2\pi i \hat{m} \cdot \hat{z}} \theta_\delta(\tilde{z} + \hat{m}^t \cdot \hat{\Omega}_t, \tilde{\Omega}_t),$$

where the sum is taken along the finite set $\{\hat{m} \in \mathbf{Z}^{\mathbf{g}-\tilde{\mathbf{g}}} / \hat{m} \cdot N \cdot \hat{m} = \delta N \cdot \hat{m}\}$ N as in (2.3) and the $c_{\hat{m}}$'s are nonzero and independent of z.

PROOF. Manoeuvring as in Example 6.4 we find for t' near t with $S_{t'}$ nonsingular

$$\theta_\delta(z, t') = \sum_{\hat{m} \in \mathbf{Z}^{\mathbf{g}-\tilde{\mathbf{g}}}} c_{\hat{m}}(t') e^{2\pi i \hat{m} \cdot \hat{z}} \theta_\delta(\tilde{z} + \hat{m}^{\cdot t} \hat{\Omega}(t'), \tilde{\Omega}(t'))$$

where

$$c_{\hat{m}}(t') = e^{\pi i\{\hat{m} \cdot \overline{\Omega}_{t'} \cdot \hat{m} - 2\hat{m} \cdot (\delta\overline{\Omega}_{t'}/2)\}}.$$

Approaching t through nonsingular curves for which $\varepsilon_i = \varepsilon$, $i = 1, \dots,$ n_{ij}; we obtain in (2.2), (2.3)

$$\text{Im } \overline{\Omega}_{t'} \approx N \cdot (1/2\pi i)\log \varepsilon, \quad \text{and}$$

$$c_{\hat{m}} \approx \varepsilon^{2(\hat{m} \cdot N \cdot \hat{m} - \delta N \cdot \hat{m})} \text{ with, by Remark 6.6,}$$

$$\hat{m} \cdot N \cdot \hat{m} - \delta N \cdot \hat{m} \geq 0.$$

Thus in the limit we are left only with the terms $\hat{m}$ satisfying $\hat{m} \cdot N \cdot \hat{m} = \delta N \cdot \hat{m}$. Finally the set of such $\hat{m}$'s is finite because the equation $x \cdot N \cdot x - \delta N \cdot x = 0$ represents a compact ellipsoid, since N is positive definite (Remark 2.5).

OBSERVATION 6.8. The finite set $\{\hat{m}/\hat{m} \cdot N \cdot \hat{m} = \delta N \cdot \hat{m}\}$ contains always the $(g - \tilde{g}) + 1$ points $\hat{m} = (0, \dots , 0), (0, \dots , 0, 1, 0, \dots , 0)$.

EXAMPLE 6.9. Let S be as in Example 2.6 then clearly

$$\theta_\delta(z, t) = \sum_{m/m_i = 0, 1} \prod_{i<j} m_i m_j \frac{(y_j - y_i)(x_j - x_i)}{(y_j - x_i)(x_j - y_i)} \prod_i e^{2\pi i m_i z_i}$$

or in Mumford's form [M1, vol. 2]

$$\theta_\delta(z, t) = \sum_{A \subset \{1, \cdot, g\}} \prod_{\substack{i<j \\ i, j \in A}} \frac{(y_j - y_i)(x_j - x_i)}{(y_j - x_i)(x_j - y_i)} \prod_A e^{2\pi i z_i}.$$

EXAMPLE 6.10. Let $t \in D$ correspond to a stable curve of genus 2 obtained by gluing two copies of $\mathbf{P}^1$ at three points giving rise to three nodes P, Q, R (see Figure 3).

One may choose a marking so that the dual forms have poles at P, R and Q, R respectively and $N = \begin{pmatrix} 2 & 1 \\ 1 & 2 \end{pmatrix}$. In this case

$$\theta_\delta(z, t) = 1 + e^{2\pi i z_1} + e^{2\pi i z_2}.$$

PROPOSITION 6.11. For each $t \in D^*$, then function $\theta_\delta(z, t)$ defines a nontrivial divisor on $J(S_t) = \mathbf{C}^g/L_t$.

PROOF. Our statement involves proving three facts,

(1) the vanishing locus of θ_δ is well defined on $J(S_t)$.
(2) θ_δ is not identically zero.
(3) θ_δ has nonempty zero set.
(1) follows from the identity

(6.2)
$$\begin{aligned}
&\theta_\delta(z + n + \tilde{m} \cdot (\widetilde{\Omega}_t, \widehat{\Omega}_t), t \\
&= e^{-\pi i(\tilde{m} \cdot \widetilde{\Omega} \cdot \tilde{m} - \tilde{m} \cdot \delta \widetilde{\Omega})} e^{-2\pi i \tilde{m} \cdot z} \theta_\delta(z, t) \\
&\hspace{6cm} \text{for } n \in \mathbf{Z}^g, \ \tilde{m} \in \mathbf{Z}^{\tilde{g}}
\end{aligned}$$

which, in view of Theorem 6.7, is itself an easy consequence of the well-known periodicity properties of theta.

To show (2) we take $\tilde{z}$, such that for some $\hat{m}$,

$$c_{\hat{m}} \theta_\delta(\tilde{z} + \hat{m} \cdot {}^t\widehat{\Omega}_t, \widetilde{\Omega}_t) \neq 0,$$

and then use the fact that the functions $e^{2\pi i \hat{m} \cdot \tilde{z}}$ are linearly independent.

To prove (3) we again take $\tilde{z}$ satisfying

$$\mu_0 = c_{\hat{m}} \theta_\delta(\tilde{z} + \hat{m} \cdot {}^t\widehat{\Omega}_t, \widetilde{\Omega}_t) \neq 0 \quad \text{for } \hat{m} = (0, \dots , 0).$$

The substitution $q_i = e^{2\pi i z_i}$ makes the function

$$(z_{\tilde{g}+1}, \dots, z_g) = \hat{z} \to \theta_\delta((\tilde{z}, \hat{z}), t)$$

into a polynomial

$$(q_{\tilde{g}+1}, \dots, q_g) = q \to \sum_{\hat{m}} \mu_{\hat{m}} q^{\hat{m}} = \mu_0 + \text{ other terms}$$

which has zeros q with all its coordinates nonzero; hence θ_δ has zeros.

REMARK 6.12. We would like to close this section with a few comments.

As the examples above show, Theorem 6.7 provides in many cases a practical rule to compute explicitly the expression for the theta function associated to stable curves.

This is so, because the monodromy matrix N on which this computation depends on, is readily obtained in terms of the number of poles encountered by the B-loops (see § 2). We observe that this matrix depends on the direction we approach our reference stable curve; in our case the diagonal direction $\varepsilon_i = \varepsilon$ for any i (see § 2.A).

In particular we see that, as far as obtaining the expression for the theta function is concerned, we could well have restricted ourselves to studying 1-dimensional families of nonsingular curves degenerating to a given stable curve. This also would have been enough for the main topics dealt with in the remainder of this paper, namely degenerate Lefschetz theory and degenerate solutions of K-P and K-dV.

It is with regard to future applications to the moduli space of stable curves, that we work with families parametrized by Bers' deformation spaces. Recall that these provide a complete collection of charts for this moduli space [Be1].

7. Lefschetz's theory for degenerate jacobians

What Proposition 6.11 says is that $\theta_\delta(z, t)$ is a section of certain theta divisor θ on $J(S_t)$; our next task is to prove that the classical Lefschetz theorem still holds.

Recall that this theorem states that $3 \cdot \theta$ is very ample i.e. we can embed $J(S_t)$ in projective space by means of products of three thetas.

One first needs a basis for $H^0(J(S_t), 3 \cdot \theta)$; in the classical case we have many choices, for instance (see [M1, vol. 1])

$$\theta[a/3; 0](3z, 3\Omega) = e^{\pi i (a/3) \cdot \Omega \cdot a} e^{2\pi i a \cdot z} \theta(3z + \Omega \cdot a, 3\Omega)$$

$a = (a_1, \dots, a_g)$ with $a_i = 0, 1, 2$.

As before it will be more convenient to replace z by $z - (\delta/2)$; we have

$$\theta[a/3; 0]\left(3z - \frac{3\delta}{2}, 3\Omega\right) = e^{\pi i (a/3) \cdot \Omega \cdot a - \pi i a \cdot \delta} e^{2\pi i a \cdot z} \theta_\delta(3z + \Omega \cdot a, 3\Omega).$$

To avoid problems of convergence we drop the first factor and take as new basis

$$(7.1) \qquad \theta_3[a](z, \Omega) = e^{2\pi i a \cdot z} \theta_\delta(3z + \Omega \cdot a, 3\Omega).$$

We are again allowed (see Observation 6.2) to write for t representing a nonsingular curve, $\theta_3[a](z, t)$ instead of $\theta_3[a](z, \Omega_t)$.

PROPOSITION 7.1. *The 3^g functions $\theta_3[a](z, t)$ extend to holomorphic functions on $\mathbf{C}^g \times D^*$.*

PROOF. The proof is similar to that of Theorem 6.4. Indeed the result is obvious for t's corresponding to stable curves of the kind considered in Example 6.3. So we only have to deal with t's corresponding to irreducible curves with just one node.

We first observe that

$$\Omega \cdot a = (\widetilde{\Omega} \cdot \tilde{a} + \widehat{\Omega} \cdot a_g, {}^t\widehat{\Omega} \cdot \tilde{a} + \Omega_{gg} a_g)$$

and then develop $\theta_\delta(3z + \Omega \cdot a, 3\Omega)$ as in Example 6.4 replacing $\widetilde{\Omega}$ by $3\widetilde{\Omega}$, z_g by $3z_g + {}^t\widehat{\Omega} \cdot \tilde{a} + \Omega_{gg} a_g$ and $m_g^2 - m_g$ by $m_g^2 - m_g + a_g(2m_g/3)$, since $\Omega_{gg} \approx (1/2\pi i)\log \varepsilon$.

We see that for $a_g = 1, 2$ only the term $m_g = 0$ remains nonzero. Hence we have $(a_g = 0)$

$$\theta_3[a](z, t)$$
$$= \theta_3[\tilde{a}](\tilde{z}, \widetilde{\Omega}) + e^{2\pi i(3z_g + {}^t\widehat{\Omega} \cdot \tilde{a})} \theta_3[\tilde{a}](\tilde{z} + \widehat{\Omega}, \widetilde{\Omega})$$

$(a_g \neq 0)$

$$\theta_3[a](z, t) = e^{2\pi i a_g z_g} \theta_3[\tilde{a}](\tilde{z} + \widehat{\Omega} \cdot (a_g/3), \widetilde{\Omega}).$$

Now the result follows again by codimensional reasons.

PROPOSITION 7.2.

$$\theta_3[a](z + n + \tilde{m} \cdot (\widetilde{\Omega}_t, \widehat{\Omega}_t), t)$$

(7.2)
$$= e^{-3\pi i(\tilde{m} \cdot \widetilde{\Omega} \cdot \tilde{m} - \tilde{m} \cdot \delta\widetilde{\Omega})} e^{-6\pi i \tilde{m} \cdot \tilde{z}} \theta_3[a](z, t)$$

$$\textit{for } n \in \mathbf{Z}^g, \ \tilde{m} \in \mathbf{Z}^{\tilde{g}}.$$

PROOF. Straightforward, see (6.2).

REMARK 7.3. By comparing (6.2) to (7.2), we see that $\theta_3[a]$ is a section of $3 \cdot \boldsymbol{\theta}$.

The key result in Lefschetz's theory is

PROPOSITION 7.4. *For each $t \in D^*$ and each $\alpha, \beta \in \mathbf{C}^g$, the functions*

$$F_{\alpha\beta}(z, t) = \theta_\delta(z - \alpha, t) \cdot \theta_\delta(z - \beta, t) \cdot \theta_\delta(z + \alpha + \beta, t)$$

are linear combinations of the $\theta_3[a](z, t)$'s.

PROOF. If t is such that S_t is either nonsingular or reducible with one separating node (Example 6.3) then its jacobian is a compact abelian variety and the result is then classical (see e.g. [**M1 or M3**]). In fact the coefficients of this linear combination are holomorphic functions of t. This can be seen as follows:

Since the 3^g functions $\theta_\delta[a]$'s are linearly independent, the function

$$(z_1, \ldots, z3^g) \rightarrow \det (\theta_3[a](z_i, t))_{a,i}$$

is $\neq 0$ (develop the det by its first row $\theta_\delta[a](z_1, t)$ varying a, and then argue by induction [1]).

Thus near such t's we can, for suitable values of $z_1, \ldots, z3^g$, make sense of the matrix $(\theta_3[a](z_i, t))_{a,i}^{-1}$ and by evaluating the linear combination $F_{\alpha\beta}(z, t) = \sum \mu_a(t)\theta_3[a](z, t)$ at these values we obtain

$$\begin{bmatrix} \mu_{a_1}(t) \\ \vdots \\ \mu_{a_3g}(t) \end{bmatrix} = (\theta_3[a](z_i, t))_{a,i}^{-1} \cdot \begin{bmatrix} F_{\alpha\beta}(t, z_1) \\ \vdots \\ F_{\alpha\beta}(t, z_3g) \end{bmatrix}.$$

Now all we need to apply our standard extension argument to the functions $\mu_a(t)$, is to have them defined up to an open set of codimension 2; in other words we have to show that a matricial identity of the kind above holds also for t's as in Example 6.4, for which in turn is enough to show that for those t's too, the functions $\theta_3[a]$'s are linearly independent.

But this is easily seen once we have (see the proof of Proposition 7.1) their explicit expression in terms of $(g - 1)$-dimensional ones for which the result is well known.

As in the classical case, the above result is all we need to prove Lefschetz's embedding theorem.

THEOREM 7.5. *For each $t \in D^*$, the map*

$$J(S_t) \rightarrow \mathbf{P^n} \quad N + 1 = 3^g$$
$$z \rightarrow (\theta_3[a](z, t))_a$$

is an embedding.

PROOF. We only have to modify the proof in the classical case, which we take from [M3].

(i) *Welldefinedness.* The $\theta_3[a]$'s cannot be simultaneously zero at a point z since this would imply $F_{\alpha\beta}(z, t) = 0$ for all α, β which is in contradiction with Proposition 6.11 so the map is well defined.

(ii) *Injectivity.*

$$\theta_3[a](u, t) = \mu\theta_3[a](v, t) \quad \text{for all } a\text{'s implies}$$
$$F_{\alpha\beta}(u, t) = \mu F_{\alpha\beta}(v, t) \quad \text{for all } \alpha, \beta \in \mathbf{C^g}.$$

Now we argue as in [M3] to get $\theta_\delta(u + \alpha) = c_0 e^{2\pi ic \cdot z}\theta_\delta(v + \alpha)$ for each α, or making $v + \alpha = z$, $\theta_\delta(z + u - v) = c_0' e^{2\pi ic \cdot z}\theta_\delta(z)$ for some constants c_0', c.

---

[1] Pointed out to me by my colleague A. Quirós.

In this formula if we substitute z by $z + e_i$, e_i the g-dimensional unitary vector, we find using (6.2) $c \cdot e_i \in \mathbf{Z}$ i.e. $c = (\tilde{c}, \hat{c}) \in \mathbf{Z}^g$.

By comparing both sides of this relation, expressed as in Theorem 6.7 and viewed as functions of $\hat{z}$, we see that $\hat{c} = 0$. Thus it becomes

$$\theta_\delta(z + u - v) = c_0' e^{2\pi i \tilde{c} \cdot \tilde{z}} \theta_\delta(z).$$

Next, substitution of z by $z + \tilde{e}_i \cdot (\tilde{\Omega}, \hat{\Omega})$ gives, again by (6.2), $v_i - u_i = \tilde{c} \cdot \tilde{\Omega} \mod \mathbf{Z}$ $i = 1, \ldots, \tilde{g}$.

Hence, we may write $v - u = n + \tilde{c} \cdot (\tilde{\Omega}, \hat{\Omega}) + w$ where w is a vector whose first $\tilde{g}$ coordinates are zero i.e. $w = (0, \hat{w})$.

Now by (6.2) the L.H.S. of our identity takes the value

$$e^{-\pi i (\tilde{c} \cdot \tilde{\Omega} \cdot \tilde{c} - \tilde{c} \cdot \delta \tilde{\Omega})} e^{2\pi i \tilde{c} \cdot \tilde{z}} \theta_\delta(z + w)$$

and, again by comparing both sides, we find that $e^{2\pi i \hat{m} \cdot \hat{w}}$ is independent of $\hat{m}$.

From here if we let $\hat{m}$ take the values $(0, \ldots, 0)$, $\hat{e}_k$, the $(g - \tilde{g})$-dimensional unitary vectors (Observation 6.8), we deduce $w \in \mathbf{Z}^g$, hence $v - u \in L_t$ as required.

(iii) *Injectivity on the tangent spaces.* Assuming the contrary, we can take $z^0 \in J(S_t)$ and a tangent vector $D = \sum a_i \partial / \partial z_i$, $i = 1, \ldots, g$, at z^0 with not all $a_i = 0$, mapped into the zero vector of the tangent space of $\mathbf{P}^N$ at the image of z^0.

There is then an a_0 such that for all α, β $a_0 F_{\alpha\beta}(z^0) + \sum a_i \partial F_{\alpha\beta} / \partial z_i(z^0)$ $= 0$ i.e. $D(\log F_{\alpha\beta})(z^0) = -a_0$.

If we put $f(z) = D(\log \theta_\delta)(z)$, this is written as

$$f(z^0 - \alpha) + f(z^0 - \beta) + f(z^0 + \alpha + \beta) = -a_0.$$

Taking derivatives with respect to each variable α_j we obtain

$$\partial f / \partial \alpha_j(z^0 - \alpha) = \partial f / \partial \alpha_j(z^0 + \alpha + \beta).$$

We see that these partial derivatives are constant functions of α hence f must be linear in a neighbourhood of α where f is defined. Integrating the equation $f(z) = D \log \theta_\delta(z)$ we obtain that there is an $a = (a_1, \ldots, a_g) \neq 0$ such that for all $\mu \in \mathbf{C}$, we have

(I)
$$\theta_g(z + \mu a) = e^{c(\mu/2)^2 + \mu f(z)} \theta_\delta(z)$$

with $c = f^0(a)$, where f^0 is the homogeneous part of the linear function f.

If, as in the earlier step, we make the substitution $z \to z + n$ we deduce $f^0 \equiv 0$ i.e. $f \equiv c$. Thus (I) becomes

(II)
$$\theta_\delta(z + \mu a) = e^{c((\mu/2)^2 + \mu)} \theta_\delta(z).$$

Next we make the substitution $z \to z + \tilde{e}_k \cdot (\widetilde{\Omega}, \widehat{\Omega}) \, k = 1, \ldots, \tilde{g}$ to obtain $\mu a_k \in \mathbf{Z}$ for any μ. Hence $a_k = 0$ for all k in other words $a = (0, \hat{a})$.

Now by comparing the expressions of both sides of (II) we finally obtain

$$e^{2\pi i (\hat{m} \cdot \hat{a})\mu} = e^{c((\mu/2)^2 + \mu)} \quad \text{for each } \mu \text{ and } \hat{m}$$

which is impossible; just let $\hat{m}$ take the value $(0, \ldots, 0)$.

REMARK 7.6. Observe that by taking the Zariski closure of Lefschetz's image of $J(S_t)$ in projective space we obtain a projective compactification $\overline{J(S_t)}$ of the jacobian of any stable curve. In fact, since we operate simultaneously at all $t \in D^*$, we have produced a compactification of the fibre space $J(D^*) \to D^*$, say $\overline{J(D^*)} \to D^*$ whose fibre over t is precisely $\overline{J(S_t)}$.

To visualize more clearly the structure of $J(S)$, consider the case when S is an irreducible curve with just one node. Accordingly $J(S)$ has just an unbounded coordinate z_g; we let it become $(-1/2\pi i) \cdot \infty$, then among all functions $\theta_3[a]$—the projective coordinates for $J(S)$—we are just left (see Proof of Proposition 7.1) with the functions $\theta_3[\tilde{a}]$—the projective coordinates for $J(\widetilde{S})$—. What this indicates is that $\overline{J(S)}$ has been obtained by adding on $J(S)$ a disjoint copy of $J(\widetilde{S})$.

In general one would have

$$\overline{J(S)} = J(S) \coprod_A J(\widetilde{S}_A)$$

where A runs among the subsets of a chosen maximal collection of nonseparating nodes and $\widetilde{S}_A$ denotes the partial normalization of S at A (see [**M1 and Ma**]).

Furthermore as above we must have something like

$$\overline{J(D^*(S))} = J(D^*(S)) \coprod_A J(D^*(\widetilde{S}_A)).$$

We do not pursue these ideas here though.

8. Solitons and stable curves

It is a well-known theorem of Kriechever that the function

$$(8.1) \qquad u(x, y, t) = 2\frac{\partial^2}{\partial x^2} \log \theta(xU + yV + tW + z) + c$$

solves the K-P equation

$$(3/4)u_{yy} = \partial/\partial x[u_t - (1/4)(6uu_x + u_{xxx})].$$

Here θ is the theta function associated to a nonsingular curve C, and

$$U = -(w_1(P), \ldots, w_g(P))$$
$$V = (-1/2)(w_1'(P), \ldots, w_g'(P))$$
$$W = (-1/3!)(w_1''(P), \ldots, w_g''(P))$$

where $w_i'(P)$ means the first derivative of the 1-form w_i at an arbitrarily chosen point $P \in C$.

As for the constant c, it is determined as follows:

Let β be the meromorphic 1-form which has P as its unique pole, has zero A-periods and is near P of the form

$$\beta = ((-1/s^2) + a(s))ds$$

s being the coordinate used to compute the vectors U, V, W, then $c = a(0)$.

Moreover when C is hyperelliptic, P a Weierstrass point and $s = \sqrt{x - x(P)}$, the square root of the hyperelliptic function, then $V = (0, \ldots, 0)$, the variable "y" disappears and (8.1) becomes

$$(8.2) \qquad u(x, t) = 2\frac{\partial^2}{\partial x^2}\log \theta(xU + tW + z) + c$$

which solves the famous K-dV equation

$$\pm u_t = (6uu_x + u_{xxx}).$$

Similarly, trigonal curves correspond to the Boussinesq equation of the nonlinear string

$$3u_{yy} = -\partial/\partial x(6uu_x + u_{xxx}).$$

The article of Dubrovin [D] is a wonderful survey on all this (see also [M1, vol. 2]).

Our fundamental observation is that, since the vector z is arbitrary, we can replace in (8.1) and (8.2) θ by θ_δ. Thus, for any $\alpha \in D^* = D^*(S)$ corresponding to a nonsingular curve, the function

$$(8.3) \qquad u_\alpha(x, y, t) = 2\frac{\partial^2}{\partial x^2}\log \theta_\delta(xU_\alpha + yV_\alpha + tW_\alpha + z_\alpha, \Omega_\alpha) + c_\alpha$$

solves the K-P equation. Furthermore the theory developed in §6 allows us to take the limit of u_α as α approaches a singular stable curve; so we have

THEOREM 8.1. *Kriechever's theorem remains valid when C is any stable curve.*

Observe that this way one may obtain solutions of extremely simple form (see for instance Example 6.10).

To illustrate how the theory can be effectively applied, we work out the following interesting case.

EXAMPLE 8.2. (compare [M1, vol. 2 and Mck]). Let S be the stable curve obtained, as in Example 2.6, by identifying in $\mathbf{P}^1$ the following g pairs of points $(b_1, -b_1), \ldots, (b_g, -b_g)$.

We first note that S is limit of hyperelliptic curves. Indeed S is isomorphic to the curve

$$C = \left\{y^2 = x\prod(x - b_i^2)^2\right\} U\{\infty\}$$

via the map

$$S \longrightarrow C$$

$$s \mapsto \begin{cases} x = s^2 \\ y = s\pi(s^2 - b_i^2). \end{cases}$$

Now C appears as the singular fibre of the family of hyperelliptic curves

$$C_\varepsilon = \left\{ y^2 = x \prod(x - b_i^2)(x - b_i^2 + \varepsilon) \right\} U\{(\infty, \varepsilon)\}.$$

The surface $V = \underset{\varepsilon}{U} C_\varepsilon$ has near $x = 0$ the usual coordinates $(\varepsilon, s) \to$ (ε, x, y) where

$$x = s^2, \qquad y = s\sqrt{\pi(s^2 - b_i^2)(s^2 - b_i^2 + \varepsilon)}$$

(replace s by $1/s$ to get coordinates near ∞).

The fact that S is limit of hyperelliptic curves has the following implication.

If for each $\varepsilon \neq 0$ we compute the vectors U_ε, V_ε, W_ε at the two Weierstrass points $x = 0$, $x = \infty$ which do not become a node for $\varepsilon = 0$, the function $u_\varepsilon(x, t)$ will be a solution of K-dV and so must therefore be $u_0(x, t)$.

In fact take a basis $\{w_1^\varepsilon, \dots, w_g^\varepsilon\}$ such that for $\varepsilon = 0$ we have (see Example 2.6)

$$w_k = (1/2\pi i)(2b_k/(s^2 - b_k^2))ds.$$

Using near ∞ the coordinate $s = 1/\sqrt{x}$, our three vectors become

$$U^0 = (1/\pi i)(b_1, \dots, b_g)$$

$$V^0 = (0, \dots, 0)$$

$$W^0 = (1/3!\pi i)(b_1^3, \dots, b_g^3).$$

Thus, by Example 6.9, we obtain

$$u_0(x, -t) = 2\frac{\partial^2}{\partial x^2}\log \left[1 + \sum_A \prod \left(\frac{b_j - b_i}{b_j + b_i} \right)^2 \prod e^{2b_i((z_i/b_i) + x - (b_i^2/3!)t)} \right] + c.$$

In fact the constant c may also be computed; to do that we observe that for $\varepsilon \neq 0$ $\beta_\varepsilon' = (1/2)x^g dx/y$ is a 1-form on C_ε with only one double pole at (ε, ∞).

To obtain the required 1-form β_ε we just normalize

$$\beta_\varepsilon = \beta_\varepsilon' - \sum_i \left(\int_{A_i(\varepsilon)} \beta_\varepsilon' \right) w_i^\varepsilon.$$

It is clear that $\beta_0' = s^{2g} ds / \prod_i (s^2 - b_i^2)$, and as in §2.

$$\int_{A_i(0)} \beta_0' = 2\pi i \cdot \text{res}_{b_i} \beta_0' = 2\pi i \cdot b_i^{2g-1}/2 \prod_{j \neq i}(b_i^2 - b_j^2).$$

Thus, replacing s by $1/s$ we obtain near ∞,

$$\beta_0 = (-1/s^2) \left\{ 1 + 0 \cdot s + \left(\sum_i \left[b_i^2 + (b_i^2)^g / \prod_{j \neq i} (b_i^2 - b_j^2) \right] \right) \cdot s^2 + \cdots \right\} ds.$$

To obtain c we just compute

$$a_0(s) = - \left\{ \left(\sum_i \left[b_i^2 + (b_i^2)^g / \prod_{j \neq i} (b_i^2 - b_j^2) \right] \right) + \text{ higher terms} \right\}$$

at $s = 0$ to obtain

$$c = - \left(\sum_i \left[b_i^2 + (b_i^2)^g / \prod_{j \neq i} (b_i^2 - b_j^2) \right] \right).$$

A rather involved Vandermonde like calculus shows that in fact $c = 0$. We summarize what we have proved

THEOREM 8.3 (Mumford). *For any distinct real numbers* $b_1, \ldots, b_g$ *and* $z_1, \ldots, z_g$ *arbitrary real numbers, the function*

$$u_0(x, -t) = 2 \frac{\partial^2}{\partial x^2} \log \left[1 + \sum_A \prod \left(\frac{b_j - b_i}{b_j + b_i} \right)^2 \prod e^{2b_i((z_i/b_i) + x - (b_i^2/3!)t)} \right]$$

where A *varies among the subsets of* $\{1, \ldots, g\}$, *is a real solution of the* K-dV *equation.*

These are the famous *soliton* solutions of K-dV, for it is clear from it explicit expression that $u_0(x, t)$ has as effective support a set of g bands $|(z_i/b_i) + x - (b_i^2/3!)t| \leq \text{small}$.

Furthermore write (8.3) in the form

$$u_\alpha(x, t) = 2D_\alpha^2 \log \theta_\delta(v, \Omega_\alpha)|_{v = xU_\alpha + tW_\alpha + z}$$

where D_α is the directional derivative D_{U_α}.

Note, using (6.2), that $D_\alpha^2 \log \theta_\delta(v)$ is a meromorphic function on $J(D)$.

Now for each $\alpha \in D$ with S_α a nonsingular hyperelliptic curve near S— e.g. the C_ε's above—consider the vectors

$$v_\alpha = xU_\alpha + tW_\alpha(1/2) \sum_{j \neq i} \int_{B_j(\alpha)} (w_1(\alpha), \ldots, w_g(\alpha))$$

$$v'_\alpha = xU_\alpha + tW_\alpha - (1/2) \sum_{j \neq i} \int_{B_j(\alpha)} (w_1(\alpha), \ldots, w_g(\alpha)).$$

Since they differ by a period, we must have

$$D_\alpha^2 \log \theta_\delta(v)|_{v = v_\alpha} = D_\alpha^2 \log \theta_\delta(v)_{v = v'_\alpha}.$$

Note that the ith coordinate of v_α (resp. v'_α) remains bounded as we approach the stable curve S whereas all the other coordinates do not. Thus, near $\varepsilon = 0$,

$$D_\alpha^2 \log \theta_\delta(v)_{|v=v_\alpha} = u_\alpha(x,t) \qquad t \gg 0,$$

$$x - (b_i^2/3!)t = z_\alpha, \text{ bounded.}$$

$$D_\alpha^2 \log \theta_\delta(v)_{|v=v'_\alpha} = u_\alpha(x,t), \qquad t \ll 0,$$

$$x - (b_i^2/3!)t = z_\alpha - \sum_{j \neq i} \int_{B_j(\alpha)} w_i(\alpha).$$

In other words, for $t \to \infty$ or $t \to -\infty$, $u_0(x,t)$ has the same shape in each band "i" except for the phase shift

$$(1/2\pi i b_i) \lim \sum_{j \neq i} \int_{B_j(\alpha)} w_i(\alpha) = (1/b_i) \sum_{j \neq i} \log \left(\frac{b_j - b_i}{b_j + b_i} \right)^2.$$

This is another well-known property of solitons.

REFERENCES

[Be1] L. Bers, *Spaces of degenerating Riemann surfaces*, Ann. of Math. Studies, no. 79, Princeton Univ. Press, Princeton, N. J., 1974, pp. 43–55.

[Be2] ____, *On spaces of Riemann surfaces with nodes*, Bull. Amer. Math. Soc. (N.S) **80** (1974), 1219–1223.

[B-P-V] W. Barth, C. Peters, and A. Van de Ven, *Compact complex surfaces*, Springer-Verlag, 1984.

[D] B. A. Dubrovin, *Theta functions and nonlinear equations*, Russian Math. Surveys **36** (1981), 11–92.

[E] C. Earle, *Families of Riemann surfaces and Jacobi varieties*, Ann. of Math. (2) 107 (1978), 255–286.

[F] J. Fay, *Theta functions on Riemann surfaces*, Lecture Notes in Math., vol. 352, Springer-Verlag, 1973.

[G-R] H. Grauert and R. Remmert, *Coherent analytic sheaves*, Springer-Verlag, 1984.

[J] Jambois, *Seminar on degeneration of algebraic varieties*, Inst. Adv. Study Princeton, 1969–1970.

[M1] D. Mumford, *Tata Lectures on theta*. I, II, Progress in Math. (Birkhäuser), **28** 1983.

[M2] ____, *On the Kordaira dimension of the Siegel modular variety*, Algebraic Geometry—Open Problems, Lecture Notes in Math., vol. 997, Springer-Verlag.

[M3] ____, *Abelian Varieties*, Oxford Univ. Press, 1970.

[Ma] A. L. Mayer, *Seminar on degeneration of algebraic varieties*, Inst. Adv. Study, Princeton, 1969–1970.

[Mck] H. P. McKean, *Theta functions, solitons, and algebraic curves*, Partial Differential Equations and Geometry, (C. I. Byrnes, ed.), Dekker, 1979.

[N] Y. Namikawa, *On the canonical holomorphic map from the moduli space of stable curves to the Igusa monoidal transform*, Nagoya Math. J. 52 (1973), 197–259.

[Wo] S. Wolpert, *On the homology of the moduli space of stable curves*. Ann. of Math. (2) **118** (1983) 491–523.

DEPARTMENTO DE MATHEMÁTICAS, UNIVERSIDAD AUTÓNOMA DE MADRID, CANTOBLANCO, 28049 MADRID, ESPAÑA

Contemporary Mathematics
Volume **136**, 1992

Geometry of the Fano Threefold of Degree 10 of the First Type

ATANAS ILIEV ILIEV

Table of Contents

1. Preliminaries

(1.1). DEFINITION. The smooth irreducible projective variety X of dimension three over the field of complex numbers is called a Fano threefold, if the anticanonical sheaf $-K_X$ (resp. the anticanonical divisor $-K_X$) is ample.

Let X be a Fano threefold and let K_X be the canonical divisor of X. From the Riemann-Roch formula and from the Kodaira vanishing theorem it follows, that:

(a) $h^i(X, O_X(-m \cdot K_X)) = 0$, for $i = 1, 2$ and for any m-integer; or, for $i > 0$, $m \geq 0$; or, for $i < 3$, $m < 0$;

(b) $h^0(X, O_X(-m \cdot K_X)) = (m(m+1)(2m+1) \cdot (K_X)^3)/12 + 2m + 1$.

1980 *Mathematics Subject Classification* (1985 *Revision*). Primary 14J30.
This paper is in final form and no version of it will be submitted for publication elsewhere.

In particular, $h^i(X, O_X) = 0$ for $i > 0$, and $h^0(X, -K_X) = \dim |-K_X| + 1 = (K_X)^3/2 + 1$.

(1.2). DEFINITION. The integer $g = g(X) = -(K_X)^3/2 + 1$ is called a genus of the Fano threefold X.

The Fano variety X is called a Fano variety of the main series, if the anticanonical divisor $-K_X$ is very ample. Therefore, the linear system $|-K_X|$ gives an embedding of X in P^{g+1} as a subvariety of degree $2 \cdot g - 2$, ($g \geq 3$). The intersections of X with codimension 2 subspaces $P^{g-1} \subset P^{g+1}$ are canonical curves of degree $2 \cdot g - 2$ and of arithmetical genus $p_a = g(X)$.

The Fano variety X is called a (Fano) variety of the 1st kind, if $\mathrm{Pic}(X) \simeq \mathbb{Z}$, where $\mathrm{Pic}(X)$ is the Picard group of X. The Fano varieties of the 1st kind (i.e., with $\mathrm{Pic} \simeq \mathbb{Z}$) are classified by V. A. Iskovskikh [I1, I2, I4] after G. Fano [F1, F2] and L. Roth [R].

In the papers [MM1, MM2] of S. Mori and S. Mukai is given a list of the Fano threefolds of rank $\mathrm{Pic} \geq 2$ (i.e., the Fano 3-folds of the 2nd kind). From a birational point of view both kinds of Fano threefolds are interesting, but nontrivial questions, connected with the problem of nonrationality, arise from the Fano 3-folds with $\mathrm{Pic} \simeq \mathbb{Z}$. Such nontrivial examples are the cubic hypersurface in P^4 [CG, B1], the quartic double solid [W, C, T1, T2, V], the complete intersection of three quadrics in P^6 [B, TN, D], the two-sheeted covering of the cone over the Veronese surface [T3, H], the quartic hypersurface in P^4 [IM, I3, L], etc.

(1.3). The Fano threefolds of the main series and of the 1st kind have genus $g \leq 12$. The 3-folds of $g = 7$ and of $g \geq 9$ are rational. The variety of $g = 8$ is interpreted as an intersection of the Grassmannian of the lines in P^5, embedded by Plücker in P^{14}, with a codimension 5 subspace $P^9 \subset P^{14}$; the last is birational to a smooth cubic hypersurface in P^4. The variety $X_8 \subset P^6$ of genus 5 is a complete intersection of three quadrics in P^6, etc. (see [I1, B2]). In particular, the Fano varieties $X_{10} \subset P^7$ of genus 6 are the last nontrivial examples of Fano threefolds for which $\mathrm{Pic} \simeq \mathbb{Z}$ and which have very ample anticanonical class. They are divided into two types (see [G]):

1st TYPE. An intersection of the Grassmannian of the lines of P^4, embedded by Plücker in P^9, with a codimension 2 subspace $P^7 \subset P^9$ and with a quadric;

2nd TYPE. An intersection of a cone (in P^7) over the threefold $X_5 \subset P^6$ with a quadric, where $X_5 \subset P^6$ is an intersection of the Grassmannian of the lines in P^4, embedded by Plücker in P^9, with a codimension 3 subspace $P^6 \subset P^9$.

It is known that the two types of Fano threefolds of genus $g = 6$ belong to one type of deformations [G, I4].

The Fano threefold of genus $g = 6$ (resp. of degree $d = 2 \cdot g - 2 = 10$)

of the first of the abovementioned types is studied in the present paper.

(1.4) The Fano threefold of degree 10 of the 1st type is represented as a 3-dimensional family of lines, lying in P^4 (see for example §(1.5) and §2). In §2 are derived some elementary incidences, which are fulfilled for the lines $l \subset P^4$, belonging to $X = X_{10}$ (see Propositions (2.2.2), (2.2.3), (2.2.4)). It is shown also the existence of a special hypersurface $R = R_6 \subset P^4$ such that the lines $l \subset P^4$, which belong to X, are at least three-tangents to R_6 (see (2.2.5) and (2.2.6)). In §3 is considered a two-dimensional analogue of the threefold X_{10}. In fact, this is the Del Pezzo surface $S = S_{2 \cdot 2}$ in P^4, represented as an intersection of the Grassmannian $G(2, 4) = G(1 : P^3)$, embedded as a Plücker quadric in P^5, with a hyperplane $P^4 \subset P^5$ and with a quadric (see Example 3.2). In this case, the lines $l \subset P^3$, which belong to S, are at least bi-tangents to a Kummer surface in P^3 (see (3.2.7)). In the rest of §3 is shown, that the propositions from §2 can be extended in an appropriate way (see Corollary (3.3.3)).

Sections 5, 6, 7 are devoted to a detailed study of the family $\mathscr{C}^3$ of rational normal cubic curves, lying on the three-fold $X = X_{10}$. First, in §5 a geometrical investigation of the possible bundles of lines in P^4, which represent the cubics $C \in \mathscr{C}^3(X)$, is made (see §5.2). It is shown also the existence of a well-defined involution $\sigma: \mathscr{C}^3 \to \mathscr{C}^3$, inducing a natural embedding of the factorfamily $\mathscr{C}_0^3 = \mathscr{C}^3/\sigma$ in the Grassmannian $G = G(2, 5)$ (see Corollary (5.3.2)). It is shown in §6, that the factorfamily $\mathscr{C}_0^3$ can be embedded in a special projective bundle over the Grassmannian $G = G(2, 5)$ as a degeneration locus (more precisely, as a second determinantal) of an appropriate net of quadrics over the Grassmannian (see §6.3, §6.4. and Proposition (6.4.4)). The essential results in §7 are the Corollaries (7.2.4) and (7.2.6), the Proposition (7.3.2) and their geometrical description in the Corollary (7.3.9). They are technical consequences from the results in §6, but under some additional open conditions on the variety X_{10} (see (6.3.5) (i), (ii) and (7.1.2) (iii)). In particular, it is shown, that the family $\mathscr{C}^3$ of rational normal cubic curves on X_{10} can be embedded naturally in the projectivized bundle $P_Z(\Sigma^*) = G_Z(4, \Sigma)$ as a zero-scheme of some section (see Proposition (7.3.2)); see also (6.3.2) and (6.3.3) for the definitions of Z and Σ.

The geometrical description of the embedding $\mathscr{C}^3 \hookrightarrow P_Z(\Sigma^*)$, given in Corollary (7.3.3) is used in the formulation and in the proof of the Tangent Bundle Theorem (T.B.T.) for the family $\mathscr{C}^3$, stated and proven in §8 (see Proposition (8.1.9) and Theorem (8.2.12)). The last describes the tangent bundle of the family $\mathscr{C}^3$ by means of some standard bundles, defined over $\mathscr{C}^3$ (see (8.2.12) and Remark (8.2.10)). In fact, the T.B.T., in its formulation presented in the paper, describes the disposition of the tangent spaces at the points of the Abel-Jacobi image of the family $\mathscr{C}^3$, inside the Intermediate Jacobian $J(X)$ of X.

In addition, some known results on the family of conics on X_{10} are followed out in §4 and in §9. The original approach to the family of conics, which inspired the author to go on further and to apply an appropriate technique to the family of rational normal cubics on X_{10}, is due to D. Logachev (see [**L1, L2**] and also (9.2.1), (9.2.5), (9.3.6), (9.3.8), (9.3.10), (9.3.11)).

The use and the possible applications of the T.B.T. to the study of the principally polarized Intermediate Jacobian of X (see for example [**CG**]) are made in the author's comments in §9.1.

(1.5) Let $V = V_5$ be the five-dimensional space $\mathbb{C}^5$ and let $G = G(2, 5) = G(2, V) = G(1 : P(V))$ be the Grassmannian of the two-dimensional subspaces of V (resp. of lines in $P(V) = P^4$). There is a standard Plücker embedding:

$$Pl: G = G(2, V) \hookrightarrow P^9 = P\left(\overset{2}{\bigwedge} V\right),$$

which can be described as follows:

Let $0 \subset L \subset V$ be a 2-dimensional subspace of V (i.e. $L \in G$), and let $L = \mathbb{C} \cdot u + \mathbb{C} \cdot v$ for some basis $\{u, v\}$ of L. Then, by definition, $Pl: L \mapsto$ (the 1-dim subspace $0 \subset \mathbb{C} \cdot u \wedge v \subset \bigwedge^2 V$), i.e. $Pl(L) \in G(1, \bigwedge^2 V) = P(\bigwedge^2 V) = P^9$. It is easy to see that the definition of $Pl(L)$ is independent of the choice of the basis $\{u, v\}$ of L.

Let $P^4 = P(V)$ has homogeneous coordinates $x_0, x_1, \ldots, x_4$, and let $e_0, e_1, \ldots, e_4$ the corresponding dual basis of vectors in V. Let $e_{ij} = e_i \wedge e_j$, $0 \ i \leq j \leq 4$ and let $\{x_{ij}\}$ be the corresponding to $\{e_{ij}\}$ dual basis. The embedded Grassmannian $G = G(2, V) \subset P(\bigwedge^2 V)$ is described as the set of all classes of decomposable 2-forms (bi-vectors) in $\bigwedge^2 V$ modulo a multiplication with an element of $\mathbb{C}^*$. Therefore, the ideal $I(G)$ of $G \subset P^9$ is degenerated by the five Plücker quadrics $Pl_m = x_{ij} \cdot x_{kl} - x_{ik} \cdot x_{jl} + x_{il} \cdot x_{jk}$, $0 \leq m \leq 4$, $0 \leq i < j < k < l \leq 4$ and $i, j, k, l \neq m$. The Plücker embedding represents the Grassmannian G as a smooth subvariety $G \subset P^4$ of degree 5 and of dimension 6.

Let $P^7 \subset P^9$ be a codimension 2 subspace of $P^9 = P(\bigwedge^2 V)$, which is in general position with the embedded Grassmannian G, let H_1 and H_2 be two hyperplanes in P^9 such that $H_1 \cap H_2 = P^7$ and let Q be a general (especially smooth) quadric in P^9 (or, equivalently, in P^7). According to the choice of the P^7 and Q, the variety $X = X_{10} = G \cap P^7 \cap Q = G \cap H_1 \cap H_2 \cap Q$ is smooth, and it is easy to see, that X is a Fano variety of dimension 3, of degree 10 and of genus 6; the anticanonical divisor $-K_X$ represents the hyperplane section of the embedded $X = X_{10} \subset P^7$.

2. Some elementary incidences, connected with the Schubert calculus on the Grassmannian of the lines in P^4

2.1. It is well known, that the integer homologies of the Grassmannian have no torsion and are generated freely by the homology classes of the Schubert cycles (see for example [**GH or F**]). In particular, the Schubert classes $\sigma_{a,b}$, $3 \geq a \geq b \geq 0$ (of real codimension $2 \cdot (a+b)$) of the Grassmannian $G = G(2, 5) = G(2, V)$ are described as follows:

Let $P^4 = P(V)$ for $V = \mathbb{C}^5$ and let $P^0 \subset P^1 \subset P^2 \subset P^3 \subset P^4 = P(V)$ be a flag in $P(V)$. Using the natural isomorphism: $G(2, V) \simeq G(1 : P(V))$, which identify the two-dimensional subspace $0 \subset L \subset V$ and the projective line $l = P(L) \subset P(V)$, we define $\sigma_{a,b} = \{l : l$ is a line in P^4 and $l \cap P^{3-a} \neq \varnothing$, $l \subset P^{4-b}\}$, where $3 \geq a \geq b \geq 0$.

The intersections of cycles in the Chow ring $A \cdot (G)$ can be described by the Pieri-formula (see for example [**GH, Chapter I, §5**]):

$$\sigma_{a,b} \cdot \sigma_{c,0} = \sum_{d+e=a+b+c, d \geq a \geq e \geq b} \sigma_{d,e} \,.$$

The intersection of general cycles $\sigma_{a,b} \cdot \sigma_{c,d}$ can be obtained from the Pieri-formula by the additive properties of the intersections of cycles (see for example the Giambelli formula in [**GH, Chapter I, §5**]).

2.2.

(2.2.1). Let $V = V_5$, $P^4 = P(V)$, $P^7 \subset P^9 = P(\wedge^2 V)$, H_1, $H_2 (H_1 \cap H_2 = P^7)$, $G = G(2, V) = G(1 : P(V))$, Q and $X = X_{10} = G \cap P^7 \cap Q = G \cap H_1 \cap H_2 \cap Q$ be as above.

Let $W = W_5 = G \cap P^7 = G \cap H_1 \cap H_2$. The fourfold W is represented by the homology class $\sigma_{1,0} \cdot \sigma_{1,0}$ in the Chow ring $A \cdot (G)$, i.e. $W = \sigma_{1,0} \cdot \sigma_{1,0} = \sigma_{2,0} + \sigma_{1,1}$ (here $=$ means the homology equivalence of cycles). In fact, the homology cycle of $\sigma_{1,0}$ represents the hyperplane section according to the Plücker embedding, so the quadrics are represented by the cycle $2 \cdot \sigma_{1,0}$.

Let x be a general point of P^4. Using the fact, that the set of the lines in P^4, which pass through x, is represented by the cycle $\sigma_{3,0}(x) = \{l \in G : x \in l\}$ (x is the P^0 in the flag, see above) and the homological equivalence $Q = 2 \cdot \sigma_{1,0}$, we derive the following:

(2.2.2). PROPOSITION. (1) *The set of the lines in* P^4, *which belong to* W *and which pass through a general point* $x \in P^4$ *is homologous to the cycle* $\sigma_{3,2}$, *which represents a plane bundle of lines through the point* $x \in P^4$ (*i.e.* $\sigma_{3,2}$ *is a Grassmann line in* G);

(2) *There are exactly two lines in* P^4, *which belongs to* X *and which pass through the general point* $x \in P^4$.

PROOF. It rests only to apply the formulae for the intersections of cycles on G.

Now let $P^3 \subset P^4$ be a general hyperplane. Using the fact, that the set (the Grassmannian) of all the lines in the chosen P^3 is represented by the cycle $\sigma_{1,1}(P^3) = \{l \in G : l \subset P^3\}$ (P^3 is the P^3 in the flag, see above), we obtain in a similar way

(2.2.3). PROPOSITION. (1) *The set of the lines, which belong to W and which lie in a fixed generally chosen hyperplane $P^3 \subset P^4$, is homologous to the cycle $\sigma_{3,1} + \sigma_{2,2}$ and represents a two-dimensional quadric $Q(P^3) = S(P^3)$, embedded in G;*

(2) *The set of the lines, which belong to X and which lie in a fixed generally chosen $P^3 \subset P^4$ is homologous to the cycle $4 \cdot \sigma_{3,2}$ and represents an elliptic curve $C(P^3)$ of degree 4 in G. The $C(P^3)$ is an intersection of the quadric $Q(P^3)$ with the quadric Q in the 3-dimensional projective space $Q(P^3) =$ Span $Q(P^3)$.*

From another point of view, the quadric surface $Q(P^3)$ is an intersection of the four-dimensional Plücker quadric $G(1 : P^3)$, embedded in $G(1 : P^4) \subset P^9$ by the natural embedding corresponding to the embedding $P^3 \subset P^4$, with the subspace $P^7 \cap P^9$. From the last it follows immediately, that, if P^3 is chosen sufficiently general, then the quadric $Q(P^3)$ and the curve $C(P^3)$ are smooth subvarieties of G. As a set of lines in $P^3 \subset P^4$, the quadric $Q(P^3)$ is described as the set of all the lines in P^3, which intersect both of a given pair of lines l_1 and l_2 in P^3, which does not intersect between them.

Let now $l \subset P^4$ be a generally chosen line, and let $\sigma_{2,0}(l) = \{m \in G : m \cap l \neq \varnothing\}$ (l is the P^1 in the flag, as above). In a similar way, we can prove the analogues of the propositions above, namely:

(2.2.4). PROPOSITION. *Let l be a general line in P^4. Then*

(1) *The set of all the elements of W, which intersect as lines in P^4 the line l, is homologous to the Grassmann cycle $2 \cdot \sigma_{3,1} + \sigma_{2,2}$ and represents a linearly normal surface S_l of degree 3; more precisely, the surface S_l is embedded as a rational normal ruled surface $S_l = \mathbb{F}_1$ in G by the linear system $|C_0 + 2 \cdot f|$, where C_0 is the (-1)-section and f is the fiber of $\mathbb{F}_1$;*

(2) *The set of the elements of X, which intersect as lines in P^4 the line l, is homologous to the cycle $6 \cdot \sigma_{3,2}$ and represents a linearly normal curve C_l of degree 6 and or (arithmetical) genus 2.*

(2.2.5). REMARKS. (1) It is clear, that the curve C_l is an intersection of the surface S_l with the quadric Q in the four-dimensional projective space $P^4(l) = \langle S_l \rangle =$ Span S_l. The hyperelliptic linear system g_2^1 on the curve C_l is described in concrete geometrical terms as follows:

Let x be a general point of l. There are exactly two lines $l'(x)$ and $l''(x)$, which are elements of X and which pass through x (see Proposition (2.2.2)), in particular $l'(x)$ and $l''(x)$ belong to C_l. The linear system $g_2^1 = |l'(x) + l''(x)|$ maps C_l twice on the projective line $l \subset P^1$; the point $x \in l$ has as preimages the two lines $l'(x)$ and $l''(x)$ as elements of C_l.

(2) As in the previous case, it is easy to see that the curve C_l is nonsingular under the general choice of the line $l \subset P^4$.

(3) Let l be chosen so that the curve C_l is smooth. Then the hyperelliptic projection g_2^1 for C_l has exactly 6 ramification points: $x_1, x_2, \ldots, x_6$; the last means, that the corresponding lines $l'(x_i)$ and $l''(x_i)$ (see (1)) coincide, $i = 1, 2, \ldots, 6$.

The arosed situation outlines the existence of a hypersurface $R = R_6 \subset P^4$ of degree 6, which is defined as the closed completion of the set of all the points x_i, $i = 1, 2, \ldots, 6$ of ramification of the hyperelliptic systems $g_2^1 = g_2^1(l)$ (see (3)) corresponding to the lines to the lines $l \in X$, for which C_l is smooth.

(2.2.6). DEFINITION-COROLLARY. The closed completion $R = \overline{R^0}$ of the set $R^0 = \{x \in P^4 :$ there is exactly one line (a "double" line), which pass through x and which belongs to X as an element of $G\}$ is a hypersurface of degree 6 in P^4. We call $R = R_6$ the ramification hypersurface for X.

3. The elements of X as three-tangents to the ramification hypersurface $R \subset P^4$

3.1. Let $l \subset P^4$ be a line, which belongs, as a point of the Grassmannian, to X. We shall show, that (when $l \in X$ is chosen sufficiently general) the line l has three points of a simple contact with the ramification hypersurface R (see (2.2.6)). The point here is not in proving this fact (by the way proving the three tangency is not so difficult), but in the suggestion that the geometrical interpretation of the elements of the subvariety X of the Grassmannian as lines, which are totally tangent to some hypersurface, is somewhat natural. The following example clarifies the situation in one more simple case.

3.2. EXAMPLE. Let $P^3 = P(\mathbb{C}^4)$ be the projective 3-space, and let $G(2, 4) = G(2, \mathbb{C}^4) = G(1 : P^3)$ be the Grassmannian of the lines in P^3, embedded as a Plücker quadric in the projective 5-space $P^5 = P(\bigwedge^2 \mathbb{C}^4)$. Let $S = G(1 : P^3) \cap H \cap Q$ be a smooth intersection of the embedded Grassmannian with the hyperplane $H = P^4 \subset P^5$ and with the quadric $Q \subset P^5$. Using, as in the case of X, the Schubert representation of S, we obtain easily:

(3.2.1). PROPOSITION. (1) *There are exactly two projective lines which pass through the general point $x \in P^3$ and which, as points of the Grassmannian $G(1 : P^3)$, belong to S.*

(2) *There are exactly two lines, which lie in the generally chosen plane $P^2 \subset P^3$ and which belong to S as points of the $G(1 : P^3)$.*

(3) *Let $l \subset P^3$ be a general line. Then the set $C_l = \{m \subset P^3 - a \text{ line}: m \in S \text{ and } m \cap l \neq \varnothing\}$, is an elliptic curve of degree 4 (the curve C_l is, in the general case, a nonsingular, though special element of the anticanonical system $|-K_S|$).*

(3.2.2). The four points of ramification x_1, x_2, x_3 and x_4 of the hyper-elliptic linear system $g_2^1 = |l'(x) + l''(x)|$ (here $l'(x)$ and $l''(x)$ are the two elements of S, which pass through the point $x \in l$, see Proposition (3.2.1) (1)), define, just as in the case of X (see Remark (2.2.5) (3)), a surface $R = R_4 \subset P^3$ of degree 4—the ramification (hyper)surface for S.

(3.2.3). Now let $P^2 \subset P^3$ be a sufficiently general plane. Let l' and l'' be the two lines in P^2, which belong to S as Grassmann points (see Proposition (3.2.1) (2)). It is easy to see, that there exists a rational map $\varphi: S \to P^2$ such that

(1) If $x \in P^2 \setminus \{l' \cup l''\}$, and $x \notin R \cap P^2$, then $\varphi^{-1}(x) = \{l'(x), l''(x)\}$, where $l'(x)$ and $l''(x)$ are as above;

(2) If $x \in P^2 \cap R$ and $x \notin l' \cup l''$, then $\varphi^{-1}(x) = l(x)$, where $l(x) = l'(x) = l''(x)$ is the unique element of S, which passes through x as a line in P^3;

(3) If $x \in l'$ (or if $x \in l''$), but $x \neq p = l' \cap l''$, then $\varphi^{-1}(x) = \{l', l''(x)\}$ (or resp. $\varphi^{-1}(x) = \{l'', l'(x)\}$);

(4) If $x = p = l' \cap l''$, then $\varphi^{-1}(x) = \{l', l''\}$.

(3.2.4). It is clear, that there exists a regularization π of the rational map φ, which makes the following diagram commutative

$$\begin{array}{ccc} & \widetilde{S} & \\ \sigma \downarrow & & \searrow \pi \\ S & \dashrightarrow^{\varphi} & P^2 \end{array}$$

(3.2.5). In the diagram above the π-preimage of $l' \cup l''$ is a union of curves $l'_1 + l'_2 + l''_1 + l''_2$ such that:

(i) If $\pi^{-1}(p) = \{p_1, p_2\}$, then $l'_1 \cap l''_1 = p_1$, $l'_2 \cap l''_2 = p_2$, $l'_1 \cap l''_2 = \varnothing$, $l''_1 \cap l'_2 = \varnothing$;

(ii) The map $\sigma: \widetilde{S} \to S$ is a product of two σ-processes, which blow down the one of the two pairs of nonintersecting lines (for example the lines l'_1 and l''_2) to a pair of nonsingular points; moreover, the lines l'_1 and l''_2 are, of course, (-1)-curves on S.

It is clear now, that the lines l' and l'' are bi-tangents to the ramification curve $R \subset P^2$ of π.

(3.2.6). There are exactly 56 lines on the surface $\widetilde{S}$, which are (-1)-curves and which are also the 56 preimages of the 28 bi-tangent lines to the plane quartic curve of ramification $R \subset P^2$; that is, the surface S is a Del Pezzo surface, which is obtained from the projective plane after blowing-up of 7 points in the general position with respect to the lines, conics and cubics on the plane. On the other hand, there are exactly 16 lines $((-1)$-curves) lying on the surface S.

Let $y_1, y_2, \ldots, y_7$ be the seven points of the projective plane $P = P^2$, after blowing-up of which is obtained the surface $\tilde{S}$ and let the lines l'_1 and l''_2 (see above) be the preimages of the points y_6 and y_7. Then the 16 lines on S correspond to: (1) the conic, which pass through the points $y_1, y_2, \ldots, y_5$; (2) the exceptional divisors over the points $y_1, y_2, \ldots, y_5$; (3) the lines through the 10 pairs of points (y_i, y_j), $1 \leq i < j \leq 5$.

(3.2.7). The interpretation of the surface S as a subvariety of the Grassmannian $G(1 : P^3)$ permits to get some interesting conclusions about the ramification surface $R \subset P^3$.

(1) Let l_0 be a line on the surface S. As a set of projective lines in P^3, the Grassmann line l_0 is described by a plane pencil of lines through a fixed point $x_0 = x(l_0)$. The plane of this pencil $P_0^2 = P^2(l_0) = (\bigcup \lambda : \lambda \subset P^3$—a line & $\lambda \in l_0)$ stands in a special position with respect to the ramification surface $R \subset P^3$. As the line $l_0 \subset S$ intersects five other lines $l_1, l_2, \ldots, l_5$ lying on the Del Pezzo surface $S = S_{2 \cdot 2} = S_4 \subset P^4$, so there are five points $x_1, x_2, \ldots, x_5$ in the plane $P_0^2 = P^2(l_0)$ each of them being the centre of the corresponding plane pencil of lines. In this case the analogue of the map φ from above $\varphi : S \to P_0^2$ is a birational morphism. More concretely, by definition, the formal preimage $\varphi^{-1}(x)$ of the point $x \in P_0^2$ is defined to be the set of all the elements of S, which pass through x as lines in P^3. The regularization π of the rational map φ can be given by the commutative diagram

$$\tilde{S}$$

$$\sigma \downarrow \quad \searrow \pi$$

$$S \overset{\varphi}{\dashrightarrow} P_0^2 = P^2(l_0),$$

where the regular map π is a composition of six blow-ups over the points x_i, $i = 0, 1, \ldots, 5$, and the map $\sigma : \tilde{S} \to S$ blows down the line $l \subset S$, which is the preimage of the conic C in P_0^2, passing through the five points $x_1, x_2, \ldots, x_5$.

(2) The plane $P_0^2 = P^2(l_0)$ is totally bi-tangent to the ramification surface R, along the conic C. It follows immediately from the description above that the projective lines, which are elements of $S \subset G(1 : P^3)$ are bi-tangents to the ramification surface R; but not every bi-tangent to R belongs to S. For example, there are 28 bi-tangent lines to R, which lie in the general plane $P^2 \subset P^3$, but only 2 of them belong to the surface $S \subset G(1 : P^3)$. There are 16 planes, corresponding to the 16 lines on S, each plane being totally tangent to the surface $R \subset P^3$; moreover, the 16 points, which are centres of the corresponding pencils of lines in P^3, are singular points for the quartic $R \subset P^3$ and there are exactly 6 singular points of R, which lie on a fixed totally tangent plane, etc.

(3.2.8). The facts, stated till now, give us a sufficient reason to claim, that the surface R is a Kummer surface in P^3; for example, there are 16 singular points on the quartic R, but the quartic surface with a maximal number of singular points is a surface of Kummer and the 16 singular points are simple nodes (see [**B3, Chapter VIII, example**]).

The family of the bi-tangent lines to a Kummer surface splits into 22 (rational) components; 16 of them are isomorphic to the projective plane, and they are the dual planes to the 16 totally tangent planes to R; the rest 6 components are isomorphic to Del Pezzo surfaces of degree 4 (as the surface S). The described set of 22 surfaces is a degeneration of the surface of the bi-tangents to a general quartic surface in P^3, the last has been studied by Welters (see [**W, part 1, §3**]) and Tikhomirov (see [**T1**]).

3.3. The elements of X as three-tangent lines to R.

(3.3.1). Let l be a line in P^4, which belongs, as a point of the Grassmannian $G = G(1 : P^4)$, to X, and let $P^2 \subset P^4$ be a sufficiently general plane through the line l. It follows from the adjunction formula, that the surface $S(P^2) = \{m \subset P^4 —\text{a line}: m \cap P^2 \neq \varnothing\}$ is a $K3$-surface of degree 10, which represents a (special) hyperplane section of the threefold $X = X_{10}$. It is easy to see, that we choose the line l and then the plane P^2 through l sufficiently general, such that the surface $S(P^2)$ be nonsingular and moreover, such that the line l be the unique element of X, which lies on the plane P^2. Let $\varphi : S = S(P^2) \to P^2$ be, as in §3.2, the natural rational map; that is, the formal preimage $\varphi^{-1}(x)$ of the point $x \in P^2$ is equal, by definition, to the set of all the lines in P^4, which pass through x and which belong, as elements of the Grassmannian, to X. In the abovedescribed case the rational φ can be regularized by the commutative diagram

where σ is a blowing-down of one of the two preimages of the line l, to a nonsingular point of S and where π is a regular two sheeted covering with a ramification curve $R \cap P^2$ of degree 6. It is clear, that the two π-preimages of the line l are (-1)-curves on $\tilde{S}$ and hence, the line l is a three-tangent to the ramification curve $R \cap P^2$; in particular, the line l is a three-tangent line to R.

(3.3.2). Let $P^5 \subset P^7$ be any subspace of codimension 2; the intersection $C = X \cap P^5$ is a canonical curve of degree 10 and of arithmetical genus $p_a(C) = 6$.

Now let $l \subset P^3$ be a pair of a line l and a hyperplane P^3 through l. The curve $C = X \cap \sigma_{2,0}(l) + X \cap \sigma_{1,1}(P^3) = C_l + C(P^3)$ (see §2.2) is an intersection of X with a special codim 2 subspace of P^7. In particular, the curve C is a can. curve of $\deg(C) = 10$ and of $p_a(C) = 6$. The Schubert calculus gives that $\deg(C_l) = 6$, $\deg(C(P^3)) = 4$ and $(C_l, C(P^3)) = 4$. As $p_a(C_m) = 2$ for the general line $m \subset P^4$ (see Proposition (2.2.4) (2)) and $p_a(C(P^3)) = 1$ for the general hyperplane $P^3 \subset P^4$ (see Proposition (2.2.3) (2)), then we derive from the numerical equalities above that $p_a(C_l) = 2$ and that $p_a(C(P^3)) = 1$ for any line $l \subset P^4$ and for any $P^3 \subset P^4$. In fact, $p_a(C_l) \geq 2$ and $p_a(C(P^3)) \geq 1$ follow from the corresponding equalities, mentioned just above. The equality $(C_l, C(P^3)) = 4$ for $l \subset P^3$ implies that $p_a(C) = p_a(C_l) + p_a(C(P^3)) + 3$. But $p_a(C) = 6$, since C is a canonical curve of degree 10. Therefore, the strong inequalities are impossible, that is $p_a(C_l) = 2$ and $p_a(C(P^3)) = 1$ for any line $l \subset P^4$ and for any hyperplane $P^3 \subset P^4$.

(3.3.3). COROLLARY. *Let $l \subset P^4$ be any line, and let $P^3 \subset P^4$ be any hyperplane. Let $C_l = \{m \in X$—a line: $m \cap l \neq \varnothing\}$, $C(P^3) = \{\min \in X$—a line: $m \subset P^3\}$ (see (2.2.3) (2) and (2.2.4) (2)). Then*

(1) C_l is a curve of degree 6 and of arith. genus 2;

(2) $C(P^3)$ is a curve of degree 4 and of arith. genus 1. Moreover, if $l \subset P^3$, then the curves C_l and $C(P^3)$ intersect between themselves in four points (with multiplicities) and the curve $C = C_l + C(P^3)$ is a canonical curve of degree 10.

(3.3.4). In particular, let the line l be a three-tangent to the ramification hypersurface $R = R_6 \subset P^4$. Then the normalization of the curve C_l is P^1 and the line l is a triple point of the curve C_l; moreover, the tangents to the three branches at the point $l \in C_l$ do not lie in one plane (see also the Corollary above).

4. A description of the conics on $X = X_{10}$

4.1. The family of the conics on the Fano threefold X_{10} of the 1st type (see (1.3)) is studied in the papers of P. Puts [P] and D. Logachev [L1, L2]. In many respects the analysis of the family $\mathscr{C}^3$ of the rational normal cubics in the present paper is similar to Logachev's study of the family of conics on X_{10}; because of that we shall give a brief exposé of some results about the family of conics on $X = X_{10}$ "in the sense of Logachev."

4.2. Let $X = G \cap H_1 \cap H_2 \cap Q = G \cap P^7 \cap Q \subset P^7 \subset P^9 = P(\wedge^2 V)$, $V = V_5$ be as before. The hypersurfaces H_1 and H_2, regarded as elements of $P((\wedge^2 V)^*) = P(\wedge^2 V^*)$ (i.e., considered as one-dimensional subspaces of the vector space $\wedge^2 V^*$), determine a 2-dimensional subspace $E \subset \wedge^2 V^*$. There exists a natural mapping $S: E \otimes V \to V^*$ defined as follows:

Let $H \in E$ and let $v \in V$; then $S(H \otimes v) = S_H(v) \in V^*$, where $S_H: V \to$

V^* is the skew-symmetric linear mapping, corresponding to the element $H \in \bigwedge^2 V^*$. The condition—the variety $W = G \cap H_1 \cap H_2$ to be smooth—means that the equality $\operatorname{rank} S_H = 4$ is valid for any $H \in E$.

Then the 2-dimensional subspace $E \subset \bigwedge^2 V^*$ defines the projective line $P(E)$ and the embedding $\psi_0 \colon P(E) \hookrightarrow P(V)$, where ψ_0 is defined by the rule: $\psi_0(H) = \operatorname{Ker} S_H$, $H \in E$. In fact, $\operatorname{Ker} S_H$ is one-dimensional subspace of V for any $H \in E$, since W is smooth (see above); hence, the definition of ψ_0 is correct. The image $C_0 = \psi_0(P(E))$ is a conic in $P^4 = P(V)$. The projective plane $P_0^2 \subset P^4$, spanned on the conic C_0 corresponds to some fixed 3-dimensional subspace $U \subset V$, such that $P_0^2 = P(U)$. The "dual" plane $(P_0^2)^* = P(U^*)$ represents the set of the projective lines, lying in the plane P_0^2, that is, $(P_0^2)^*$ is a ρ-plane, lying in the fourfold $W \subset P^7 = H_1 \cap H_2$; moreover, the $(P_0^2)^*$ is the unique ρ-plane in W (see **[P or L1]**).

Let $l_0 = P((V/U)^*) \subset P(V^*)$ be the "orthogonal" line to the plane $P_0^2 = P(U) \subset P(V)$, that is, $l_0 = \{ V_4 \subset V$—a four-dimensional subspace: $U \subset V_4 \subset V \}$. There exists a natural isomorphism $s \colon C_0 \xrightarrow{\sim} l_0$, defined by the rule: $s(v) = S(E \otimes v) \in P(V^*)$, for $v \in l_0$.

The element $v \in C_0$ defines the Schubert cycle $\sigma(v) = \sigma_{3,1}(v, s(v)) = \{ l \in G = G(1 : P^4) \colon v \in l \subset s(v) \}$. The Schubert cycle $\sigma(v)$ represents the set of all the projective lines, passing through the point $v \in C_0 \subset P(V)$ and lying in the three-dimensional embedded subspace $P^3(v) = s(v) \subset P^4 = P(V)$, that is, $\sigma(v)$ is a σ-plane in the Grassmannian G (ibid.).

(4.2.1). CLAIM (Logachev, **[L1]**). The set $\{ \sigma(v) \colon v \in C_0 \}$ describes all the σ-planes, lying in the fourfold $W = G \cap H_1 \cap H_2$. Moreover, for any $v \in C_0$, the set $l(v) = \sigma(v) \cap (P_0^2)^* = \sigma(v) \cap (\operatorname{Span} C_0)^*$ is a line, which is tangent to the dual conic $C_0^* \subset (P_0^2)^*$. For any $V_4 \in l_0$ the surface $S(P(V_4)) = W \cap \sigma_{1,1}(P(V_4))$ is a degenerate quadric, which splits into two planes: $S(P(V_4)) = (P_0^2)^* \cup \sigma(s^{-1}(V_4))$. The opposite is also true, namely:

If $S(P(V_4))$ splits into two planes, then $V_4 \in l_0$ (for the definition of $S(P(V_4))$ see (2.2.3)).

4.3. Now let $q \subset X$ be a conic, lying on the Fano three-fold $X = G \cap H_1 \cap H_2 \cap Q$. As a subvariety of the Grassmannian $G = G(1 : P(V))$, the conic q represents a one-dimensional bundle of projective lines in $P^4 = P(V)$, which is of degree 2 with respect to the cycle $\sigma_{1,0}$ corresponding to the hyperplane section of the embedded by Plücker-Grassmannian $G \subset P^9$. Consequently, the surface $S_q = (\bigcup l, \ l$—a line in $P^4 \ \& \ l \in q) \subset P^4$ is one of the following:

(1) $S_q \subset \operatorname{Span} S_q = P_q^3 \subset P^4$ is a nonsingular quadric surface and q corresponds to one of the two bundles of lines on the quadric S_q, that is, q is a τ-conic on X (see **[P]**);

(2) $S_q \subset \operatorname{Span} S_q = P_q^3 \subset P^4$ is a quadratic cone and q corresponds to

the one-dimensional family of lines, lying on the cone S_q, that is, q is a σ-conic on X (ibid.);

(3) Span $S_q = P_q^2 \subset P^4$ is a plane and q corresponds to the set of all the tangent lines to a fixed conic in P_q^2, that is, q is a p-conic on X (ibid.).

Since $X = W \cap Q$ is a general intersection of W with a quadric $Q \subset P^7 = H_1 \cap H_2$, we can suppose that the above cases of possible conics, lying on X, are realized "in general position" that is, in particular, we can suppose, that the p-conics (resp., the σ-conics) on $X = W \cap Q$ are obtained as intersections of the p-planes (resp., of the σ-planes) on W with the quadric Q. In the last context, the opposite is also true (see [**P or L1**]). In fact, the general conic on X is of type τ; the σ-conics on X correspond to the points of the curve C_0 (see Claim (4.2.1)), and there is a unique p-conic on X, which corresponds to the unique p-plane $(P_0^2)^*$ on W.

4.4. Now let q be a τ-conic or a σ-conic on X (that is, $q \neq$ the unique p-conic $q_0 = (P_0^2)^* \cap Q$ on X); in particular, $q \subset G(1: P_q^3 = \text{Span} S_q)$. As we know (see (2.2.3) and (3.3.3)), the surface $S(P_q^3) = W \cap \sigma_{1,1}(P_q^3)$ is a quadric and the curve $C(P_q^3) = S(P_q^3) \cap Q \cap X$ is a space curve of degree 4 and of arithmetical genus 1. As q is a component of the curve $C(P_q^3)$, there exists an additional conic $\bar{q} \subset X$, such that $C(P_q^3) = q + \bar{q}$ and $(q, \bar{q}) = 2$. In particular, $P_q^3 = P_{\bar{q}}^3 = P^3(q, \bar{q})$. The conic $\bar{q}$ can be τ-conic, σ-conic, or p-conic. In fact, if $\bar{q} = q_0$, then the symbol $P_{\bar{q}}^3$ does not make sense. Then, under $P_{\bar{q}}^3$ we mean the 3-space P_q^3. The incidence $\bar{q} = q_0$ is possible *iff* q is a σ-conic (see (4.2.1)). From the Claim (4.2.1) we derive also that in the case when q is a σ-conic, the quadric $S(P_q^3)$ splits into two planes: $S(P_q^3) = (P_q^2)^* + \sigma(x_0(q))$ (here $x_0(q) \in C_0$ is the center of the cone $(\bigcup l : l \subset P^4$—a line & $l \in q)$, i.e. $\bar{q} = q_0$ is the unique p-conic on X. In the case, when q is a τ-conic, $\bar{q}$ is also a τ-conic and the surface $S(P_q^3) = S(P_{\bar{q}}^3)$ is a nonsingular quadric surface in W.

(4.4.1). DEFINITION. The set of conics

$$F_c = F_c(X) = \{q \text{—a conic}: q \subset X\}$$

is called a surface of the geometrical conics on X.

The adjective "geometrical" is necessive in the context of the following investigations; see, for example, the difficulties, arising from the attempts to define correctly the natural involution on the surface of the conics on X.

(4.4.2). DEFINITION. We call the set of pairs: $F = F(X) = \{q, V_4\}: q$ is a conic on X, and V_4 is a subspace of dim 4 of V, such that $S_q \subset P(V_4)\}$ an extended surface of conics on X (or, simply, a surface of conics for X).

4.5. The mapping $q \mapsto \bar{q}$, $q \neq q_0$ defines a birational isomorphism of F_c, which can be completed to an involution i on the surface $F = F(X)$ in view of the following considerations:

(4.5.1). It follows immediately from the definitions of F and F_c that the projection $(q, V_4) \mapsto q$ on the first factor defines naturally a morphism $r_F : F \to F_c$, which is one-to-one over $F_c \backslash \{q_0\}$. The curve $r_F^{-1}(q_0) \subset F$ is described by the set of all the four-dimensional subspaces $V_4 \subset V$, for which $P(V_4)$ contains the projective plane $P_{q_0}^2 = P_0^2$. Evidently, the curve $r_F^{-1}(q_0)$ is naturally isomorphic to the curve of σ-conics (= $\{q : q$ is a σ-conic on $X\} \subset F_c$); moreover, the morphism r_F is a σ-process over the point $q_0 \in F_0$. In this way the mapping $\bar{\cdot} : F_c \to F_c$ defines correctly the involution $i : F(X) \to F(X)$.

(4.5.2). DEFINITION. We define $F_0 = F_0(X)$ to be the factorsurface F/i of F under the involution i; that is F_0 is identified with the set of all (nonordered) pairs of involutive elements of F.

4.6. The projection on the second factor $(q, V_4) \mapsto V_4$ defines a morphism $\varphi : F \to (P^4)^* := P(V^*)$.

Let $p_F : F \to F_0$ be the natural two-sheeted covering, defined by the involution i. Clearly, there exists a commutative diagram:

$$
\begin{array}{ccc}
F & & \\
\downarrow{\scriptstyle p_F} & \searrow{\scriptstyle \varphi} & \\
F_0 & \xrightarrow{\varphi_0} & P(V^*),
\end{array}
$$

because the involutive elements of F have same φ-images.

It is easy to see also that the mapping φ_0 provides an embedding of the factorsurface F_0 in the projective space $P(V^*)$.

5. The family $\mathscr{C}^3$ of rational normal cubic curves on X

5.1. PROPOSITION (see [G, §2, Propositions 2.1 and 2.2]). ($X = X_{10}$ is, as usual, a general Fano threefold of degree 10 and of the 1st type, see (1.3))

(1) Let C be a rational normal cubic curve on X. Then the normal sheaf $N_{C/X}$ is one of the following:

$$
N_{C/X} = \begin{cases} O_C \oplus O_C(1), \\ O_C(-1) \oplus O_C(2), \\ O_C(-2) \oplus O_C(3); \end{cases}
$$

(2) There are rational normal cubic curves on X;

(3) Let T be an irreducible reduced component of the family $\mathscr{C}^3$ of all the rational normal cubic curves on X, and let x be a sufficiently general point of X. Then

(a) $\dim\{C \in T : C$ passes through $x\} = 1$,

(b) The normal sheaf $N_{C/X}$ of the general $C \in T$ is isomorphic to $O_C \oplus O_C(1)$.

5.2. A geometrical representation of the rational normal cubics on X as cubic scrolls.

(5.2.1). Let $C \subset X$ be a rational normal cubic, lying on the Fano threefold $X = G \cap H_1 \cap H_2 \cap Q$. The points of the curve C, regarded as lines in $P^4 = P(V)$, sweep out a surface $S = S_C$ of degree 3 in P^4. The ruled surface $S_C \subset P^4$ is one of the following:

CASE 1. $S_C \subset \operatorname{Span} S_C = P^4$, and,

(1.a) $S = \mathbb{F}_1$ is a rational normal cubic scroll embedded in P^4 by means of the linear system $|s + 2 \cdot f|$, where s is the (-1)-section and f is the fiber of the rules surface $\mathbb{F}_l = P(O_{P^1} \oplus O_{P^1}(1))$,

(1.b) S_C is a cone over a rational normal cubic curve Z of degree 3, with a center outside of the "ambient" space $\operatorname{Span} Z$;

CASE 2. $S_C \subset \operatorname{Span} S_C = P^3 \subset P^4$, and,

(2.a) S_C is a projection of a cubic scroll $\mathbb{F}_l$ (see (1.a)) on $P^3 \subset P^4$,

(2.b) S_C is a projection of a rational cubic cone (see (1.b)) on P^3.

(5.2.2) CLAIM. The cases (1.b) and (2.b) do not occur for the general X.

PROOF. We shall consider simultaneously the cases (1.b) and (2.b).

Let $S = S_C$ be a rational cubic cone (in P^4, or resp. in some $P^3 \subset P^4$), corresponding to some rational normal cubic $C \subset X$. As we know (see 4.2) the set of all σ-conics on X describes a rational curve on the surface of geometrical conics F_c (see Definition (4.4.1)). The lines, which are points of such a conic, sweep out a quadratic cone in P^4. The rational curve on F_c, which points corresponds to the σ-conics on X, is naturally isomorphic to the conic $C_0 = \psi_0(P(E))$ (see Claim (4.2.1)). Therefore, the set of the centers of the corresponding quadratic cones describes some rational curve $\widehat{C}_0$ in P^4. In fact, the curve $\widehat{C}_0$ is a nonsingular conic, lying on the plane P_0^2 (see §4.2 and §4.3); it is not hard to see also that the curve $\widehat{C}_0$ coincides with the conic C_0 (see Claim (4.2.1)), but we do not need such a precision.

Let x_0 be the center of the cubic cone, corresponding to the cubic $C \subset X$ and let y be a center of some quadratic cone, corresponding to some σ-conic q on X. Let $l = \langle x_0, y \rangle$ be the line through the points x_0 and y. As we know, the curve $C_l = \sigma_{2,0}(l) \cap X$ has a degree 6 and an arithmetical genus 2, independently of the choice of the line $l \subset P^4$ (see Corollary (3.3.3) (1)). Consequently, the curve C_l has an additional component L of degree 1, such that $C_l = C + q + L$, where C and q are the cubic and the conic from above (all the lines of C pass through $x_0 \in l$, and all the lines of q pass through $y \in l$). As the curve C_l is connected and all the projective lines, corresponding to the points of the "Grassmann" line L, intersect the line $l \subset P^4$, the center z of the plane pencil of lines, which describes L, lies on the line $l = \langle x_0, y \rangle$. Obviously, the point z is the unique point on the line $\langle x_0, y \rangle$ with the property, that z is a center of some plane pencil of lines in P^4, corresponding to some line, lying on X.

The last considerations show, that the existence of a cubic $C \subset X$ of the type (1.b) or (2.b) implies an existing of an isomorphism between the rational curve C_0 and some component of the curve Γ (see [**P or M**]), parametrizing the lines, lying on the Fano threefold X. But, as it is known [**P**, §8], if X is general, then the family Γ has only one component; moreover, the geometrical genus of Γ is 71, which leads us into a contradiction. Therefore, the cases (1.b) and (2.b) do not occur, if X is sufficiently general.

(5.2.3). COROLLARY. *Let C be a rational normal cubic curve, lying on X. Then, the corresponding to C bundle of lines in P^4 describes either the ruled surface $\mathbb{F}_1$, realized as a cubic scroll in P^4, or some projection of a cubic scroll $\mathbb{F}_1$ on some $P^3 \subset P^4$.*

(5.2.4). In both the cases (1.a) and (2.a) there exists a unique line $l = l(C)$ in P^4, which coincides geometrically with the (-1)-section of the corresponding $\mathbb{F}_1$ (or of its projection) and such that the cubic curve C is a component of the curve $C_l = \sigma_{2,0}(l) \cap X$.

(5.2.5). On the other hand, $\deg C_l = 6$ and $p_a(C_l) = 2$. The last implies that there exists an additional rational normal cubic $\overline{C}$ on X, such that $C_l = C + \overline{C}$ and $(C, \overline{C}) = 3$. It is possible for curve $\overline{C}$ to be singular, which in our particular case means "a degenerated rational normal cubic"; because of this, as in the case of the family of conics, we introduce the following refinement of the definition of the family of rational normal cubics on X:

(5.2.6). DEFINITION. We call the set

$$\mathscr{C}^3 = \mathscr{C}^3(X) = \{C \subset X : C \text{ is a rational normal}$$

$$\text{(possibly, degenerate) cubic curve}\}$$

a family of the rational normal cubic curves (or, simply r.n. cubics) on X.

5.3. The rational mapping $C \mapsto \overline{C}$ defines an involution on the family of rational normal cubics on X. To be precise, we need to consider separately the special cases of "degenerate" rat. normal cubics on X.

Let $C \in \mathscr{C}^3$ be a degenerate r.n. cubic on X. We can separate the following possible cases for C:

(i) $C = q + L$, where $q \in F_c(X)$ is a conic on X and $L \subset X$ is a line, such that $(q, L) = 1$;

(ii) $C = L_1 + L_2 + L_3$, where L_1, L_2 and L_3 are lines on X and $(L_1, L_2) = (L_2, L_3) = 1$, $(L_1, L_3) = 0$.

We shall find the involutive curve $\overline{C}$ in the cases (i) and (ii).

CASE (i'). Let q be a τ-conic. Then the surface $S_q = (\bigcup l, l \subset P^4 -$ a line & $l \in q$) is a nonsingular quadric and the set of lines $\{l: l \in q\}$ describes one of the two pencils of lines on S_q. Let P_L^2 be the plane $(\bigcup l, l \subset P^4 -$ a line & $l \in L)$, and let $y \in P_L^2$ be the center of the plane pencil of lines $\{l: l \in L\}$. The equality $(q, L) = 1$ means that there exists a line $l_0 \subset P^4$, which belongs to both of the pencils; evidently, $y \in l_0$. Let $l' \subset S_q$ be the line from the second pencil of lines on S_q, which passes through the point y. Then the conic q and the line L are components of the curve $C_{l'} = \sigma_{2,0}(l') \cap X$. Since $\deg C_{l'} = 6$ and $p_a(C_{l'}) = 2$, then the residue component $\overline{C} = C_{l'} - q - L$ belongs to the family $\mathscr{C}^3(X)$; moreover $(q + L, \overline{C}) = 3$, i.e. $\overline{C} = \overline{q + L}$ is the involutive of $C = q + L$.

CASE (i''). Let q be a σ-conic. Then the surface $S_q = (\bigcup l, \ l \subset P^4$—a line & $l \in q)$ is a quadratic cone and the equality $(q, L) = 1$ means that there is a line l_0, which is common for the pencils $\{l: l \in q\}$ and $\{l: l \in L\}$. Let x_0 be the singular point (the center) of the cone S_q and let y be the center of the plane pencil of lines, corresponding to L (see (i')). Since $(q, L) = 1$, then the plane of the second pencil P_L^2 does not lie in the subspace $\operatorname{Span} S_q \subset P^4$. In the opposite case the plane P_L^2 intersects the cone S_q along a pair of lines l_0 and l_0'. Both l_0 and l_0' are common members of the pencils $\{l: l \in q\}$ and $\{l: l \in L\}$, i.e. $(q, L) = 2$—a contradiction. Moreover, if the points y and x_0 coincide, then the configuration, $S_q + P_L^2$ corresponds to a degeneration of the case (1.b) (see (5.2.1)), which does not occur (see Claim (5.2.2)); the proof of the "degenerate" variant of the Claim (5.2.2) is, obviously, the same. Consequently, the points x_0 and y determine the line $l_0 = \langle x_0, y \rangle$. As in the case (i'), $C_{l_0} = q + L + \overline{C}$, where $(q + L, \overline{C}) = 3$, and the involutive curve $\overline{C} = \overline{q + L}$ belongs to $\mathscr{C}^3$.

CASE (i'''). Let $q = q_0$ be the unique p-conic on X. Then, the center y of the bundle $\{l: l \in L\}$ lies on the plane $P_0^2 = \operatorname{Span} C_0$ (see §4.2). The points $v \in C_0$ are centers of the bundles of lines in P^4 corresponding to the Schubert cycles (σ-planes) $\sigma(v) = \sigma_{3,1}(v, s(v))$ (see §4.2) on the fourfold $W = G \cap H_1 \cap H_2 \subset H_1 \cap H_2 = P^7$. The "Grassmann" quadric Q intersects each σ-plane, $\sigma(v)$ in the σ-conic $q(v) = \sigma(v) \cap Q$; the center of the corresponding cone $S_{q(v)}$ coincides with the point v, i.e. $x_0(v) = v$, $v \in C_0$. Now let $l_0 \subset P_0^2$ be a line through the center y of the bundle $\{l: l \in L\}$. The line l_0 intersects the conic C_0 in two points: v_1 and v_2; the last are the centers of the cones, corresponding to the σ-conics $q(v_1)$ and $q(v_2)$. Therefore, the curve C_{l_0} has as components the conics $q(v_1)$, $q(v_2)$, q_0 and the line L (the centers v_1, v_2 and y lie on l_0 and l_0 lies in the plane P_0^2 of the dual conic q_0), hence, $6 = \deg C_{l_0} \geq \deg q(v_1) + \deg q(v_2) + \deg q_0 + \deg L = 7$—a contradiction. Therefore, the case (i''') does not occur.

(5.3.1). REMARK. As $q_0 = (P_0^2)^* \cap Q$, the "Grassmann" conic q_0 is represented by the set of all lines in P_0^2 which are lines of intersection of the cones $S_{q(v)}$, $v \in C_0$ and the plane P_0^2. For every $v \in C_0$ the cone $S_{q(v)}$ intersects P_0^2 in a pair of lines through the point v; the set of all these lines coincides with the set of the tangent lines to some fixed conic $C_0' \subset P_0^2$. In particular, every σ-conic $q(v)$, $v \in C_0$ intersects its "involutive" $\overline{q(v)} = q_0$ in a pair of points (or, in a double point, if v is a point of an intersection of C_0 and C_0'), which correspond to the pair of tangents to the conic C_0' through the point $v \in C_0$.

CASE (ii). Let the degenerate rational normal cubic C splits into a chain of three lines: $C = L_1 + L_2 + L_3$, $(L_1, L_2) = (L_2, L_3) = 1$, $(L_1, L_3) = 0$. Let P_1^2, P_2^2 and P_3^2 be the planes, and y_1, y_2 and y_3 be the centers of the corresponding to the "Grassmann" lines L_1, L_2 and L_3 plane pencils of lines in P^4. The conditions for the intersections give, that the points y_1 and y_3 lie in the plane P_2^2. The case $y_1 = y_2 = y_3$ represents a special degeneration of the case (1.b) or (2.b) (see (5.2.1)), which does not occur for the general X (see Claim (5.2.2)). If $y_1 = y_3 \neq y_2$ or if y_1, y_2 and y_3 are collinear, then the line $l = \langle y_1, y_2 \rangle = \langle y_3, y_2 \rangle$ is a common point of the "Grassmann" lines L_1 and L_3; but the last two do not intersect between them—a contradiction. Therefore, the line $l_0 = \langle y_1, y_3 \rangle$ is correctly defined and moreover l_0 does not pass through the point y_2. Just as in the case (i), $C_{l_0} = L_1 + L_2 + L_3 + \overline{C}$, where the curve $\overline{C}$ belongs to the family $\mathscr{C}^3$, and $(L_1 + L_2 + L_3, \overline{C}) = 3$, i.e. $\overline{C} = \overline{L_1 + L_2 + L_3}$ is the correctly defined involutive of $C = L_1 + L_2 + L_3$.

(5.3.2). COROLLARY. *Let $\mathscr{C}^3 = \mathscr{C}^3(X)$ be the family of the rational normal cubics on X. Then, there is a correctly defined involution*: $\sigma: \mathscr{C}^3 \to \mathscr{C}^3$, *such that* $\sigma(C) = \overline{C}$, $C \in \mathscr{C}^3$. *The factorscheme* $\mathscr{C}_0^3 = \mathscr{C}^3/\sigma =$ {*the nonordered pairs* $(C, \overline{C})$ *of involutive elements of* $\mathscr{C}^3$} *is naturally embedded in the Grassmannian* $G = G(1: P(V))$ *by the rule*:

$$(C, \overline{C}) \mapsto l,$$

where the line $l \subset P^4 = P(V)$ *is defined as above (see cases (5.2.1) (1.a), (2.a) and §5.3 (i), (ii)); i.e. l is the unique line in $P^4 = P(V)$, such that $C_l = \sigma_{2,0}(l) \cap X = C + \overline{C}$. The involutive pair of rational normal cubics C and $\overline{C}$ is defined equivalently by the numerical condition $(C, \overline{C})_X = 3$.*

To prove the Corollary, it remains to note that the uniqueness of the line l is demonstrated separately at every of the cases above; from the last it follows immediately that $\overline{\overline{C}} = C$ for every $C \in \mathscr{C}^3(X)$.

5.4. Comments.

(5.4.1). Let $(C, \overline{C})$ be an involutive pair of $\mathscr{C}_0^3$ and let $C \cdot \overline{C} = t_1 + t_2 + t_3$, where t_1, t_2 and t_3 are points of $P^7 = H_1 \cap H_2$, possibly with multiplicities. As C and $\overline{C}$ are rational normal cubic curves, the intersection $C \cdot \overline{C}$ determines correctly a unique plane (the plane, "spanned on the intersection points of C and $\overline{C}$").

Let $l \subset P^4$ be the line, such that $C_l = C + \overline{C}$, and let $p_i = t_i \cap l$, $i = 1, 2, 3$ (the points t_1, t_2 and t_3 are regarded here as lines in P^4). It is evident that the line l is a three-tangent to the ramification hypersurface $R = R_6 \subset P^4$ (see Definition-Corollary (2.2.6)), that is $l \cdot R = 2 \cdot p_1 + 2 \cdot p_2 + 2 \cdot p_3$.

(5.4.2). The involutive pairs of the factorfamily $\mathscr{C}_0^3$ as sections of the rational cubic scrolls $S_l \subset W$ with the quadric Q.

As above, let $W = G \cap H_1 \cap H_2$ and let l be a line in P^4. The intersection $S_l = \sigma_{2,0}(l) \cdot W$ is homologous to the cycle $2 \cdot \sigma_{3,1} + \sigma_{2,2}$ and represents a surface of degree 3; moreover, for the general line $l \subset P^4$, the surface S_l coincides with a surface $\mathbb{F}_1 = P(O_{P^1} \oplus O_{P^1}(1))$, embedded in some four-dimensional subspace of $P^7 = H_1 \cap H_2$ as a rational normal cubic scroll (see Proposition (2.2.4) (1)). As $X = W \cap Q$, where Q is a (sufficiently general) quadric, the curve $C_l = \sigma_{2,0}(l) \cdot X$ is an intersection of the surface S_l with Q; in the notations of Proposition (2.2.4), the curve C_l belongs to the linear system $|2 \cdot C_0 + 4 \cdot f|$ (C_0 is the (-1)-section, f is the fiber of $\mathbb{F}_1$). The splitting $C_l = C + \overline{C}$ means that the restriction of the quadric Q on the surface S_l ($= \mathbb{F}_1$, in the general case) can be represented as a sum of two hyperplane sections (elements of the linear system $|C_0 + 2 \cdot f|$) of the cubic scroll $\mathbb{F}_1$. In the degenerate cases the surface S_q is a degeneration of a family of $\mathbb{F}_1$'s; we need to be careful with the symbols, because the surfaces S_C in the cases (1.a) and (2.a) (see §5.2) and also the surfaces $S_q + P_L^2$ and $P_1^2 + P_2^2 + P_3^2$ (see §5.3), which are cubic scrolls or their degenerations, represent cubic surfaces in P^4, which are swept out by lines in P^4 corresponding to the points of the curves $C \in \mathscr{C}^3$. We shall study the conditions for the splitting $C_l = C + \overline{C}$ in detail in the next paragraph.

6. The factorfamily $\mathscr{C}_0^3$ as a determinantal variety

6.1. Tautological sequences on $P(V^*)$ and on $G(2, V)$.

In the present section we need some details, concerning the well-known standard tautological sequences on the Grassmannians.

(6.1.1). We shall regard the "dual" projective space $(P^4)^* = P(V^*)$ as a Grassmannian: $P(V^*) = G(4, V) = \{V_4 : V_4$—four-dimensinal subspace of $V\}$. In this interpretation the standard tautological sequence on $P(V^*)$:

$$0 \to \tau_{4,V} \to P(V^*) \times V \to \tau^*_{1,V^*} \to 0$$

parametrizes the family of embeddings

$$\{0 \to V_4 \to V : V_4 \in G(4, V) = P(V^*)\}.$$

Let $f = {}^t(f_1, \ldots, f_4)$ and $g = {}^t(g_1, \ldots, g_4)$ be the vector-columns of two given bases of the subspace $V_4 \subset V$ and let $f = A \cdot g$ be a change of the basis f by the basis g. Then $\bigwedge^4 f = f_1 \wedge f_2 \wedge f_3 \wedge f_4 = \det A \cdot g_1 \wedge g_2 \wedge g_3 \wedge g_4 = \det A \cdot \bigwedge^4 g$. On the other hand, the Plücker embedding $Pl(\bigwedge^4): G(4, V) \to P^4$ provides the natural isomorphism: $G(4, V) \simeq P(V^*)$; $Pl(\bigwedge^4): V_4 \to \bigwedge^4 f$ (mod proportionally by elements of $\mathbb{C}^*$), $V_4 \in G(4, V)$. From here we get immediately, that $\det \tau_{4,V} = O_{G(4,V)}(-1)$, according to the embedding above. Keeping in mind the last, we replace $G(4, V)$ with $P(V^*)$; in particular, we write $\det(\tau_{1,V^*}^*) = \tau_{1,V^*}^* = O_{G(4,V)}(1) = O_{(V^*)}(1)$.

(6.1.2). Let now $G = G(2, V) = \{V_2 : V_2 \text{ is a 2-dim subspace of } V\}$ be a Grassmannian and let

$$0 \to \tau_{2,V} \to G(2, V) \times V \to \tau_{3,V^*}^* \to 0$$

be the standard tautological sequence on G. Over the element $V_2 \subset V$ of $G(2, V)$ the embedding of the left side coincides with the natural embedding $0 \to V_2 \to V$. If we perform a base change $f = A \cdot g$ in the fiber $V_2(f = {}^t(f_1, f_2)$ and $g = {}^t(g_1, g_2)$ are two bases of V_2, as above), we obtain a change of the second exterior powers $\bigwedge^2 f = \deg A \cdot \bigwedge^2 g$.

On the other hand, the Plücker embedding

$$Pl\left(\overset{2}{\bigwedge}\right) : G(2, V) \to P\left(\overset{2}{\bigwedge} V\right) = P^9$$

maps the element $V_2 \in G$ to the class $\bigwedge^2 f$ (mod proportionality by elements of $\mathbb{C}^*$). In view of the last, the former change means that $\det(\tau_{2,V}) = O_{G(2,V)}(-1)$, according to the Plücker embedding $Pl(\bigwedge^2)$.

6.2. Pfaff ideals on $P(V^*)$ and $G(2, V)$.

(6.2.1). We shall define a subbundle Pf of the bundle of quadrics $S^2 \bigwedge^2 \tau_{4,V}^*$ over $P(V^*) = G(4, V)$.

Let $V_4 \subset V$ be an element of $P(V^*) = G(4, V)$, let $f = {}^t(f_1, \ldots, f_4)$ and $g = {}^t(g_1, \ldots, g_4)$ be, as above, bases of V_4 and let $y = (y_1, \ldots, y_4)$ and $z = (z_1, \ldots, z_4)$ be the corresponding bases of coordinates on V_4. Assigned to the base f, we define the fibre of Pf to be: $Pf(V_4)_f =$ the Plücker quadric $y_{12} \cdot y_{34} - y_{13} \cdot y_{24} + y_{14} \cdot y_{23}$, where y_{ij}, $1 \leq i < j \leq 4$ are the coordinates, which correspond to the basis $f_{ij} = f_i \wedge f_j$, $1 \leq i < j \leq 4$ of $\bigwedge^2 V_4$ (the same for the coordinates z_{ij} and for the base vectors g_{ij} of $\bigwedge^2 V_4$). Let $f = A \cdot g$ be a base change of V_4; the change of the coordinates is $y \cdot A = z$, i.e. $y = B \cdot z$, where $B = A^{-1}$. The last allows to compute the change of

$$Pf: Pf(V_4)_f = y_{12}y_{34} - y_{13} \cdot y_{24} + y_{14} \cdot y_{23}$$
$$= \cdots = \det B \cdot (z_{12} \cdot z_{34} - z_{13} \cdot z_{24} + z_{14} \cdot z_{23})$$
$$= (\det A)^{-1} \cdot Pf(V_4)_g;$$

but A changes the bases of the bundle $\tau_{4,V}$ (see (6.1.1)), hence $Pf = (\det \tau_{4,V})^{-1} = O_{P(V^*)}(1)$. Here we shall describe one geometrical interpretation of the bundle of quadrics Pf.

Taking the second exterior power of the embedding $0 \to \tau_{4,V} \to P(V^*) \times V$

from the tautological sequence, we obtain the embedding

$$0 \to \bigwedge^2 \tau_{4,V} \to P(V^*) \times \bigwedge^2 V,$$

which parametrizes the family of embeddings $\bigwedge^2 V_4 \hookrightarrow \bigwedge^2 V$, $V_4 \in P(V^*)$. The fiber $Pf(V_4)$ coincides with the one-dimensional vector space, spanned on the equation of the embedded Grassmannian $G(2, V_4) = \sigma_{1,1}(P(V_4)) \subset$ Span $\sigma_{1,1}(P(V_4)) = P(\bigwedge^2 V_4)$ as a subvariety of $G(2, V) \subset P(\bigwedge^2 V) = P^9$.

(6.2.2). We take the exterior product of the embedding $0 \to \tau_{2,V} \to G(2, V) \times V$ with the constant bundle $G(2, V) \times V$. The obtained embedding

$$0 \to \tau_{2,V} \wedge V \to G(2, V) \times \bigwedge^2 V$$

can be interpreted geometrically as follows:

Let $V_2 \subset V$ be an element of $G(2, V)$ and let $\sigma_{2,0}(P(V_2))$ be the Schubert cycle $\{l \subset P(V)$—a line: $l \cap P(V_2) \neq \varnothing\} = \{L \subset V$—a subspace of dim = 2: $\dim(L \cap V_2) \geq 1\}$, embedded in $P(\bigwedge^2 V)$ as a subvariety of the embedded Grassmannian $G(2, V)$. We can check directly that Span $\sigma_{2,0}(P(V_2)) = P(V_2 \wedge V)$. Now we can define the "ideal" subbundle $\mathbb{I} \subset S^2(V_2 \wedge V)$, namely:

$\mathbb{I}(V_2) =$ [the set of all the quadrics in $P(V_2 \wedge V)$

(regarded as elements of $S^2(V_2 \wedge V)^*$) which vanish on

the subvariety $\sigma_{2,0}(P(V_2)) = H^0(P(V_2 \wedge V))$,

$$O(2 - \sigma_{2,0}(P(V_2))))] \subset S^2(V_2 \wedge V)^*.$$

As in (6.2.1) we compute the cocycle of the base changes of the bundle $\mathbb{I}$; as a result we obtain that $\mathbb{I}$ is isomorphic to the bundle $\tau_{3,V}^*$.

(6.2.3). COROLLARY. (i) *Let* $Pf \subset S^2 \bigwedge^2 \tau_{4,V}^*$ *be the sheaf of quadrics over* $P(V^*)$ *with a fiber* $Pf(V_4) = H^0(P(\bigwedge^2 V_4), O(2 - \sigma_{1,1}(P(V_4))))$ *over the element* $V_4 \subset V$ *of* $P(V^*) = G(4, V)$. *Then there is a natural isomorphism*:

$$Pf \simeq O_{P(V^*)}(1);$$

(ii) *Let* $\mathbb{I} \subset S^2(\tau_{2,V} \wedge V)$ *be the bundle of quadrics over* $G = G(2, V)$ *with a fiber* $\mathbb{I}(V_2) = H^0(P(V_2 \wedge V), O(2 - \sigma_{2,0}(P(V_2))))$ *over the element* $V_2 \subset V$ *of* $G(2, V)$. *Then there is a natural isomorphism*

$$\mathbb{I} \simeq \tau_{3,V^*}^*,$$

where τ_{2,V^*}^* *is the tautological factorbundle over* $G(2, V)$.

(6.1-2). COMMENTS. The bundles of quadrics $Pf \to P(V^*)$ and $\mathbb{I} \to G(2, V)$ are the components of degree 2 in the graded bundles of ideals of the families of the embedded Schubert cycles

$$\left\{ \sigma_{1,1}(P(V_4)) \subset P\left(\bigwedge^2 V_4 \right) : V_4 \in P(V^*) \right\}$$

and

$$\{\sigma_{2,0}(P(V_2)) \subset P(V_2 \wedge V): V_2 \in G(2, V)\}$$

respectively. Obviously, the components Pf and $\mathbb{I}$ generate the corresponding ideals.

6.3. The factorfamily $\mathscr{C}_0^3$ as a set of quadrics.

(6.3.1). Let $Q \subset S^2(\tau_{2,V} \wedge V)^*$ be the bundle of quadrics over $G = G(2, V)$, which parametrizes the family of restrictions of the quadric $Q \in S^2(\wedge^2 V)^*$ on the subspaces of the form $V_2 \wedge V$, $V_2 \in G(2, V)$. As the quadric Q is sufficiently general, the corresponding bundle of restrictions $Q(V_2) = $ [the $\mathbb{C}^*$-class of the equation of the restriction of the quadric hypersurface $Q = 0$ in $P(V_2 \wedge V)$] is correctly defined.

(∗) NOTE. We use the same symbol Q for the quadric $Q \in S^2 \wedge^2 V^*$, for the surface $(Q = 0)$ in $P(\wedge^2 V)$ and for the bundle "of quadrics" Q over $G(2, V)$.

Obviously, the bundle Q, defined above, is trivial, i.e. the corresponding sheaf Q is isomorphic to the structure sheaf $O_{G(2,V)}$ over $G = G(2, V)$.

(6.3.2). PROPOSITION. *Let* $Z \subset P_G(S^2(\tau_{2,V} \wedge V)^*)$ *be the set*:

$$Z = P\{(V_2, q): V_2 \in G, q \in H^0(P(V_2 \wedge V), O(2 - \sigma_{2,0}(P(V_2)) \cdot X)).$$

Then Z *is naturally isomorphic to the projectivizated bundle*

$$P_G(\tau_{3,V^*}^* \oplus O_G)$$

over $G = G(2, V)$.

PROOF. The proposition follows immediately from Corollary (6.2.3) (ii) and from (6.3.1). It remains to see that the fiber of the natural projection $Z \to G(2, V)$ over the "point" $V_2 \in G(2, V)$ coincides with the space

$$\mathrm{Span}\{\mathbb{I}(V_2) \cup Q(V_2)\} = \mathbb{I}(V_2) \oplus Q(V_2).$$

(6.3.3). Now let $X = G(2, V) \cap H_1 \cap H_2 \cap Q$ and let $V_8 \subset V_{10} = \wedge^2 V$ be the subspace, such that $P^7 = H_1 \cap H_2 = P(V_8)$. We define a bundle Σ over $G(2, V)$ with fibers $\Sigma(V_2) = (V_2 \wedge V) \cap V_8$, $V_2 \in G(2, V)$. We can suppose that the subspace $P^7 = P(V_8)$ is chosen a "sufficiently general" in such a way, that all the intersections $(V_2 \wedge V) \cap V_8$ are transversal (and, hence, are vector spaces of dimension 5) for every $V_2 \in G(2, V)$. Considerations on the level "intersection of Schubert cycles" give that the last requirement is fulfilled for the elements V_8 of an open subset of the Grassmannian $G(8, V_{10})$.

(6.3.4). Let $V_2 \in G(2, V)$. As we know,

$$S_{P(V_2)} = \sigma_{2,0}(P(V_2)) \cap P(V_8)$$

$$= \{l \subset P^4 \text{—a line}: l \in W, \ l \cap P(V_2) \neq \varnothing\}$$

(see Proposition (2.2.4) (1) and (5.4.2). Let $V_8 \subset V_{10} = \bigwedge^2 V$ be chosen general. Then, taking into account the considerations above, we have that all the restriction maps

$$H^0(P(V_2 \wedge V), \ O(2 - \sigma_{2,0}(P(V_2))))$$

$$\to H^0(P(\Sigma(V_2)), \ O(2 - S_{P(V_2)})),$$

are isomorphisms. So we obtained

(6.3.5). COROLLARY. *Let* $V_8 \subset V_{10} = \bigwedge^2 V$ *and* Q *are chosen sufficiently general, such that*

(i) *The variety* $X = G \cap Q \cap P(V_8)$ *is smooth*;

(ii) $\dim(V_8 \cap (V_2 \wedge V)) = 5$, *for every* $V_2 \in G = G(2, V)$.

Let $\mathbb{I} \to G$ *and* $Q \to G$ *be the bundles, as in (6.2.3) (ii) and (6.3.1). Then we can consider that* $\mathbb{I}$ *and* Q *are embedded in* $S^2 \Sigma^*$ *in such a way that*

$$\mathbb{I}(V_2) = H^0(P(\Sigma(V_2)), \ O(2 - S_{P(V_2)})),$$

$$Q(V_2) = H^0(P(\Sigma(V_2)), \ O_{P(\Sigma(V_2))}) \otimes \mathbb{C} \cdot Q,$$

where $V_2 \in G(2, V)$; *here* $S_{P(V_2)} = \sigma_{2,0}(P(V_2)) \cdot W$, $V_2 \in G$.

(6.3.6). Now let $C \subset X$ be a rational normal cubic curve. As we know (see Corollary 5.3.2)), there exists a line $l \subset P^4 = P(V)$ (resp., a subspace $V_2 \subset V$, such that $P(V_2) = l$), such that $C_l = \sigma_{2,0}(l) \cdot X = C + \overline{C}(\overline{C}$ is the involutive of C). But $C_l = S_l \cdot Q$, where S_l, $l = P(V_2)$, is as in the Corollary above. The last means that there exists a quadric q, which is a point of the space $\mathbb{I}(V_2) \oplus Q(V_2) \subset S^2 \Sigma^*(V_2)$ and which splits, as a subvariety of $P(\Sigma(V_2))$ into two hyperplanes: $(q = 0) = \langle C \rangle + \langle \overline{C} \rangle \subset P(\Sigma(V_2))$. The last is equivalent, in the interpretation of the Corollary (6.3.5), to the existence of a quadric of rank ≤ 2 in the vector-space "of quadrics" $(\mathbb{I} \oplus Q)(V_2)$. Obviously, the opposite is also true: the existence of a quadric $q \in (\mathbb{I} \oplus Q)(V_2)$ of rank ≤ 2, $q = H \cdot \overline{H}$ means that the curves $C = S_{P(V_2)} \cdot P(H)$ and $C = S_{P(V_2)} \cdot P(\overline{H})$ are an involutive pair of rational normal cubics such that $C_{P(V_2)} = C + \overline{C}$.

(6.3.7). COROLLARY. *The factorfamily* $\mathscr{C}_0^3 = \mathscr{C}^3/\sigma$ *(see (5.3.2)) is embedded naturally in the projectivized vector bundle* $P_G(\mathbb{I} \oplus Q) \subset G(2, V) \times P(S^2 \Sigma^*)$ *as the set*

$$D_2 = P\{(V_2, q) : V_2 \in G, \ q \in (\mathbb{I} \oplus Q)(V_2) \subset S^2 \Sigma^*(V_2), \ \mathrm{rank}(q) \leq 2\}.$$

Postponing the comments of the Corollary till later on, we shall explain in brief some facts about the degeneration loci (determinantals) (see [**F, Chapter 14**]).

6.4. The factorfamily $\mathscr{C}_0^3$ as a determinantal.

(6.4.1). DEFINITION. (i) Let $\varphi: E \to F$ be a homomorphism of vector bundles (a vector bundle map) over the variety Y; let $\mathrm{rank}\, E = e$, $\mathrm{rank}\, F =$

f. Let k be a nonnegative integer, such that $k \leq \min(e, f)$. The locus $D_k = D_k(\varphi) = \{y \in Y \colon \operatorname{rank}\varphi(y) \leq k\} \subset Y$ is called the kth degeneration locus (the kth determinantal) of φ.

(ii) In particular, let $F = E^* \otimes L$ for some invertible sheaf L on Y. Multiplying by L^{-1}, we obtain the map $\varphi \otimes L^{-1} \colon E \otimes L^{-1} \to E^*$. Taking the dual $(\varphi \otimes L^{-1})^* \colon E^{**} = E \to E^* \otimes L$, we obtain another vector bundle map from E to $F = E^* \otimes L$. The vector bundle map $\varphi \colon E \to E^* \otimes L$ is called symmetric, if φ is "selfdual," i.e. if $\varphi = (\varphi \otimes L^{-1})^*$. Using the natural correspondence between so defined symmetric maps and the "nets of quadrics" $\hat{\varphi} \colon L^{-1} \to S^2 E^*$, we define the kth degeneration locus (the kth determinantal) of the net of quadrics $\hat{\varphi} \colon L^{-1} \to S^2 E^*$ to be the kth determinantal of the corresponding (symmetric) vector bundle map

$$\varphi \colon E \to E^* \otimes L.$$

(6.4.2). As we know, $\mathbb{I} \oplus Q \simeq \tau_{3, V^*}^* \oplus O_{G(2, V)}$, that is

$$Z = P_{G(2, V)}(\tau_{3, V^*}^* \oplus O_{G(2, V)}) \simeq P_{G(2, V)}(\mathbb{I} \oplus Q)$$

(see Corollary (6.2.3)).

Let $\pi \colon Z = P_G(\tau_{3, V^*}^* \oplus O_G) \to G = G(2, V)$ be the natural projection and let

$$\tau_{3, V^*}^* \oplus O_{G(2, V)} \xrightarrow{\sim} \mathbb{I} \oplus Q \hookrightarrow S^2 \Sigma^*$$

be the abovedescribed embedding of vector bundles. Taking the tautological sequence for the projectivized vector bundle $\pi \colon Z = P_G(\tau_{3, V^*}^* \oplus O_G) \to G(2, V)$,

$$0 \to O_\pi(-1) \xrightarrow{j} \pi^*(\tau_{3, V^*}^* \oplus O_G) \to R \to 0,$$

we obtain the composition of natural embeddings,

$$0 \to O_\pi(-1) \xrightarrow{j} \pi^*(\tau_{3, V^*}^* \oplus O_G) \xrightarrow{i} \pi^*(S^2 \Sigma^*)$$

of vector bundles over Z.

(6.4.3). By the construction of the tautological sequences, the projectivization of the embedding of the sheaf $O_\pi(-1)$ ($=$ the relative $O(-1)$, or the tautological subbundle):

$$P(O_\pi(-1)) \hookrightarrow P(\pi^*(\tau_{3, V^*}^* \oplus O_G))$$

represents the embedding of the points of the fibers of $P_G(\tau_{3, V^*}^* \oplus O_G)$, regarded, in view of the isomorphism $\tau_{3, V^*}^* \oplus O_G \simeq \mathbb{I} \oplus Q$, as quadratic hypersurfaces in the corresponding projective spaces $P(\Sigma(V_2))$, $V_2 \in G(2, V)$, lifted to Z. From the last and from Corollary (6.3.7) we obtain

(6.4.4). PROPOSITION. *The factorfamily $\mathscr{C}_0^3 = \mathscr{C}^3/\sigma$ is embedded naturally in $Z = P_{G(2, V)}(\tau_{3, V^*}^* \oplus O_{G(2, V)})$ as a second determinantal of the net of quadrics:*

$$0 \to O_\pi(-1) \xrightarrow{j} \pi^*(\tau_{3, V^*}^* \oplus O_{G(2, V)}) \xrightarrow{i} \pi^*(S^2 \Sigma^*);$$

i.e. $\mathscr{C}_0^3 \simeq D_2(\hat{\varphi})$, *where* $\hat{\varphi} = i \cdot j$ *(see (6.4.2))*.

7. The families $\mathscr{C}^3$ and $\mathscr{C}_0^3$ as zero-schemes

7.1. We shall use the same symbol q for the quadratic form $q \in S^2\Sigma^*(V_2)$ and for the corresponding symmetric operator $q \colon \Sigma(V_2) \to \Sigma^*(V_2)$. Then Proposition (6.4.4) means that $\mathscr{C}_0^3 = \{(V_2, q) \in Z \colon \mathrm{rank}(q) \le 2\} = D_2(\hat{\varphi})$, (the q in the pair is regarded as a form, the q after the colon is regarded as an operator). The condition $\mathrm{rank}(q) = k$ means that $\dim \mathrm{Im}(q) = k$ ($\mathrm{Im}(q)$ is a subspace of $\Sigma^*(V_2)$).

Let $0 \le k \le 5 = \dim \Sigma(V_2)$ and let

$$\tilde{D}_k = D_k(\hat{\varphi}) = (z\,; U_k) \in G_Z(k, \Sigma^*) \colon \mathrm{Im}(q) \subseteq U_k, \text{ where } z \ni z = (V_2, q)\}.$$

The natural projection $\pi_k \colon (z\,; U_k) \mapsto z$ provides a map: $\pi_k \colon \tilde{D}_k \to D_k$, which is an isomorphism outside the locus $D_{k-1} \subset D_k$, $1 \le k \le 5 = \dim(\text{fibers of } \Sigma)$. The natural embedding $\tilde{D}_k \hookrightarrow G_Z(k, \Sigma^*)$ can be included in the commutative diagram

$$
\begin{array}{ccc}
\tilde{D}_k & \hookrightarrow & G_Z(k, \Sigma^*) \\
\pi_k \downarrow & & \downarrow \rho \\
D_k & \hookrightarrow & Z
\end{array}
$$

(7.1.1)

where $\rho \colon G_Z(k, \Sigma^*) \to Z$ is the natural projection.

(7.1.2). REMARK. Let $Q \in S^2 V_8^*$ (resp. $Q = 0) \subset P(V_8))$ be a quadric. It is not hard to check that the set of all the quadrics Q in $P(V_8)$ (resp. $Q \in S^2 V_8^*$), such that some of the spaces $\mathbb{I}(V_2) \oplus Q(V_2)$ contains a quadric of rank 1 (a double hyperplane in the corresponding $P(\Sigma(V_2))$ (resp., of $S^2\Sigma^*(V_2)$)), has a codimension > 1 inside the space of quadrics in $P(V_8)$ (resp., of $S^2 V_8^*$). Therefore, for the general quadric Q in V_8 the first determinantal $D_1(\hat{\varphi})$ vanishes. Because of that we can suppose in addition that the quadric Q is chosen such that the open condition

(iii) $D_1(\hat{\varphi}) = \varnothing$ is fulfilled (see also the conditions (i) and (ii) in the Corollary (6.3.5)).

Taking into account (iii) from the Remark, we obtain that the projection $\pi_2 \colon \tilde{D}_2 \to D_2$ is an isomorphism.

7.2.

(7.2.1). Let

$$0 \to \tau_{2,\Sigma^*} \xrightarrow{\alpha} G_Z(2, \Sigma^*) \times \Sigma^* \xrightarrow{\beta} \tau_{3,\Sigma}^* \to 0$$

be the standard tautological sequence on the grassmannization $\rho \colon G_Z(2, \Sigma^*) \to Z$. The embedding $\alpha \colon \tau_{2,\Sigma^*} \hookrightarrow G_Z(2, \Sigma^*) \times \Sigma^*$ defines an embedding

$S^2\alpha\colon S^2\tau_{2,\Sigma^*} \hookrightarrow G_Z(2, \Sigma^*) \times S^2\Sigma^*$. The embedding $S^2\alpha$ can be included in the exact sequence:

$$0 \to S^2\tau_{2,\Sigma^*} \xrightarrow{S^2\alpha} G_Z(2, \Sigma^*) \times S^2\Sigma^* \xrightarrow{\gamma} \mathrm{Coker}(2) \to 0,$$

where $\mathrm{Coker}(2)$ is the factor-bundle $S^2\Sigma^*/S^2\alpha(S^2\tau_{2,\Sigma^*})$ on $G_Z(2, \Sigma^*)$.

(7.2.2). On the other hand, the net of quadrics $\hat{\varphi}$ defines the embedding $0 \to O_\pi(-1) \xrightarrow{\hat{\varphi}} S^2\Sigma^*$ of bundles over Z. Taking into account the projection $p\colon G_Z(2, \Sigma^*) \to Z$ we can look at the sheaves $0_\pi(-1)$ and Σ as sheaves on $G_Z(2, \Sigma^*)$ (here we omit, for simplicity, the symbol ρ^*).

Composing the morphisms in the sequences above, we obtain the sequence

$$0 \to O_\pi(-1) \xrightarrow{\hat{\varphi}} G_Z(2, \Sigma^*) \times S^2\Sigma^* \xrightarrow{\gamma} \mathrm{Coker}(2)$$

of natural maps of sheaves (resp., of bundles) over $G_Z(2, \Sigma^*)$. Multiplying by the sheaf $(O_\pi(-1))^* = O_\pi(1)$, we obtain the composition:

$$0 \to O_{G(2,\Sigma^*)} \xrightarrow{\hat{\varphi}\otimes O_\pi(1)} O_\pi(1) \otimes S^2\Sigma^*$$
$$\xrightarrow{\gamma\otimes O_\pi(1)} O_\pi(1) \otimes \mathrm{Coker}(2);$$

the last defines a section s_0 of the sheaf $O_\pi(1) \otimes \mathrm{Coker}(2)$ over the variety $G_Z(2, \Sigma^*)$, $s_0 = (\gamma \otimes O_\pi(1)) \circ (\hat{\varphi} \otimes O_\pi(1))$.

(7.2.3). By definition the zero scheme $Z(s_0) \subset G_Z(2, \Sigma^*)$ of the section $s_0 \in H^0(G_Z(2, \Sigma^*), O_\pi(1) \otimes \mathrm{Coker}(2))$ coincides with the 0th determinantal $D_0(s_0)$ of the section s_0, regarded as a vector bundle map $s_0\colon O_{G(2,\Sigma^*)} \to O_\pi(1) \otimes \mathrm{Coker}(2)$. Consequently,

$$Z(s_0) = D_0(s_0)$$
$$= \{(z; U_2) \in Z \times G(2, \Sigma^*(V_2)):$$
$$z = (V_2, q \ \& \ s_0(z; U_2) = 0\}.$$

So we have

(1) $\hat{\varphi} = i \cdot j$, where the embedding $j\colon O_\pi(-1) \to \pi^*(\tau_{3,V^*}^* \oplus O_{G(2,V)})$ represents the closed embedding of the "point" q in the fiber $\pi^{-1}(V_2) \subset Z$ and the embedding $i\colon \pi^*(\tau_{3,V^*}^* \oplus O_{G(2,V)}) \hookrightarrow S^2\Sigma^*$ corresponds to the natural representation of the points in the fibers of $\tau_{3,V^*}^* \oplus O_{G(2,V)}$ as quadrics in the fibers $\Sigma(V_2)$ of Σ (here we use the isomorphism $\tau_{3,V^*}^* \oplus O_{G(2,V)} \widetilde{\to} \mathbb{I} \oplus Q$, see (6.2.3) (ii) and (6.3.1));

(2) The maps

$$\gamma(z) = \gamma(V_2, q)\colon S^2\Sigma^*(V_2, q; U_2) \to \mathrm{Coker}(2),$$

as factor-maps, send the quadrics $q \in S^2\Sigma^*(V_2)$ to the corresponding classes $q(\mathrm{mod}\, S^2 U_2)$, over the elements $(z; U_2) = (V_2, q; U_2) \in G_Z(2, \Sigma^*)$.

From (1) and (2) we derive, that

$$Z(s_0) = \{V_2, q; U_2) \in G_Z(2, \Sigma^*): q(\mathrm{mod}\, S^2 U_2) = 0;$$

$$q \text{ regarded as an element of } S^2\Sigma^*(V_2)\}$$

$$= \{(V_2, q; U_2) \in G_Z(2, \Sigma^*): U_2 \subseteq \mathrm{Image}(q) \subset \Sigma^*(V_2);$$

$$q \text{ regard as an operator } q \in \mathrm{Sym}(\Sigma(V_2), \Sigma^*(V_2))\}.$$

Consequently, $Z(s_0) = \tilde{D}_2(\hat{\varphi}) \subset G_Z(2, \Sigma^*)$, (see 7.1 and (7.1.1)).

But, by the condition (iii) (see Remark (7.1.2.)), $D_1(\hat{\varphi}) = \varnothing$, that is, there are no double hyperplanes in the fibers $H^0(P(\Sigma(V_2)), O(2 - \sigma_{2,0}(P(\Sigma(V_2)) \cdot X_{10})), V_2 \in G(2, V)$ of the bundle "of quadrics" $\mathbb{I} \oplus Q$.

Consequently, the natural projection $\pi_2: \tilde{D}_2(\hat{\varphi}) \to D_2(\hat{\varphi})$ (cf. (7.1.2)) is an isomorphism.

Summing up the obtained incidences, we derive:

(7.2.4). COROLLARY. *Let* $V_8 \subset V_{10} = \wedge^2 V$ *and* Q *fulfill the open conditions* (i), (ii) *from the Corollary* (6.3.5) *and* (iii) *from the Remark* (7.1.2). *Let* $\mathscr{C}_0^3 = \mathscr{C}^3/\sigma$ *be the factorfamily of the family of rational normal cubics on* $X = X_{10} = G(2, V) \cap P(V_8) \cap Q$ *and let* $s_0 \in H^0(G_Z(2, \Sigma^*), O_\pi(1) \otimes \mathrm{Coker}(2))$ *be the section, described in* (7.2.2). *Then* $\mathscr{C}_0^3$ *embeds naturally in* $G_Z(2, \Sigma^*)$ *as a zero scheme of the section* s_0, *that is:*

$$\mathscr{C}_0^3 \overset{\sim}{\to} Z(s_0) \subset G_Z(2, \Sigma^*).$$

(7.2.5). NOTE. By convention, the symbol ρ^*, where $\rho: G_Z(2, \Sigma^*) \to Z$, is dropped, i.e. with some abuse of the notations we write $\rho^* O_\pi(1) \otimes \mathrm{Coker}(2) = O_\pi(1) \otimes \mathrm{Coker}(2)$.

There exists a natural isomorphism of duality

$$\delta: G_Z(2, \Sigma^*) \overset{\sim}{\to} G_Z(3, \Sigma),$$

induced by the natural duality of the fibers

$$\delta(V_2): G(2, \Sigma^*(V_2))$$
$$\overset{\sim}{\to} G(3, \Sigma(V_2)), \qquad V_2 \in G(2, V).$$

The composition $s_0^* = \delta \cdot s_0$ defines a section of (the isomorphic δ-preimage of) the sheaf $O_\pi(1) \otimes \mathrm{Coker}(2)$ on $G_Z(3, \Sigma)$. So we obtained:

(7.2.6). COROLLARY. *In the conditions of the Corollary* (7.2.4) *there is a natural embedding* $\mathscr{C}_0^3 \hookrightarrow G_Z(3, \Sigma)$, *such that*

(1) $\mathscr{C}_0^3 \overset{\sim}{\to} Z(s_0^*) \subset G_Z(3, \Sigma)$, *where* $s_0^* = \delta \cdot s_0$ *is the section, defined just above*;

(2) *Let* $(C, \overline{C})$ *be an element of* $\mathscr{C}_0^3$ (*i.e.* $(C, \overline{C})$ *is a pair of involutive rational normal cubics on* $X = X_{10}$). *Let* $V_2 \subset V$ *be the subspace (of* $\dim = 2$), *such that* $C_{P(V_2)} = \sigma_{2,0}(P(V_2)) \cdot X_{10} = C + \overline{C}$ (*see Corollary* (5.3.2)) *and let* H_C *and* $H_{\overline{C}} \in \Sigma^*(V_2)$ *be the hyperplanes in* $\Sigma(V_2)$ *such that* $\langle C \rangle =$

$\mathrm{Span}(C) = P(H_C)$ *and* $\langle \overline{C} \rangle = \mathrm{Span}(\overline{C}) = P(H_{\overline{C}})$. *Then the embedding*
$\mathscr{C}_0^3 \hookrightarrow G_Z(3, \Sigma)$ *sends the pair* $(C, \overline{C})$ *to the triple* $(V_2, q = H_C \cdot H_{\overline{C}}; H_C \cap H_{\overline{C}}) \in (V_2, q = H_C \cdot H_{\overline{C}}) \times G(3, \Sigma(V_2)) \subset G_Z(3, \Sigma)$.

PROOF. Point (1) follows immediately from the definition of the section $s_0^* = \delta \cdot s_0$ and from the previous Corollary. To prove (2) it is sufficient to see that the third member $H_C \cap H_{\overline{C}}$ of the triple corresponds, by the duality δ, to the subspace $U_2 := \mathrm{Image}(q) \subset \Sigma^*(V_2)$, where $q = H_C \cdot H_{\overline{C}}$ is regarded as an operator from $\mathrm{Sym}(\Sigma(V_2), \Sigma^*(V_2))$. Moreover, $\dim(\mathrm{Image}(q = H_C \cdot H_{\overline{C}})) = 2$, because of the fact, that $D_1(\hat{\varphi}) = \varnothing$. In particular, H_C and $H_{\overline{C}}$ are not proportional, that is $U_2 = \mathrm{Image}(q = H_C \cdot H_{\overline{C}})$ belongs to the Grassmannian $G(2, \Sigma(V_2))$.

7.3.
(7.3.1). Let

$$0 \to \tau_{4,\Sigma} \to P_Z(\Sigma^*) \times \Sigma \to \tau_{1,\Sigma^*}^* \to 0$$

be the standard tautological sequence on the projectivization $P_Z(\Sigma^*) = G_Z(4, \Sigma)$. Taking the dual, we obtain the natural surjection $P_Z(\Sigma^*) \times \Sigma^* \to \tau_{4,\Sigma}^* \to 0$, hence, the surjection

$$\varepsilon \colon P_Z(\Sigma^*) \times S^2\Sigma^* \to S^2\tau_{4,\Sigma}^* \to 0.$$

Let now $\tau \colon P_Z(\Sigma^*) \to Z$ be the natural projection. Taking the τ-preimages of the sheaves, we lift the map $\hat{\varphi}$ onto $P_Z(\Sigma^*)$,

$$0 \to O_\pi(-1) \xrightarrow{\hat{\varphi}} P_Z(\Sigma^*) \times S^2\Sigma^*;$$

as usual the symbol τ^* is dropped.

Just as in (7.2.2), after multiplying by $O_\pi(1)$, we obtain the composite map

$$0 \to O_{P_Z(\Sigma^*)} \xrightarrow{\hat{\varphi} \otimes O_\pi(1)} P_Z(\Sigma^*) \times O_\pi(1) \otimes S^2\Sigma^*$$

$$\xrightarrow{\varepsilon \otimes O_\pi(1)} O_\pi(1) \otimes S^2\tau_{4,\Sigma}^*,$$

hence, we obtain a well-defined section s of the sheaf $O_\pi(1) \otimes S^2\tau_{4,\Sigma}^*$ over $P_Z(\Sigma^*)$, $s = (\varepsilon \otimes O_\pi(1)) \circ (\hat{\varphi} \otimes O_\pi(1))$.

As in §7.2 we shall prove the following

(7.3.2). PROPOSITION. *In the conditions of the Corollary* (7.2.4), *the family* $\mathscr{C}^3$ *of the rational normal cubics on* X *is embedded naturally in the projectivized bundle* $P_Z(\Sigma^*)$ *as a zero scheme of the section* s, *that is*

$$\mathscr{C}^3 \xrightarrow{\sim} Z(s) \subset P_Z(\Sigma^*).$$

PROOF (for more details see the proof of the Corollary (7.2.4)). As in (7.2.3) by the definition of the tautological sequence the map $\varepsilon \colon P_Z(\Sigma^*) \times S^2\Sigma^* \to S^2\tau_{4,\Sigma}^*$ sends the element $(V_2, q; Y_4)$ of the fiber $\tau^{-1}(V_2, q)$ to the element $(V_2, q; q|_{Y_4}) \in S^2 \tau^*_{4,\Sigma}$ (here $q|_{Y_4}$ is the restriction of the quadric $q \in S^2 \Sigma^*(V_2)$ to the subspace $Y_4 \subset \Sigma(V_2)$, $\dim(Y_4) = 4$). Consequently, the elements of the zero scheme of s are defined by the condition $Y_4 \subseteq$ Kernel (q); here the quadric q is regarded as an operator from

$$\mathrm{Sym}(\Sigma(V_2), \Sigma^*(V_2)).$$

In other words, the zero scheme

$$Z(s) = \{(V_2, q; Y_4): (V_2, q) \in Z,$$

$Y_4 \subset \Sigma(V_2)$ is an element of $G(4, \Sigma(V_2))$ such that $Y_4 \subseteq$ Kernel $(q)\}$.

But the symmetric operator $q: \Sigma(V_2) \to \Sigma^*(V_2)$, $\dim \Sigma(V_2) = 5$, contains four-dimensional subspaces in its kernel, if and only if $\mathrm{rank}(q) \leq 2$. By the way, there are no quadrics q of rank $= 1$, since $D_1(\hat{\varphi}) = \varnothing$ (see for example (7.2.3)—the proof of the Corollary (7.2.4)). Because of that $\mathrm{rank}(q) \leq 2$ *iff* $\mathrm{rank}(q) = 2$, that is *iff* $q = H \cdot \overline{H}$, for some nonproportional linear functions (noncoincident hyperplanes) H and $\overline{H}$ of $\Sigma^*(V_2)$. But the projective hyperplanes $P(H)$ and $P(\overline{H})$ cut out on X (in the described situation) a pair of involutive rational normal cubics $C = X \cap P(H)$ and $\overline{C} = X \cap P(\overline{H})$. Because of that the zero scheme $Z(s)$ describes the set of the components of the quadrics q of rank $= 2$, q being the second element of the pair $(V_2, q) \in D_2(\hat{\varphi})$ (see Proposition (6.4.4), §7.1 and §7.2). Equivalently, the zero scheme $Z(s)$ describes the set of the rational normal cubics (the intersections of $X = X_{10}$ with the corresponding components $P(H)$ and $P(\overline{H})$ of the quadrics $q = H \cdot \overline{H}$ of rank $= 2$, as above). The Proposition is proved.

From the proof of the Proposition (7.3.2), just as for the Corollary (7.2.6) (2), we obtain the following geometrical description of the embedding $\mathscr{C}^3 \hookrightarrow P_Z(\Sigma^*) = G_Z(4, \Sigma)$:

(7.3.3). COROLLARY. *Let $C \in \mathscr{C}^3$ be a rational normal cubic curve on $X = X_{10}$. Let $V_2 \subset V$ be the subspace (of $\dim = 2$), such that $C_{P(V_2)} = \sigma_{2,0}(P(V_2)) \cdot X = C + \overline{C}$, where $\overline{C}$ is the involutive of C (see Corollary (5.3.2)) and let H_C and $H_{\overline{C}} \in \Sigma^*(V_2)$ be the hyperplanes in $\Sigma(V_2)$, such that $\langle C \rangle = \mathrm{Span}(C) = P(H_C)$ and $\langle \overline{C} \rangle = \mathrm{Span}(\overline{C}) = P(H_{\overline{C}})$. Then the embedding $\mathscr{C}^2 \hookrightarrow P_Z(\Sigma^*) = G_Z(4, \Sigma)$ sends the element $C \in \mathscr{C}^3$ to the triple $(V_2, q = H_C \cdot H_{\overline{C}}; H_C) \in (V_2, q = H_C \cdot H_{\overline{C}}) \times G(4, \Sigma(V_2)) \subset G_Z(4, \Sigma) = P_Z(\Sigma^*)$.*

8. Tangent Bundle Theorem for the family of rational normal cubics $\mathscr{C}^4$

8.1. Before formulating and proving the Tangent Bundle Theorem (T.B.T.), we shall carry out some constructions, which clarify the local case. From now on we shall suppose that the curve $C \in \mathscr{C}^3$ is chosen sufficiently general, in particular, we can suppose that the curve C and its involutive $\overline{C}$ are smooth rational normal cubics with normal bundles $N_{C/X}$ and $N_{\overline{C}/X}$, isomorphic to $O_{P^1} \oplus O_{P^1}(1)$ (see Proposition 5.1), etc.

(8.1.1). Let C and $\overline{C}$ be a pair of involutive rational normal cubics on X, let $V_2 \subset V$ be the corresponding subspace of $\dim = 2$, such that $C_l = C + \overline{C}$, where $l = P(V_2)$ (see Corollary (5.3.2)). Let $W = G \cdot P(V_8)$, $X = X_{10} = W \cdot Q = G \cdot P(V_8) \cdot Q$ and $S_l = \sigma_{2,0}(l) \cdot W$ are defined as before. Examine the pairs of successive embeddings:

$$C \subset X \subset W \quad \text{and} \quad C \subset S_l \subset W$$

and the corresponding exact sequences of normal sheaves:

(1) $$0 \to N_{C/X} \to N_{C/W} \to N_{X/W} \otimes O_C \to 0$$

and

(2) $$0 \to N_{C/S_l} \to N_{C/W} \to N_{S_l/W} \otimes O_C \to 0.$$

Multiplying the sequences (1) and (2) by the sheaf $O(-1) = O_{P(V_8)}(-1)$ and identifying the middle members, we obtain the diagonal map

(3) $$\varphi_C \colon N_{C/S_l} \otimes O(-1) \to N_{C/W} \otimes O(-1) \to N_{X/W} \otimes O(-1) \otimes O_C.$$

Since $\deg(C) = 3$, then $O(-1) \otimes O_C \simeq O_C(-3)$, $C \simeq P^1$. From $X = W \cdot Q$, where Q is a quadric in $P^7 = P(V_8)$, it follows that $N_{X/W} = O_X(2)$. Therefore, $N_{X/W} \otimes O(-1) \otimes O_C \simeq O(1) \otimes O_C \simeq O_C(3)$.

(8.1.2). On the other hand, $N_{C/S_l} = O(1) \otimes O_C$, since C is a hyperplane section of the rat. norm cubic scroll $S_l \subset \mathrm{Span}(S_l) = P(\Sigma(V_2))$, $1 = P(V_2)$; (see (5.4.2)). Therefore

(3′) $$\varphi_C \colon O_C \to O(1) \otimes O_C = O_C(3).$$

Taking into account the identification above, the sequence $(2) \otimes O(-1)$ takes the form

(2′) $$0 \to O_C \to N_{C/W} \otimes O(-1) \to N_{S_l/W} \otimes O(-1) \otimes O_C \to 0.$$

(8.1.3). **LEMMA.** *Let $C \in \mathscr{C}^3(X)$ be a general r.n. cubic. Then $N_{S_l/W} \otimes O_C \simeq O_C(2) \oplus O_C(2)$.*

The proof of the lemma will be given later on (see (8.1.10)).

(8.1.4). From (8.1.3) we derive the identification

(4) $$N_{S_l/W} \otimes O(-1) \otimes O_C \simeq O_C(-1) \oplus O_C(-1),$$

therefore the long exact sequence of cohomologies, associated to the sequence $(2′)$, defines the natural isomorphisms

(5) $$H^i(C, O_C) \xrightarrow{\sim} H^i(C, N_{C/W} \otimes O(-1)), \qquad i = 0, 1, \dots.$$

In fact, all the cohomology groups of the sheaf $N_{S_l/W} \otimes O(-1) \otimes O_C$ vanish.

(8.1.5). Now let us look at the sequence $(1) \otimes O(-1)$. Since $N_{C/X} \otimes O(-1) \simeq O_C(-3) \oplus O_C(-2)$, then $H^0(C, N_{C/X} \otimes O(-1)) = 0$.

From $H^1(C, N_{C/X}) = H^1(C, O_C \oplus O_C(1)) = 0$, we obtain that the tangent space $(\mathcal{T}_{\mathscr{C}^3}(C)$ is isomorphic to $H^0(C, N_{C/X})$.

Let $\Omega_{\mathscr{C}^3} = \mathcal{T}_{\mathscr{C}^3}^*$ be the cotangent sheaf of and let ω_X be the sheaf of higher differential forms on X (the canonical sheaf of X). Using the Serre duality, we can write the following sequence of identities:

$$\Omega_{\mathscr{C}^3}(C) = \mathcal{T}_{\mathscr{C}^3}^*(C) = H^0(C, N_{C/X})^*$$

$$= H^1(C, N_{C/X}^* \otimes \omega_C) = H^1(C, N_{C/X} \otimes \det N_{C/X}^* \otimes \omega_C)$$

$$= H^1(C, N_{C/X} \otimes O_C(-1) \otimes \omega_C) = H^1(C, N_{C/X} \otimes O(-1))$$

$$= H^1(C, N_{C/X} \otimes \omega_X), \text{ since } \omega_X = O(-1) \otimes O_X.$$

From the last we obtain that the long exact cohomology sequence, associated to the exact sequence $(1) \otimes O(-1)$ takes the form

$$0 \to (H^0(C, N_{C/X} \otimes \omega_X) = 0) \to H^0(C, N_{C/W} \otimes O(-1))$$

(1')
$$\to H^0(C, O(1) \otimes O_C) \to (H^1(C, N_{C/X} \otimes \omega_X) = \Omega_{\mathscr{C}^3}(C))$$

$$\to H^1(C, N_{C/W} \otimes O(-1)) \to \cdots.$$

Using (5) we obtain the sequence (see also (8.1.2) (3')),

(1'')
$$0 \to H^0(C, O_C) \xrightarrow{H^0(\varphi_C)} H^0(C, O(1) \otimes O_C)$$

$$\to \Omega_{\mathscr{C}^3}(C) \to (H^1(C, O_C) = 0).$$

(8.1.6). If we suppose for a while that we have performed a globalization of the constructions above, simultaneously for all the elements C of the family of rational normal cubics $\mathscr{C}^3 = \mathscr{C}^3(X)$, then the members of the sequence (1'') will be induced by restriction on the fibers from sheaves, defined globally on the family $\mathscr{C}^3$ (as the fiber $\Omega_{\mathscr{C}^3}(C)$).

(6) REMARK. In fact, the diagonal map $\varphi_C: O_C \to O(1) \otimes O_C$ (see (3')) corresponds to the choice of the divisor $C \cap \overline{C}$ on C; the last divisor represents the hyperplane section of C in the projective space $\langle C \rangle = \mathrm{Span}(C) \subset P^3$ (see (5.4.1) and (5.4.2)).

(8.1.7). On the other hand, as we know, the family $\mathscr{C}^3$ is embedded in $P_Z(\Sigma^*) = G_Z(4, \Sigma)$ as a zero scheme of the section s, and the factorfamily $\mathscr{C}_0^3 = \mathscr{C}^3/\sigma$ is embedded in $G_Z(3, \Sigma)$ as a zero scheme of the section s_0^* (see Corollary (7.2.6), Proposition (7.3.2) and Corollary (7.3.3)). From the last we conclude, that the tautological sheaves $\tau_{3, \Sigma}$ and τ_{2, Σ^*}^* from the standard tautological sequence

(a)
$$0 \to \tau_{3, \Sigma} \to G_Z(3, \Sigma) \times \Sigma \to \tau_{2, \Sigma^*}^* \to 0$$

are well defined on the family $\mathscr{C}_0^3$, and the tautological sheaves $\tau_{4, \Sigma}$ and τ_{1, Σ^*}^* from the sequence

(b)
$$0 \to \tau_{4, \Sigma} \to P_Z(\Sigma^*) \times \Sigma \to \tau_{1, \Sigma}^* \to 0$$

are well defined on the family $\mathscr{C}^3$.

But the existence of the natural 2-sheeted covering $\pi\colon \mathscr{C}^3 \to \mathscr{C}_0^3$, induced from the involution $\sigma\colon \mathscr{C}^3 \to \mathscr{C}^3 (\sigma(C) = \overline{C}, \sigma^2 = \mathrm{id})$ shows that the (π-preimages of the) sheaves $\tau_{3,\Sigma}$ and τ_{2,Σ^*}^* are defined also on the family $\mathscr{C}^3$; as usual we write $\tau_{3,\Sigma}$ and τ_{2,Σ^*}^* instead of $\pi^*\tau_{3,\Sigma}$ and $\pi^*\tau_{2,\Sigma^*}^*$. Then, the (lifted by π) exact sequence (a) on, together with the exact sequence (b), define an embedding $j\colon \tau_{3,\Sigma} \hookrightarrow \tau_{4,\Sigma}$ of sheaves on $\mathscr{C}^3$ (see also (7.2.6), (7.3.2) and (7.2.6)).

Let $R = \tau_{4,\Sigma}/\tau_{3,\Sigma}$ be the factor-sheaf of the embedding j. Hence, we obtain the exact sequence

(c) $$0 \to \tau_{3,\Sigma} \xrightarrow{\ j\ } \tau_{4,\Sigma} \xrightarrow{\ \beta\ } R \to 0$$

of sheaves on the family $\mathscr{C}^3$.

(8.1.8). From the cited corollaries and proposition we derive also that the embedding $j\colon \tau_{3,\Sigma} \hookrightarrow \tau_{4,\Sigma}$ corresponds to the embeddings $H_C \cap H_{\overline{C}} \hookrightarrow H_C$ of subspaces of $\Sigma(V_2)$; remember that $P(H_C) = \mathrm{Span}(C) \subset P(\Sigma(V_2))$, $P(H_{\overline{C}}) = \mathrm{Span}(\overline{C}) \subset P(\Sigma(V_2))$, and $C + \overline{C} = C_l = \sigma_{2,0}(l) \cdot X_{10} \subset \mathrm{Span}(C_l) = P(\Sigma(V_2))$, where $l = P(V_2)$)(ibid.).

But the last means exactly that the epimorphism $\beta\colon \tau_{4,\Sigma} \to R = \tau_{4,\Sigma}/\tau_{3,\Sigma}$ coincides with the dual of the map

$$H^0(\varphi_C)\colon H^0(C, O_C) \to H^0(C, O(1) \otimes O_C)$$
$$\simeq H^0(C, O_{P(\Sigma(V_2))}(1) \otimes O_C)$$

(see the sequence $(1'')$ and the description of the map φ_C in the beginning of the Remark). In fact, the description of the epimorphism $\beta\colon \tau_{4,\Sigma} \to R$ corresponds to the description of the global sections of the linear system $|P(H_C \cap H_{\overline{C}})|_C| = |C \cap \overline{C}|$ on C, defined by the map φ_C. Therefore, the exact sequence $(1'')$ takes the form

$(1''')$
$$0 \to (\tau_{4,\Sigma}/\tau_{3,\Sigma})^*(C) \xrightarrow{\ H^0(\varphi_C)=\beta^*\ } \tau_{4,\Sigma}(C)$$
$$\to \Omega_{\mathscr{C}^3}(C) \to 0,$$

therefore

$$\Omega_{\mathscr{C}^3}(C) \simeq \tau_{3,\Sigma}^*(C).$$

So, we formulate the following

(8.1.9). PROPOSITION (The open Tangent Bundle Theorem for the family $\mathscr{C}^3$ of rational normal cubic curves on X). *Let* $U = \{C \in \mathscr{C}^3\colon C$ *and* $\overline{C}$ *are smooth rational normal curves on* $X = X_{10} = G \cap P(V_8) \cap Q$, *such that* $N_{C/X}$ *and* $N_{\overline{C}/X}$ *are isomorphic to* $O_{P^1} \oplus O_{P^1}(1)$ *(see* §5.1*), and also, such that the surfaces* S_l *and* S_C *are rational normal cubic scrolls (see* (2.2.4) (1) *and* §5.2*) and* $N_{S_l/W} \otimes O_C \simeq N_{S_l/W} \otimes O_{\overline{C}} \simeq O_{P^1}(2) \oplus O_{P^1}(2)\} \subset \mathscr{C}^3$.

Then, there exists a naturally defined isomorphism

$$\tau_{3,\Sigma}^*|_U \rightarrow \Omega_{\mathscr{C}^3}|_U.$$

To prove the Tangent Bundle Theorem (8.1.9) we need (1) to prove Lemma (8.1.3), and (2) to globalize the local constructions, leading to the sequence $(1''')$.

(8.1.10). PROOF OF LEMMA (8.1.3). We shall suppose in addition that the surface $S_C = (\bigcup l: l$—a line in P^4 & $l \in C)$ is a rational normal cubic scroll (see (5.2.1); by the way, the surface S_C cannot be a cone (see Corollary (5.2.3))). Though this superfluous restriction will be not used in the proof the author supposes that the normal sheaves $N_{C/Y}$ for the curves C of the type (5.2.1) (2.a) (see also Corollary (5.2.3)), where $Y = Y_5$ is the threefold defined below (see (1), (5) and (7)), are isomorphic to $O_{P^1}(1) \oplus O_{P^1}(3)$. In any case, the last does not prevent us from proving (8.1.3).

(1) It is well known (see [F, (B.5.8)]) that the tangent bundle $T(G(2, V))$ is isomorphic to the bundle of homomorphisms $\operatorname{Hom}(\tau_{2,V}, \tau_{3,V^*}^*) = \tau_{2,V}^* \otimes \tau_{3,V^*}^*$.

Let $P^3 = P(V_4)$, where $0 \subset V_4 \subset V$. Then, the embedded cycle $\sigma_{1,1}(P^3)$ carries out, in practice, an embedding of the Grassmannian $G(2, V_4)$ into $G(2, V)$; in particular, we have the exact sequence:

$$0 \rightarrow \operatorname{Hom}(\tau_{2,V_4}, \tau_{2,V_4^*}^*) \rightarrow \operatorname{Hom}(\tau_{2,V}, \tau_{3,V^*}^*) \otimes O_{\sigma_{1,1}}(P^3)$$

$$\rightarrow N(\sigma_{1,1}(P^3), G) \rightarrow 0.$$

We can derive from here that the normal sheaf $N(\sigma_{1,1}(P^3), G)$ equals to the restriction $\tau_{2,V}^* \otimes O_{\sigma_{1,1}}(P^3)$.

Now let $Y = Y_5 = G \cap P^6$, and let $P^3 \subset P^4$ is chosen in such a way that $\sigma_{1,1}(P^3)$ intersects Y transversely; we shall suppose also that Y is chosen to be a smooth intersection. The intersection $q = \sigma_{1,1}(P^3) \cap Y$ is a conic; we shall suppose in addition that the surface $S_q = (\bigcup l: l$—a line in P^4 & $l \in q)$ is a smooth quadric (see §4.3 (1)). Then we shall have $N_{q/Y} =: N(q, Y) = N(\sigma_{1,1}(P^3) \cap P^6, G \cap P^6) = N(\sigma_{1,1}(P^3), G) \otimes O_q = \tau_{2,V}^* \otimes O_q = O_q(1) \oplus O_q(1)$.

(2) Let now l be a line lying on $Y = Y_5$. It is easy to see that $N_{l/Y}$ is one of the sheaves $O_l \oplus O_l$ or $O_l(-1) \oplus O_l(1)$. In [I1, (5.2)] (see also [I4, (7.2)]) was proven that $N_{l/Y} = O_l \oplus O_l$ for the general line $l \subset Y_5$ (see also [FN]).

(3) Now let q and l be a conic and a line, lying on $Y = Y_5$ and let q and l intersect in a point. Let $H = P^5 \subset P^6$ be a sufficiently general hyperplane through $q \cup l$, such that the surface $S = S_5 = Y \cap H$ is smooth. The surface S is a Del Pezzo surface of degree 5 in $P^5 = H \subset P^6$. The surface $S = S_5$ is obtained from the projective plane P^2 after blowing-up

of four points x_1, x_2, x_3 and $x_4 \in P^2$. As S_5 is chosen to be smooth, then no three of the points x_1, x_2, x_3 and x_4 are collinear.

(4) Now let $z \in P^2$ be a point, which does not lie on the union of the lines $l_{ij} = \langle x_i, x_j \rangle$, $1 \le i < j \le 4$. Let $\sigma = \prod_{1 \le k \le 4} \sigma(x_k): S \to P^2$ be the map described in (3). The inverse of σ is the rational map φ given by the noncomplete linear system

$$|O_{P^2}(3 - x_1 - x_2 - x_3 - x_4)|: P^2 \to P^5.$$

Now let Q be a conic in P^2 through the points x_1, x_2, x_3 and z. It is easy to see that the proper σ-preimage $C \subset S = S_5$ of the conic $(Q = 0) \subset P^2$, $Q \in H^0(O_{P^2}(2 - x_1 - x_2 - x_3 - z))$ is a rational normal cubic curve.

The line l (see (3)) is one of the ten lines, which lie on the Del Pezzo surface $S = S_5 \subset P^5$. Obviously, we can suppose that the morphism σ is chosen in such a way that $l = \sigma^{-1}(x_4) \subset S_5$.

Let $(t) = (t_1 : t_2)$ be the homogeneous coordinates of

$$P^1 = |O_{P^2}(2 - x_1 - x_2 - x_3 - z)|.$$

Evidently, the proper preimage $C_{(t)}$ of the conic $Q_{(t)} \in P^1 = P^1(t)$ splits *iff* $Q_{(t)}$ passes through the point x_4. Let $(1) = (1 : 1)$, and let $x_4 \in Q_{(1)}$. then $C_{(1)} = q \cup l$, where q is the conic and l is the line $l = \sigma^{-1}(x_4)$; obviously, $(q, l)_S = 1$.

(5) Now let $\beta_{(t)}: Y_{(t)} \to Y = Y_5$, $(t) \ne (1)$, be the blowing-up of the curve $C_{(t)}(\subset S_5) \subset Y = Y_5$; let also $S_{(t)} = \beta_{(1)}^{-1}(C_{(t)})$ be the ruled surface over the curve $C_{(t)}$, $(t) \ne (1)$. It is easy to see that the normal sequence for $C_{(t)} \subset S_5 \subset Y$ is

$$0 \to O_{C_{(t)}}(1) \to N_{C_{(t)}/Y} \to O_{C_{(t)}}(3) \to 0.$$

Therefore, $N_{C_{(t)}/Y} = O_{C_{(t)}}(a) \oplus O_{C_{(t)}}(b)$, where $a + b = 4$ and $1 \le a \le b \le 3$.

Let

$$\mathscr{S} = \overline{\bigcup_{(t) \ne (1)} S_{(t)}} \subset \overline{\bigcup_{(t) \ne (1)} Y_{(t)}} = \mathscr{Y}.$$

Obviously, $\mathscr{S}$ (resp. $\mathscr{Y}$) is a bundle of rational surfaces (resp. of rational threefolds) over the projective line $P^1 = P^1(t)$.

(6) Let the general rational normal cubic $C \subset Y_5$ has a normal bundle $O_C(1) \oplus O_C(3)$. Then we can choose the hyperplane H (and, hence, the surface $S_5 = Y_5 \cap H$) through the given degenerate rational normal cubic $C_{(1)} = q \cup l$ (see (3)), and the point $z \in P^2$ (see (4)) in such a way that $N_{C_{(t)}/Y} = O_{C_{(t)}}(1) \oplus O_{C_{(t)}}(3)$ for any $(t) \in P^1(t)$ in some neighborhood of $(t) = (1)$. We can also suppose that the normal sheaves of q and l are as in (1) and (2), i.e. $N_{q/Y} = O_q(1) \oplus O_q(1)$ and $N_{l/Y} = O_l \oplus O_l$. The exceptional surfaces $S_{(t)} = \beta_{(t)}^{-1}(C_{(t)}) \subset Y_{(t)}$ are isomorphic to the ruled surface $\mathbb{F}_2 =$

$P(N_{C_{(t)}/Y}) = P(O(1) \oplus O(3))$, for all $(t) \neq (1)$, around (1). The surface $S_{(1)} = \mathscr{S}\setminus[\bigcup_{(t)\neq(1)} S_{(t)}]$ is a degeneration of $\mathbb{F}_2$; on the other hand, $S_{(1)}$ is a union of the two quadrics—the quadric A over q and the quadric B over l.

The smooth quadrics A and B intersect along the common generator $A \cap B$, because $(q, l)_S = 1$. But then the union $A \cup B$ cannot be a degeneration of the family of $\mathbb{F}_2$'s, described above. Consequently, the general rational normal cubic curve $C \subset Y = Y_5$ has a normal bundle $N_{C/Y} = O_C(2) \oplus O_C(2)$ (see (5)).

(7) Let now $C \subset X_{10}$ be a general rational normal cubic curve. In particular, C is embedded in the rational normal cubic scroll $S_1 \subset W$ as a hyperplane section (see Proposition (2.2.4) (1) and Corollary (5.3.2)).

Let $H = P^6 \subset P^7$ be a sufficiently general hyperplane through the curve C, such that the intersection $Y = Y_5 = W \cap H$ is smooth. In particular, let H intersect S_1 transversely along the curve C. Then $N_{S_1/W} \otimes O_C = N_{(S_1 \cap H)/(W \cap H)} \otimes O_C = N_{C/Y}$.

Clearly, we can also suppose that the curve C is sufficiently general in $Y = Y_5$, i.e. $N_{C/Y} = O_C(2) \oplus O_C(2)$. But then $N_{S_1/W} \otimes O_C = N_{C/Y} = O_C(2) \oplus O_C(2)$. Lemma (8.1.3) is proved.

8.2. Proof of the Tangent Bundle Theorem (Proposition (8.1.9)).

(8.2.1). The proof of the T.B.T. consists in the repeating of the local constructions from the beginning globally over the whole family $\mathscr{C}^3 = \mathscr{C}^3(X)$. For this purpose we introduce the following subvarieties of the projectivized bundle $P_{\mathscr{C}^3}(\Sigma)$ on the planes of the curve $C \subset X = X_{10}$ and of the surface $S_l \subset W$:

$$D_{\mathscr{C}^3} = \{(C, x) \in \mathscr{C}^3 \times X : x \in C\}$$

and

$$D_{\mathscr{C}^3}(G) = \{(C, x) \in \mathscr{C}^3 \times W : x \in S_1\};$$

remind that $C + \overline{C} = C_l = S_l \cdot Q \subset S_l \subset P(\Sigma(V_2))$, where $l = P(V_2)$.

(8.2.2). The pairs of successive embeddings

$$D_{\mathscr{C}^3} \subset \mathscr{C}^3 \times X \subset \mathscr{C}^3 \times W$$

and

$$D_{\mathscr{C}^3} \subset D_{\mathscr{C}^3}(G) \subset \mathscr{C}^3 \times W$$

globalize respectively the pairs of embeddings $C \subset X \subset W$ and $C \subset S_l \subset W$. Correspondingly to the normal sequences (8.1.1) (1) and (8.1.1) (2) they induce the sequences of normal sheaves

$$(1, \mathscr{C})\quad 0 \to N(D_{\mathscr{C}^3}, \mathscr{C}^3 \times X) \to N(D_{\mathscr{C}^3}, \mathscr{C}^3 \times W)$$

$$\to N(\mathscr{C}^3 \times X, \mathscr{C}^3 \times W) \otimes O_{D_{\mathscr{C}^3}} \to 0$$

and

$$(2, \mathscr{C})0 \to (D_{\mathscr{C}^3}, D_{\mathscr{C}^3}(G)) \to N(D_{\mathscr{C}^3}, \mathscr{C}^3 \times W)$$
$$\to N(D_{\mathscr{C}^3}(G), \mathscr{C}^3 \times W) \otimes O_{D_{\mathscr{C}^3}} \to 0.$$

(8.2.3). Let $O(1)$ be the antitautological sheaf on the projective bundle $P_{G(2,V)}(\Sigma) \to G = G(2, V)$. According to the definition of the bundle Σ (see (6.3.3)), the sheaf Σ on G is embedded in a natural way in the sheaf $V_8 \otimes O_G$ (corresponding to the constant bundle $G \times V_8$ over $G = G(2, V)$). In fact, for every $V_2 \in G = G(2, V)$, the embedding of the corresponding fibers $\Sigma(V_2) \subset (V_8 \otimes O_G)(V_2)$ coincides with the natural embedding $(V_2 \wedge V) \cap V_8 \subset V_8$. Therefore, the sheaf $O(1)$ coincides with the restriction of the sheaf $O_{P(V_8)}(1) = O_{P(\wedge^2 V)} \otimes O_{P(V_8)}$ on the subvariety $P_G(\Sigma) \subset P_G(V_8 \otimes O_G) = G \times P(V_8)$, (the "old" $O(1)$, see above).

(8.2.4). As in the local case we multiply the sequence $(2, \mathscr{C})$ by the sheaf $O(-1)$:

$$(2', \mathscr{C})0 \to N(D_{\mathscr{C}^3}(G)) \otimes O(-1)$$
$$\to N(D_{\mathscr{C}^3}, \mathscr{C}^3 \times W) \otimes O(-1)$$
$$\to N(D_{\mathscr{C}^3}(G), \mathscr{C}^3 \times W) \otimes O_{\mathscr{C}^3} \otimes O(-1) \to 0.$$

Similarly, we obtain the sequence $(1, \mathscr{C}) \otimes O(-1)$:

$$(1', \mathscr{C}) \qquad 0 \to N(D_{\mathscr{C}^3}, \mathscr{C}^3 \times X) \otimes O(-1)$$
$$\to N(D_{\mathscr{C}^3}, \mathscr{C}^3 \times W) \otimes O(-1) \to O(1) \to 0,$$

since $N(\mathscr{C}^3 \times X, \mathscr{C}^3 \times W) = O(2)$.

(8.2.5). Let, for the brevity, denote by B the sheaf $N(D_{\mathscr{C}^3}, \mathscr{C}^3 \times W) \otimes O(-1)$ and let α and β be the natural projections for the Fano family $D_{\mathscr{C}^3}$:

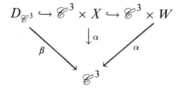

As before, using the equality $R_\beta^1 N(D_{\mathscr{C}^3}, \mathscr{C}^3 \times X) = 0$, which can be obtained on the level of fibers from the local equality $H^1(C, N_{C/X}) = 0$, and also the relative duality of Serre, we obtain

$$\Omega_{\mathscr{C}^3} = R_\beta^1(N(D_{\mathscr{C}^3}, \mathscr{C}^3 \times X) \otimes O(-1))$$

(here $R_.^i$ is the operation of taking the ith direct image).

(8.2.6). In a similar way, using the local equalities:

$$H^i(N_{S_l/W} \otimes O_C \otimes O(-1)) = 0, \qquad i = 1, 2, \ldots$$

we derive the identities

$$(5, \mathscr{C}) \qquad R^i_\beta(N(D_{\mathscr{C}^3}(G), \mathscr{C}^3 \times W)) \otimes O_{D_{\mathscr{C}^3}} \otimes O(-1) = 0,$$

$$i = 1, 2, \ldots.$$

(8.2.7). As is already clear, $N(\mathscr{C}^3 \times X, \mathscr{C}^3 \times W) = O(2)$.

Using the obtained just above and taking the direct β-images of the sequences $(1', \mathscr{C})$ and $(2', \mathscr{C})$, we obtain

$$(2'', \mathscr{C}) \qquad \begin{aligned} &0 \to \beta_*(N(D_{\mathscr{C}^3}, D_{\mathscr{C}^3}(G)) \otimes O(-1)) \to \beta_* B \to 0, \\ &0 \to R^1_\beta(N(D_{\mathscr{C}^3}, D_{\mathscr{C}^3}(G)) \otimes O(-1)) \to R^1_\beta B \to 0 \end{aligned}$$

and

$$(1'', \mathscr{C}) \qquad \begin{aligned} &0 \to \beta_*(N(D_{\mathscr{C}^3}, \mathscr{C}^3 \times X) \otimes O(-1)) \to \beta_* B \\ &\to \beta_* O(1) \to \Omega_{\mathscr{C}^3} \to R^1_\beta B \to R^1_\beta O(1) \to 0, \end{aligned}$$

where $\beta_* = R^0_\beta$.

(8.2.8). Now, we introduce the variety

$$D_{\mathscr{C}^3}(\langle \cdot \rangle) = \{(C, x): C \in \mathscr{C}^3, \ x \in \langle C \rangle = \operatorname{Span}(C)\}.$$

As $\operatorname{Span}(C) \subset P(\Sigma(C)) = P(\Sigma(V_2))$, then $D_{\mathscr{C}^3}(\langle \cdot \rangle)$ is embedded naturally in the projectivized bundle $P_{\mathscr{C}^3}(\Sigma)$. In fact, the bundle Σ is defined, originally on the Grassmannian $G = G(2, V)$, but it is defined also on the subsets of $G(2, V)$ and on the preimages of these subsets under morphisms; as usual, we drop the symbols of the maps. From the Remark (6) (see (8.1.6)), (8.1.7) and (8.1.8), we conclude that

$$D_{\mathscr{C}^3}(\langle \cdot \rangle) = P_{\mathscr{C}^3}(\tau_{4, \Sigma}),$$

(see also the Corollary (7.3.3)).

The Fano family $D_{\mathscr{C}^3}$ is embedded identically in the projectivized bundle $D_{\mathscr{C}^3}(\langle \cdot \rangle) = P_{\mathscr{C}^3}(\tau_{4, \Sigma})$; in particular, the projection $\beta: D_{\mathscr{C}^3} \to \mathscr{C}^3$ is a restriction of the projection $\alpha: D_{\mathscr{C}^3}(\langle \cdot \rangle) = P_{\mathscr{C}^3}(\tau_{4, \Sigma}) \to \mathscr{C}^3$. From the properties of the antitautological sheaf $O(1)$ on $P_{\mathscr{C}^3}(\tau_{4, \Sigma})$ we have

$$\beta_* O(1) = \tau^*_{4, \Sigma} \quad \text{and} \quad R^1_\beta O(1) = 0.$$

(8.2.9). Let us denote the sheaf $N(D_{\mathscr{C}^3}, D_{\mathscr{C}^3}(G)) \otimes O(-1)$ by A. Taking into account the identity

$$\beta_*(N(D_{\mathscr{C}^3}, \mathscr{C}^3 \times X) \otimes O(-1)) = 0,$$

which follows from the corresponding local equality, we conclude, just as in the local case, that the exact sequence $(1'', \mathscr{C})$ can be written in the form

$$(1''', \mathscr{C}) \qquad 0 \to \beta_* A \to \tau^*_{4, \Sigma} \to \Omega_{\mathscr{C}^3} \to R^1_\beta A \to 0,$$

(see the "local" sequences (8.1.5) $(1'')$ and (8.1.8) $(1''')$).

According to the comments in the Remark (8.1.6) (6), the sheaf $\beta_* A$ co-incides with the factorsheaf $R = (\tau_{4,\Sigma}/\tau_{3,\Sigma})^*$. If, moreover, the curve C $(C \in \mathscr{C}^3)$ is general (in the sense of the conditions of Proposition (8.1.9)), then the first direct image $R^1_\beta A$ vanishes over C, since the fiber of $R^1_\beta A$ over $C \in U \subset \mathscr{C}^3$ is $H^1(N_{C/S_1} \otimes O(-1)) = H^1(C, O_C) = 0$ $(C \simeq P^1)$. The open T.B.T. (Proposition (8.1.9)) is proved.

(8.2.10). REMARK. The correct formulation of the T.B.T. must take into account a joint geometrical interpretation of the sequences $(1, \mathscr{C})$ and $(2, \mathscr{C})$, taking into consideration also the disposition of the locus $\mathscr{C}^3_{ex} \subset \mathscr{C}^3$, where the local constructions fail; it is already clear that $\mathscr{C}^3_{ex} \subseteq \mathscr{C}^3 \setminus U$ (see (8.1.4) (5) and (8.1.3)).

(8.2.11). The geometry of the Fano threefold $X = X_{10}$ can provide an additional useful information by means of the geometrical analogue of the just proven open T.B.T. First, the exact sequence $(1''', \mathscr{C})$ induces a natural isomorphism $\psi: \tau^*_{3,\Sigma}|_U \to \Omega_{\mathscr{C}^3}|_U$ over the open subset $U \subset \mathscr{C}^3$. The definition of $H^0(\varphi_C)$ in (8.1.5) $(1'')$ shows that the fiber $\tau_{3,\Sigma}(C)$ coincides with the embedded plane $H_C \cap H_{\overline{C}} \subset \Sigma(C)(= \Sigma(V_2) = \Sigma(\overline{C}))$; see also the Corollary (7.2.6) (2). Here, as usual, $C + C = C_{P(V_2)}$, $P(H_C) = \langle C \rangle$, $P(H_{\overline{C}}) = \langle \overline{C} \rangle$.

The dual map Ψ^* defines the isomorphism:

$$\Psi^*: \mathscr{T}_{\mathscr{C}^3}|_U \to \tau_{3,\Sigma}|_U \quad \text{over } U \subset \mathscr{C}^3$$

(here $\mathscr{T}_{\mathscr{C}^3} = \Omega^*_{\mathscr{C}^3}$ is the tangent sheaf of the family $\mathscr{C}^3$).

So, we can formulate the following:

(8.2.12). THEOREM (Geometric (open) T.B.T. for $\mathscr{C}^3$). *There is a naturally defined isomorphism (see (8.1.9))*

$$\Psi^*: \mathscr{T}_{\mathscr{C}^3}|_U \to \tau_{3,\Sigma}|_U,$$

such that Ψ^* *coincides with the natural map*

$$\Psi^*(C): \mathscr{T}_{\mathscr{C}^3}(C) \widetilde{\to} H_C \cap H_{\overline{C}} \subset \Sigma(V_2) \subset V_8$$

on the fiber $\mathscr{T}_{\mathscr{C}^3}(C)$; *(see Corollary (5.3.2) and Corollary (7.2.6) for the definitions of* V_2, $\overline{C}$, H_C *and* $H_{\overline{C}}$).

9. Tangent Bundle Theorem for the family of conics on the Fano threefold $X = X_{10}$

9.1. In the present section we shall follow some results of the thesis of D. Logachev [L1] (see also [L2 and P]) on the family of conics on $X = X_{10} = G \cap P(V_8) \cap Q$. The original idea of representing the family of conics on X_{10} as a degeneration locus, resp. as a zero scheme of a section of some vector bundle, and the corresponding statement of the tangent bundle theorem, belongs to Logachev.

The main observations in the present work are

(1) The use of the simple fact that the conics and the rational normal cubics on X_{10} correspond to the standard splittings of the codim 2-cycles on X_{10} of the types $\sigma_{1,1} \cap X_{10}$ and $\sigma_{2,0} \cap X_{10}$ respectively; the last generate the subring of the ring of cycles of codimension 2 on $X = X_{10}$, which are induced from cycles on the Grassmannian $G(2, V)$;

(2) The possibility to perform the constructions of Logachev also to the family of the rational normal cubic curves on X_{10}.

The two marked families are, probably, the only families of curves on X_{10}, for which the mentioned methods and representations are adequate.

In particular, (1), together with the geometrical interpretations of the tangent bundle theorems for both families (see the T.B.T. (8.2.12) and the T.B.T.$^{-s}$ (9.4.1, 3, 5) below), grounds us to carry out a geometrical approach to the Abel-Jacobi maps for some families of curves, naturally arising on X (see for example [**CG**]). Let, for example

$$\mathscr{C}_1^5 = \{C - \text{a curve} : C \subset X, \ p_a(C) = 1, \ \deg(C) = 5\}$$

be the family of the linearly normal elliptic quintics on X; the last corresponds to the set of components of the degenerate canonical curves of degree 10 on X_{10} into pairs of elliptic curves. The T.B.T.$^{-s}$ for the families $F = \{\text{conics on } X_{10}\}$ and $\mathscr{C}^3 = \{\text{rat.norm cubics on } X_{10}\}$, enables us to perform a geometrical approach to the Abel-Jacobi map of the family $\mathscr{W} = \mathscr{C}^3 + F = \{C + q : C \in \mathscr{C}^3, \ q \in F\}$.

Let $\mathscr{C} = \mathscr{C}_1^5 \cap \mathscr{W} = \{C + q : C \in \mathscr{C}^3, \ q \in F, \ (C, q) = 2\}$ be the threedimensional family, the elements of which are, by definition, degenerate elliptic quintics on X, which split into a sum of a rat.norm cubic and a conic, intersecting between them in two points. The results in the paper [**V**] of Claire Voisin on the Quartic Double Solids give us a hope to suppose that the general points z of the two-dimensional subvariety Z of the Intermediate Jacobian $J(X_{10})$ of X_{10}, which is the Abel-Jacobi image of the "diagonal" family $\mathscr{C}$, are double points of the theta-divisor $\Theta(X_{10}) \subset J(X_{10})$. The author's suggestion is that the tangent cones $TC_z(\Theta)$ to the theta divisor Θ at the singular points z of Θ, are quadrics of rank $= 6$. The last will provide us with a geometrical proof of the nonrationality of the Fano threefold $X = X_{10} = G(2, 5) \cap P^7 \cap Q$ (see [**V**, §4]).

9.2. In (6.2.1) was defined the bundle $Pf \to P(V^*)$; the fiber $Pf(V_4)$ over the element $V_4 \in P(V^*)$ was defined to be the Plücker quadric, corresponding to the 4-space V_4 (for more details, see §6.2). It was proven that the corresponding invertible sheaf Pf over $P(V^*)$ is isomorphic to the sheaf $(\det \tau_{4,V})^{-1} = O_{P(V^*)}(1)$.

Now let V_8, $V_8 \subset V_{10} = \bigwedge^2 V$ be chosen sufficiently general, such that
(i) the variety $X = G \cap P(V_8) \cap Q$ is smooth;
(ii) $\dim(V_8 \cap (\bigwedge^2 V_4)) = 4$, for every $V_4 \in P(V^*) = G(4, V)$.

Then we can define correctly a bundle M over $P(V)$ with 4-dimensional fibers $M(V_4)$ such that

$$M(V_4) = \bigwedge^2 V_4 \cap V_8, \quad V_4 \in P(V^*) = G(4, V).$$

On the other hand, the restriction of the Plücker bundle $Pf \to P(V^*)$ (regarded as a bundle of quadrics on $\bigwedge^2 \tau_{4,V} \to P(V^*)$) to the subbundle $M = \bigwedge^2 \tau_{4,V} \cap V_8 \subset \bigwedge^2 \tau_{4,V}$, represents the bundle Pf as a subbundle of $S^2 M^* \to P(V^*)$; moreover Pf has fibers

$$Pf(V_4) = H^0 \left(P\left(\bigwedge^2 V_4 \cap V_8\right), O(2 - (\sigma_{1,1}(P(V_4)) \cap P(V_8))) \right)$$

(see Corollary (6.2.3)).

We can also define an analogue to the bundle $Q \to G(2, V)$, which was regarded as a subbundle of $S^2 \Sigma^* \to G(2, V)$, namely

Let $X = G \cap P(V_8) \cap Q$, where Q is a quadric. We define the bundle (the invertible sheaf) Q over $P(V)$ such that

(1) Q is a subbundle of $S^2(\bigwedge^2 \tau_{4,V})^*$;

(2) $Q(V_4) = \mathbb{C} \cdot Q|_{P(\wedge^2 V_4)} \subset H^0(P(\bigwedge^2 V_4), O(2))$, i.e. $Q(V_4)$ is the restriction of the equation of the quadric Q, etc.

From the conditions (i) and (ii) we derive that we can look at the bundle $Q \to P(V^*)$ also as a bundle of quadrics (a "quadric") on the bundle M. Moreover, it is evident that the corresponding sheaf Q is isomorphic to the structure sheaf $O_{P(V^*)}$. Just as in §6.3 (see Proposition (6.3.2)) we can prove the following:

(9.2.1).   PROPOSITION (Logachev, [L1]). *Let* $Z \subset P_{P(V^*)}(S^2(\bigwedge^2 \tau_{4,V})^*)$ *be the set*

$$Z = P\{(V_4, q) : V_4 \in P(V^*), q \in H^0\left(P\left(\bigwedge^2 V_4\right), O(2 - Q \cap \sigma_{1,1}(P(V_4)))\right).$$

Then Z is naturally isomorphic to the projectivized bundle

$$P_{P(V^*)}(O_{P(V^*)}(1) \oplus O_{P(V^*)}).$$

Moreover, from (i) and (ii) (see above), follows, that

$$Z = P\Big\{(V_4, q) : V_4 \in P(V^*), \; q \in H^0\Big(P\Big(\bigwedge^2 V_4 \cap V_8\Big),$$
$$O(2 - \sigma_{1,1}(P(V_4)) \cap P(V_8) \cap Q)\Big)\Big\}$$
$$= P\{(V_4, q) : V_4 \in P(V^*), \; q \in H^0(P(M(V_4)),$$
$$O(2 - \sigma_{1,1}(P(V_4)) \cap X))\}$$
$$\subset P_{P(V^*)}(S^2 M^*) \text{ (see Proposition (6.3.2))}.$$

(9.2.2). From the definition of the variety Z follows that the fiber $\eta^{-1}(V_4)$ $= Z(V_4)$ of the projective bundle $\eta: Z \to P(V^*)$ is isomorphic to the projective line $\langle Pf(V_4), Q(V_4) \rangle = \mathrm{Span}\{Pf(V_4) \cup Q(V_4)\}$ in the projective space $P(S^2 M^*(V_4))$ of quadrics in $P(M(V_4))$.

The surface F_c of the geometrical conics on $X = X_{10}$, the surface F (of "conics on X") and the factorsurface $F_0 = F/i = \{(q, \bar{q}; V_4): q \in F_c, (\bar{q}, V_4) = i(q, V_4)\}$ were defined in §§4.4 and in 4.5 (here i is the involution, which is correctly defined on F). The noncorrectness of the defining the rational map $(\cdot)^{-1}: F_c \dashrightarrow F_c$ arises from the fact that the linear Span of the set $S_{q_0} = (\bigcup l: l \subset P(V)$—a line, $l \in q_0)$, where q_0 is the unique ρ-conic on X, is contained in a one-dimensional family of P^3's; the last coincides with the line $P((V/U)^*) = \{P^3: P^3 \subset P^4 = P(V)$ & (the plane) $P(U) = S_{q_0} \subset P^3\}$ (see §§4.2 and 4.3).

(9.2.3). Let now $V_4 \in P(V^*)$ and let W denotes, as usual, the intersection $W = G \cap P(V_8) \subset P(V_8) \subset P(\bigwedge^2 V)$. As we know (see Proposition (2.2.3) (2)) the cycle $Q(P(V_4)) = \sigma_{1,1}(P(V_4)) \cap P(V_8) \subset W$ is a two-dimensional quadric. It follows from the definition of the bundle $M \to P(V^*)$ (see the introduction of §9.2) that $\langle Q(P(V_4)) \rangle = \mathrm{Span}(Q(P(V_4))) = P(M(V_4))$. By definition, $W \cap Q = X$. Hence, the elliptic quartic $C(P(V_4))$ (see Corollary (3.3.3)), which is, by definition, an intersection of the cycle $\sigma_{1,1}(P(V_4))$ with X, coincides with the intersection of the quadrics $Q(P(V_4))$ and Q in the space $P(M(V_4)) = P(\bigwedge^2 V_4 \cap V_8)$.

The quadric $Q(P(V_4))$ defines a section of the Pfaff bundle $Pf \to P(V^*)$; the restrictions of the quadric Q on the fibers of the bundle $M \to P(V^*)$ define a section of the bundle $Q \to P(V)$, defined above. Therefore, the condition of splitting of the curve $C(P(V_4))$ into a pair of involutive conics is equivalent to the existence of a quadric of rank ≤ 2 on the line $\langle Pf(V_4) \cup Q(V_4) \rangle$ in the space of quadrics in $P(M(V_4))$. The last arguments are correct on the subset $F_c \subset \{q_0\} \subset F_c$ (q_0 is the unique ρ-conic on X), but it is not hard to see that the "exceptional" quadrics of F, which lie on the exceptional line $r_F^{-1}(q_0)$ (see (4.5.1)) correspond to degenerate quadrics in the bundle $Pf \oplus Q$ (see for example §4.3, §4.4, Claim (4.2.1) and (4.5.1)). The involution i on F identifies the elements of the exceptional curve $r_F^{-1}(q_0)$ with the elements of the curve of the σ-conics on X (ibid.).

(9.2.4). The tautological sequence

$$0 \to O_\eta(-1) \to \eta^*(O_{P(V^*)}(1) \oplus O_{P(V^*)}) \to O_Z(1) \to 0$$

on the projective bundle

$$\eta: Z = P(O_{P(V^*)}(1) \oplus O_{P(V)}) \to P(V^*)$$

defines, just as in (6.4.4) (see also (6.4.2)), a net of quadrics on Z. It follows from the properties of the tautological sequence that the projectivization of the embedding

$$0 \to O_\eta(-1) \xrightarrow{j} \eta^*(O_{P(V^*)}(1) \oplus O_{P(V^*)})$$

represents the "embedded" points of the fibers of the projective bundle Z; these points are interpreted also as quadrics in the corresponding spaces $P(M(V_4))$. Because of that, as in §6, Proposition (6.4.4), we can conclude the following:

(9.2.5). CLAIM (Logachev, [L1]). The factorfamily (of pairs of involutive conics) $F_0 = F/i$ is embedded naturally in the projective bundle

$$Z = P_{P(V^*)}(O_{P(V^*)}(1) \oplus O_{P(V^*)})$$

as a second determinantal of the net of quadrics:

$$0 \to O_\eta(-1) \xrightarrow{j} \eta^*(O_{(V^*)}(1) \oplus O_{P(V^*)}) \xrightarrow{i} \eta^*(S^2 M^*),$$

i.e. $F_0 = D_2(\hat\varphi)$, where $\hat\varphi = i \cdot j$.

9.3. The remaining constructions and results are carried out parallel to the corresponding ones for the family of the rational normal cubics on X (see §7).

(9.3.1). Let Z and $\hat\varphi \colon O_\eta(-1) \to \eta^*(S^2 M^*)$ be as above. We define

$$\begin{aligned}
\widetilde{D}_k = \widetilde{D}_k(\hat\varphi) &= \{(z\,;\,U_k) \\
&= (V_4,\, q\,;\,U_k) \in G_Z(k,\, M^*)\colon \operatorname{Im}(q) \subseteq U_k\} \\
&\subset G_Z(k,\, M^*),\, 0 \le k \le 4.
\end{aligned}$$

(9.3.2). It is not hard to prove that the first determinantal $D_1(\hat\varphi) \subset Z$ vanishes for the general intersection $X = G \cap P(V_8) \cap Q$, hence, the natural projection $\pi_2 \colon \widetilde{D}_2(\hat\varphi) \to D_2(\varphi)$ (see (7.1.1)) is an isomorphism for the general X (see Remark (7.1.2)).

(9.3.3). From Corollary (3.3.3) (2) we have that the curve $C(P(V_4)) \subset X$ is a linearly normal curve of arithmetical genus 1 and of degree 4; the linear normality follows from the fact that the curve $C(P(V_4))$ is a component of the canonical curve $C(P(V_4)) + C_l$, $l = P(V_2)$ (see (3.3.2) and Corollary (3.3.3)), where $V_2 \in G(2, V)$ is chosen in such a way that $V_2 \subset V_4$. In particular, in the case, when $C(P(V_4))$ splits into a sum of two plane sections: $C(P(V_4)) = q + \overline{q}$, we obtain that $(q, \overline{q}) = 2$, hence, the intersection $\langle q \cap \overline{q} \rangle$ of the spaces $\langle q \rangle = \operatorname{Span}(q)$ and $\langle \overline{q} \rangle = \operatorname{Span}(\overline{q})$ is a line in $P(V_8)$. In the case when the two points of intersection of q and $\overline{q}$ coincide, the line $\langle q \cap \overline{q} \rangle$ is equal, by definition, to their common tangent line (see also (5.4.1) and (5.4.2)).

(9.3.4). The tautological sequence on $G_Z(2, M)$, together with the injection $\hat\varphi \colon O_\eta(-1) \hookrightarrow \eta^* S^2 M^*$, define, as in (7.2.2), a section s_0 of the sheaf $O_\eta(1) \otimes \operatorname{Coker}(2)$ over $G_Z(2, M)$, where $\operatorname{Coker}(2) = \operatorname{Coker}(S^2 \tau_{2, M^*} \to \eta^* S^2 M^*)$.

(9.3.5). The same arguments, as in Corollary (7.2.4), give us a ground to impose the following open conditions on the subspace

$$V_8 \subset V_{10} = \overset{2}{\bigwedge} V (\dim(V_8) = 8)$$

and on the quadric Q, where $X = G \cap P(V_8) \cap Q$ (see also Corollary (6.3.5) and Remark (7.1.2)):

(i) $X = G \cap P(V_8) \cap Q$ is smooth;

(ii) $\dim(V_8 \cap (\bigwedge^2 V_4)) = 4$, for every $V_4 \in P(V^*) = G(4, V)$;

(iii) $D_1(\hat{\varphi}) = \varnothing$.

(9.3.6). CLAIM (Logachev, [L1]). Let $X = G \cap P(V_8) \cap Q$, where the subspace $V_8 \subset \bigwedge^2 V$ and the quadric Q fulfill the open conditions (i), (ii) and (iii). Let Z, M, $\hat{\varphi}$, s_0 and $F_0 = F/i$, where F is the surface of conics on X (see Definition (4.4.2)), be as above.

Then F_0 is embedded in $G_Z(2, M^*)$ as a zero scheme of the section s_0, i.e.

$$F_0 \overset{\sim}{\to} Z(s_0) \subset G_Z(2, M^*).$$

(9.3.7). The section s_0, together with the duality $G_Z(2, M^*) \simeq G_Z(2, M)$ define a section s_0^* with zeros on $G_Z(2, M)$ (see (7.2.5)). So we can formulate

(9.3.8). COROLLARY. *In the conditions of the Claim* (9.3.6) *the "dual" natural embedding* $F_0 \hookrightarrow G_Z(2, M)$ *represents the isomorphic image of* F_0 *in* $G_Z(2, M)$ *as a zero scheme of the section* s_0^*, *i.e.*

$$F_0 \overset{\sim}{\to} Z(s_0^*) \subset G_Z(2, M).$$

Moreover, this embedding can be described as follows:

Let (q, V_4) *and* $(\overline{q}, V_4) = i(q, V_4)$ *be a pair of involutive elements (conics) of* F. *Let* H_q *and* $H_{\overline{q}} \in M^*(V_4)$ *be the hyperplanes in* $M(V_4)$, *such that* $\langle q \rangle = \mathrm{Span}(q) = P(H_q)$ *and* $\langle \overline{q} \rangle = \mathrm{Span}(\overline{q}) = P(H_{\overline{q}})$. *Then the embedding* $F_0 \hookrightarrow G_Z(2, M)$ *sends to the pair* $((q, V_4), (\overline{q}, V_4))$ *to the triple*

$$(V_4, H_q \cdot H_{\overline{q}}; H_q \cap H_{\overline{q}}) \in G_Z(2, M).$$

(9.3.3). REMARK. Note that $H_q \cap H_{\overline{q}}$ is always a plane since $P(H_q \cap H_{\overline{q}})$ coincides with the projective line $\langle q \cap \overline{q} \rangle$ (see (9.3.3)).

Similarly, as in the "rational normal cubic" case (see (7.3.1)), the tautological sequence on $P_Z(M^*)$ together with the injection $\hat{\varphi}$ define a section s of the sheaf $O_\eta(1) \otimes S^2 \tau_{4, M}^*$ over $P_Z(M^*)$. The same arguments, as in (7.3.1) and (7.3.2), give

(9.3.10). CLAIM (Logachev, [L1]). In the conditions of the Claim (9.3.6), the family F of the conics for the Fano threefold $X = X_{10} = G(2, V) \cap P(V_8) \cap Q$ is embedded naturally in the projective bundle $P_Z(M^*)$ as a zero scheme of the section s, defined just above, that is

$$F \overset{\sim}{\to} Z(s) \subset P_Z(M^*).$$

Using the duality $P_Z(M^*) \simeq G_Z(3, M)$, we can formulate the following:

(9.3.11). COROLLARY. *The embedding* $F \hookrightarrow P_Z(M^*) \simeq G_Z(3, M)$ *sends the element* $(q, V_4) \in F$ *to the triple* $V_4, H_q \cdot H_{\bar{q}}; H_q) \in G_Z(3, M)$ *(the hyperplanes* H_q *and* $H_{\bar{q}}$ *are defined as in the Corollary* (9.3.8)).

9.4. Now let $p_F \colon F \to F_0 = F/i$ be the two-sheeted covering, which corresponds to the involution $i \colon F \to F$ (see (4.5.1) and (4.5.2)). The embeddings $F_0 \hookrightarrow G_Z(2, M)$ and $F \hookrightarrow P_Z(M^*) = G_Z(3, M)$, together with p_F define natural maps $F \to F_0 \to G_Z(2, M)$ and $F \to G_Z(3, M)$. Therefore, the standard tautological bundles $\tau_{2, M}$ on $G_Z(2, M)$ and $\tau_{3, M}$ on $G_Z(3, M) = P_Z(M^*)$ can be lifted on the family F. Just as in the "rational normal cubic" case, there is a representation of the tangent bundle $\mathcal{T}_F$ by means of some bundle on F, which can be described in geometrical terms.

(9.4.1). CLAIM (Logachev), (the open Tangent Bundle Theorem for the family F of "conics" for X; see [**L1**]). Let $l_\sigma = \{(q, V_4) \colon q$ is a σ-conic on $X\} \subset F$ be the smooth rational curve of the σ-conics in the family of conics F (see §4.2); let also $Z, M, G_Z(2, M), P_Z(M^*)$ and the bundles $\tau_{2, M}$ and $\tau_{3, M}$ be as above. Let $\mathcal{R}_F$ be the cotangent sheaf (resp., the cotangent bundle) of F.

Let $U = F \backslash (l_\sigma \cup i(l_\sigma)) \subset F$. Then there exists a naturally defined isomorphism of sheaves over the open subset $U \subset F$,

$$\Psi^* \colon \tau_{2, M}^* |_U \to \Omega_F |_U .$$

(9.4.2). REMARK. The curve $i(l_\sigma)$ coincides with the exceptional curve $r_F^{-1}(q_0)$ (the curve of the exceptional "conics" of F (see §4.2, or the comments (9.2.2)). The curves l_σ and $i(l_\sigma)$ are mutually disjoint (ibid.). Moreover, there is an exact global version of the T.B.T. for the family of conics, proven by Logachev:

(9.4.3). THEOREM (Logachev, [**L1**]). *There is a naturally defined exact sequence of sheaves over the surface of conics* F *for* X,

$$0 \to \tau_{2, M}^* \to r^* \Omega_{F_m} \to O_{l_\sigma} \vee O_{i(l_\sigma)} \to 0 ,$$

where $r \colon F \to F_m$ *is the morphism onto the minimal model* F_m *of* F. *In fact, the morphism* r *is a composition of* $r_F \colon F \to F_c$ *(which blows-down the curve* $r_F^{-1}(q_0)$ *of the exceptional conics in* F), *and of the blowing-down of the* r_F-*image of its involutive curve (the curve of the* σ-*conics on* X *in* F_c). *The both contractions blow-down* (-1)-*curves to (nonsingular) points of the smooth minimal surface* F_m *(see also* [**P**]).

(9.4.4). COMMENTS.

(1) Note that the statement of the global T.B.T. (9.4.3) is, in fact, a correct treatment of the analogue of the sequence $(1''', \mathscr{C})$ for the family F (see (8.2.9), seq. $(1''', \mathscr{C})$ and Remark (8.2.10)).

(2) The factor-sheaf $O_{l_\sigma} \vee O_{i(l_\sigma)}$ has a support on the closet subset $l_\sigma \cup i(l_\sigma)$ of codim $= 1$ in F.

(9.4.5). COROLLARY (see (8.2.12)). *Let* $\Psi^* : \mathscr{T}_F|_U \to \tau_{2,M}|_U$ *be the dual of* Ψ *(see Claim* (9.4.1)*). Then* Ψ^* *coincides with the natural map*

$$\Psi^*(q, V_4) : \mathscr{T}_F(q, V_4) \xrightarrow{\sim} H_q \cap H_{\bar{q}} \subset M(V_4) \subset V_8$$

on the fiber $\mathscr{T}_F(q, V_4)$, *where* $(q, V_4) \in U$ *and the hyperplanes* H_q *and* $H_{\bar{q}}$ *are defined as in Corollary* (9.3.8).

Acknowledgment

In the end I would like to express my gratitude to Professors V. A. Iskovskikh, V. Kanev and K. Ranestad for lending their comprehensive support.

REFERENCES

[B1] A. Beauville, *Les singularités du diviseur theta de la Jacobienne intermédiaire de l'hypersurface cubique dans* P^4, Lecture Notes in Math., vol. 947, Springer-Verlag, 1982, pp. 190–208.

[B2] ____, *Variétés de Prym et jacobienne intermédiaires*, Ann. Sci. École Norm Sup. (4) **10** (1977), 309–391.

[B3] ____, *Complex algebraic surfaces*, London Math. Soc. Lecture Note Series no. 68, Cambridge Univ. Press, Cambridge, 1983.

[C] C. H. Clemens, *Double solids*, Adv. in Math. **47**: 2 (1983), 107–230.

[CG] C. H. Clemens and Ph. A. Griffiths, *The intermediate Jacobian of the cubic threefold*, Ann. of Math. (2) **95** (1972), 281–356.

[D] O. Debarre, *Le théorème de Torelli pour les intersections de trois quadriques*, Invent. Math. **95**: 3 (1989), 507–528.

[F] W. Fulton, *Intersection theory*, Ergebn. der Math. und ihrer Grenzgebiete: 3 Folge, Band **2**, A Series of Modern Survey in Math., Springer-Verlag, Berlin-Heidelberg, 1984.

[F1] G. Fano, *Sulle varietè a tre dimensioni a curve-sezioni canoniche*, Mem. R. Acad. d'Italia **8** (1937), 27–37.

[F2] ____, *Su alcune varietà algebriche a tre dimensioni razionali e aventi curve-sezioni canoniche*, Comment. Math. Helv. **14** (1942), 202–211.

[FN] M. Furushima and N. Nakayama, *The family of lines on the Fano threefold* V_5, Nagoya Math. J. **116** (1989), 111–122.

[G] N. P. Gushel', *On Fano varieties of genus 6*, Izv. A. N. SSSR Ser. Math. **46**: 6 (1982), 1159–1174 (Russian), English transl. in Math. USSR Izv. **21** (1983), 445–459.

[GH] Ph. A. Griffiths and J. Harris, *Principles of algebraic geometry*, John Wiley & Sons, N.Y., 1978.

[H] S. I. Hashin, *Birational automorphisms of the double Veronese cone of dimension three*, Vest. Mosk. Univ. Ser. I, No. 1 (1984), 13–16 (Russian); English transl. in Moscow Univ. Math. Bull. **39**: 1 (1984).

[I1] V. A. Iskovskikh, *Fano threefolds. I*, Izv. A.N. SSSR Ser. Mat. **41**: 3 (1977), 516–562 (Russian); English transl. in Math. USSR Izv. **11**: 3 (1977), 485–527.

[I2] ____, *Fano threefolds. II*, Izv. A.N. SSSR Ser. Mat. **42**: 3 (1978), 506–549 (Russian); English transl. in Math. USSR Izv. **12**:3 (1978), 469–506.

[I3] ____, *Birational automorphisms of three-dimensional algebraic varieties*, Itogi Nauki i Tekhniki: Sovr. Probl. Mat.-Vol. 12, VINITI, Moscow (1979), 159–236 (Russian); English transl. in J. Soviet. Math. **13** No. 6 (1980), 815–868.

[I4] ____, *Lectures on algebraic threefolds: Fano threefolds*, Moscow Univ. Publ., Moscow, 1988.

[IM] V. A. Iskovskikh and Yu. I. Manin, *Three-dimensional quartics and counterexamples to the Lüroth problem*, Mat. Sb. **86** (1971), 140–166 (Russian); English transl. in Math. USSR Sb. **15** (1971), 141–166.

[L] M. Letizia, *The Abel-Jacobi mapping for the quartic threefold*, Invent. Math. **75** (1984), 477–492.

[L1] Dm. Yu. Logachev, *Abel-Jacobi isogeny for the Fano threefold of genus 6*, Yaroslavl' Gos. Ped. Inst.—K. D. Ushinskij (1982)—Thesis (Russian).

[L2] ____, *Abel-Jacobi isogeny for the Fano threefold of genus 6*, Sb. Constructive Algebraic Geometry, Yaroslavl', **200** (1982), 67–76.

[M] Dm. G. Markushevich, *Numerical invariants of the families of lines of some Fano varieties*, Mat. Sb. **116**: 2 (1981), 265–288 (Russian); English transl. in Math. USSR Sb. **45** (1981).

[MM1] S. Mori and S. Mukai, *Classification of Fano threefolds with $B_2 \geq 2$*, I, Algebraic and Topological Theories, Tokyo and Amsterdam: Kinokuniya Company LTD 1985, pp. 496–548.

[MM2] ____, *Classification of Fano threefolds with $B_2 \geq 2$*, Manuscripta Math. **36**: 2 (1981), 147–162.

[P] P. J. Puts, *On some Fano threefolds that are sections of Grassmannians*, Indag. Math. **44** (1982), 77-90.

[R] L. Roth, *Sulle V_3 algebraiche su cui l'aggiunzione si estingue*, Rend. Acc. Lincei (8) **9** (1950), 246–249.

[T1] Al. S. Tikhomirov, *Geometry of the Fano surface of the double space of P^3 with ramification on a quartic*, Izv. A.N. SSSR Ser. Math. **44**: 2 (1980), 415–442 (Russian); English transl. in Math. USSR Izv. **16** (1981).

[T2] ____, *The Abel-Jacobi map of sextics of genus 3 on double spaces of P^3 of index two*, Dokl. A.N. SSSR **286**: 4 (1986), 821–824 (Russian); English transl. in Soviet Math. Dokl. **33**: 1 (1986), 204–206.

[T3] ____, *The Fano surface of a double Veronese cone*, Izv. A.N. SSSR Ser. Mat. **45**: 5 (1981), 1121–1197 (Russian); English transl. in Math. USSR Izv. **19** (1982)

[TN] A. N. Tjurin, *On the intersection of quadrics*. I, II, Usp. Mat. Nauk **30**: 6 (1975), 51–99 (Russian); English transl. in Russian Math. Surveys **30**: 6 (1975), 51–105.

[V] C. Voisin, *Sur la jacobienne intermédiaire du double solide d'indice deux*, Duke Math. J. **57**: 2 (1988), 629–646.

[W] G. E. Welters, *Abel-Jacobi isogenies for certain types of Fano threefolds*, Math. Centre Tracts 141, Math. Centrum, Amsterdam (1981)—Thesis.

DEPARTMENT OF ALGEBRA, INSTITUTE OF MATHEMATICS, BULGARIAN ACADEMY OF SCIENCES, 1090 SOFIA, BULGARIA

Contemporary Mathematics
Volume **136**, 1992

Some Uses of Analytic Torsion in the Study of Weierstrass Points

Jay Jorgenson

Abstract. *In his foundational study of Arakelov theory for arithmetic surfaces, Faltings [Fa] was led to define a volume on determinant lines associated to holomorphic line sheaves over a compact Riemann surface. The Faltings construction used the Riemann theta function. In [Q], Quillen presented another construction using analytic torsion (determinants of the Laplacian). Combining the two methods, one obtains an interesting and rich interplay between the holomorphic theory of the Riemann theta function and the spectral theory of analytic torsion. In this note we will show how one can apply these techniques to the study of Weierstrass points and codimension one Weierstrass points associated to subspaces of the space of holomorphic one-forms. We shall study limits of Weierstrass points on certain degenerating families of algebraic curves. Our results are closely related to theorems due to Diaz [Dz] and Eisenbud-Harris [EH].*

§1. Notation and Definitions

Let X denote a non-singular algebraic curve of genus g defined over $\mathbf{C}$, and let $\mathcal{K}$ denote the canonical sheaf on X. Let P be a point of X and

$$\psi_P : X \longrightarrow J(X)$$

1980 Mathematics Subject Classification (1985 Revision) 30F10, 14H15, 32G05, 34B25
This paper is in final form and no version of it will be submitted for publication elsewhere.

be the Abel-Jacobi map of X into its Jacobian which sends P to the origin. Let

$$W_{g-1,P} = W_{g-1}$$

denote the divisor of $J(\mathbf{X})$ which is equal to the sum of $\psi_P(X)$ taken $g-1$ times. Consider the ensuing principal theta polarization corresponding to the hermitian form $\mathbf{H}_X$. The principal theta polarization gives rise to the Riemann theta function and its theta divisor, which we denote by Θ. There is a well-defined choice of a sheaf $\mathcal{S}$ of degree $g-1$ such that $\mathcal{S}^2$ is isomorphic to the canonical sheaf $\mathcal{K}$ on X. The sheaf $\mathcal{S}$ is characterized as follows. Up to linear equivalence, there is a unique divisor D of degree $g-1$ such that

$$\Theta = W_{g-1,P} - \psi_P(D).$$

Let $\mathcal{S}$ be the unique, up to isomorphism, line sheaf such that $\mathcal{S} \cong \mathcal{O}(D)$. Using Riemann's vanishing theorem, one can show that $\mathcal{S}^2 \cong \mathcal{K}$. Proofs can be found in [FK], [F2], [La1], [M1] or [M2].

A classical description of the above set-up is as follows. Choose a canonical basis of $H_1(X, \mathbf{Z})$ which we denote by $A_1, B_1, \ldots, A_g, B_g$. Let $\zeta_1, \ldots, \zeta_g$ denote a basis of $H^0(X, \mathcal{K})$ dual to the marking and set

$$\zeta = (\zeta_1, \ldots, \zeta_g).$$

Let Ω denote the g by g matrix defined by

$$\Omega = \left(\int_{B_j} \zeta_k \right).$$

The Jacobian $J(X)$ can be realized as the complex torus obtained by $\mathbf{C}^g$ modulo the lattice $L(\Omega)$ generated by the period matrix (I_g, Ω). Let M_X denote the matrix associated to the hermitian form $\mathbf{H}_X$. Then,

$$M_X = (\mathrm{Im}\,\Omega)^{-1}.$$

This is proved in Chapter 13, Proposition 3.1 of [La1]. By definition,

$$\det M_X = \det \mathbf{H}_X.$$

The sheaf $\mathcal{S}$ corresponds to the degree $g-1$ divisor class determined by the vector of Riemann constants (see [FK] page 298, or [F2] page 10). The **Riemann theta function** is defined for $z \in \mathbf{C}^g$ by the series

$$(1.1) \qquad \theta(z, \Omega) = \sum_{n \in \mathbf{Z}^g} \exp(\pi i{}^t n \Omega n + 2\pi i{}^t nz).$$

Let us introduce the notation $||\theta||^2$ to denote

$$(1.2) \quad ||\theta||^2(z, \Omega) = (\det(\mathrm{Im}(\Omega)))^{\frac{1}{2}} \exp\left(- 2\pi {}^t y \mathrm{Im}(\Omega) y\right) \Big|\theta(z, \Omega)\Big|^2$$

where $y = \mathrm{Im}\, z$. The function $||\theta||^2(z, \Omega)$ is well-defined for all $z \in J(X)$. Basically, the Riemann theta function (1.1) is a section of the bundle $\mathcal{O}(-\Theta)$ over $J(X)$, and $||\theta||^2$ is the norm of that section with respect to a certain metric. Basic properties of the theta function will be assumed.

Throughout we require that X be of genus $g \geq 2$. A **Weierstrass point** is a point P on X for which there exists a holomorphic one-form on X that vanishes at P to order at least g. Let $\mathcal{A}$ denote a codimension one subspace of $H^0(X, \mathcal{K})$. A **Weierstrass point relative to $\mathcal{A}$**, or a **codimension one Weierstrass point**, is a point P on X for which there exists a holomorphic one-form in $\mathcal{A}$ that vanishes at P to order at least $g - 1$.

One can view the Jacobian variety $J(X)$ as the space of isomorphism classes of degree zero line sheaves. The Picard variety $\mathrm{Pic}_{g-1}(X)$ is the space of isomorphism classes of degree $g - 1$ line sheaves ([Ha]). We have a canonical isomorphism

$$(1.3) \qquad J(X) \to \mathrm{Pic}_{g-1}(X)$$

given by

$$\mathcal{L} \to \mathcal{L} \otimes \mathcal{S}.$$

Let us use $[\mathcal{L}]$ to denote the point in $J(X)$ corresponding to the degree zero line sheaf $\mathcal{L}$. In particular, $[\mathcal{L}]$ can be used as an argument of the Riemann theta function (1.1). Classically, the argument $[\mathcal{L}]$ is written as follows. Let K_P denote the vector of Riemann constants relative to the base point P of the Abel-Jacobi map ψ_P. Let $\mathcal{L}$ denote a degree zero line sheaf isomorphic to

$$\mathcal{L} \cong \mathcal{O}(P_1 + \cdots + P_d - R_1 - \cdots - R_d).$$

Then

$$[\mathcal{L}] = \sum_{j=1}^{d} \left(\int_P^{P_j} \zeta - \int_P^{R_j} \zeta \right) \mod L(\Omega).$$

Let $\mathcal{M}$ denote the degree $g-1$ sheaf

$$\mathcal{M} = \mathcal{O}(P_1 + \cdots + P_{g-1}).$$

The point $[\mathcal{M}^{-1} \otimes \mathcal{S}]$ in $J(X)$ is classically given by

$$[\mathcal{M}^{-1} \otimes \mathcal{S}] = \sum_{j=1}^{g-1} \int_P^{P_j} \zeta + K_P \mod L(\Omega).$$

A metric on a line sheaf $\mathcal{E}$ of rank r is defined on page 94 of [La3]. Briefly, the definition is as follows. Let $\{U_j\}$ denote an open cover of X such that $\mathcal{E}|U_j$ has a trivialization. Let g_{ij} denote the transition functions of $\mathcal{E}$ with respect to the open cover ([GH], page 67). For each i, let ρ_i denote a smooth map of U_i into the space of positive real numbers such that on $U_i \cap U_j$ we have

$$\rho_i = |g_{ij}|^2 \rho_j.$$

Let s_1, s_2 denote sections of $\mathcal{E}$ over some U_i, viewed as complex valued functions on U_i. Define the pointwise inner product $< s_1, s_2 >_P$ at the point P by

$$< s_1, s_2 > (P) = \frac{s_1(P)\bar{s}_2(P)}{\rho_j(P)}.$$

The value on the right-hand side is independent of the open set U_i as can be seen by the transformation laws of the metric and the sections. Given a positive $(1,1)$ form μ on X, one obtains an inner product of sections of $\mathcal{E}$ by integrating the pointwise inner product with respect to μ. If $\mathcal{L}$ is a degree zero line sheaf, there exists a trivialization of $\mathcal{L}$ for which the transition functions g_{ij} are constants. Hence, one can choose a metric such that the corresponding functions ρ_i are constants. Such a metric is called a **flat metric**. For the purposes of this paper, we assume that all metrics on degree zero line sheaves are flat metrics.

A metric ρ on $\mathcal{K}$ is determined by a positive $(1,1)$ form μ on X. The relation between ρ and μ in terms of the local coordinate z on X is given by the expression

$$\mu(z) = \rho(z)^{-1} \frac{i}{2\pi} dz \wedge d\bar{z}.$$

By square root, there is an induced metric $\rho^{\frac{1}{2}}$ on the line sheaf $\mathcal{S}$. The **first Chern form** $c_1(\mu)$ is locally given by

$$c_1(\mu) = dd^c \log \rho.$$

The associated Griffiths function $G(\mu) = G(\rho)$ is the function defined by

$$G(\mu)\mu = c_1(\mu).$$

Classically, $-G$ is the Gauss curvature of the metric μ (see [La3] page 100).

Given a line sheaf $\mathcal{E}$ with metric ρ, let d denote the operator associated to the deRham complex of forms with values in $\mathcal{E}$ (see [GH] page 106 or [RS]). Let μ be a given positive $(1,1)$ form on X. Associated to this data we take our Laplacian $\Delta_{\mathcal{E},\rho,\mu}$ to be d-Laplacian

$$(1.4) \qquad\qquad \Delta_{\mathcal{E},d} = dd^* + d^*d.$$

Let $\bar{\partial}$ denote the operator associated to the Dolbeault complex. With this operator, one obtains a $\bar{\partial}$-Laplacian defined by $\Delta_{\mathcal{E},\bar{\partial}} = \bar{\partial}\bar{\partial}^* + \bar{\partial}^*\bar{\partial}$. Since all metrics we consider are Kähler, the Laplacians are related by

$$\Delta_{\mathcal{E},d} = 2\Delta_{\mathcal{E},\bar{\partial}}.$$

As an example, let X be a genus one curve realized as $\mathbf{C}$ modulo the lattice generated by 1 and τ. Let $b = \mathrm{Im}(\tau)$ and z denote the standard holomorphic coordinate on $\mathbf{C}$. Associated to the metric

$$\mu = \frac{i}{2b}dz \wedge d\bar{z}$$

is the Laplacian

$$\Delta_{\mathcal{O},\mu} = -4b\frac{\partial^2}{\partial z\partial\bar{z}},$$

which acts on smooth functions (sections of the trivial sheaf $\mathcal{O}$) on X.

Fix a positive $(1,1)$ form μ on X and a hermitian metric ρ on a holomorphic line sheaf $\mathcal{E}$ of rank r over X. Associated to this data is a **Laplacian** $\Delta_{\mathcal{E},\rho,\mu}$ (often written as $\Delta_{\mathcal{E}}$) that acts on the space of smooth sections of $\mathcal{E}$. The (unbounded) operator $\Delta_{\mathcal{E}}$ is positive and self-adjoint on the L^2 span of smooth sections of $\mathcal{E}$, and $\Delta_{\mathcal{E}}$ has a purely discrete spectrum. We shall denote the non-zero eigenvalues by

$$0 < \lambda_1(\mathcal{E}) \le \lambda_2(\mathcal{E})\ldots.$$

By Weyl's Law, we can define the **spectral zeta function** $\zeta_{\mathcal{E}}(s)$ for $\mathrm{Re}(s)$ sufficiently large by the series

$$\zeta_{\mathcal{E}}(s) = \sum_{j=1}^{\infty} \lambda_j^{-s}(\mathcal{E}).$$

The spectral zeta function can be written in the following form.

$$(1.5) \qquad \zeta_{\mathcal{E}}(s) = \frac{1}{\Gamma(s)} \int_0^{\infty} \left(\sum_{j=1}^{\infty} e^{-t\lambda_j(\mathcal{E})} \right) t^s \frac{dt}{t}.$$

The exponential sum in (1.5) is the trace of the heat kernel associated to the Laplacian $\Delta_{\mathcal{E}}$ minus the integer $h^0(X, \mathcal{E})$. Recall that $h^0(X, \mathcal{E})$ is equal to the number of zero eigenvalues of $\Delta_{\mathcal{E}}$.

By Seeley's theorem [Se] (i.e., small time asymptotics of the heat kernel), $\zeta_{\mathcal{E}}(s)$ has a meromorphic continuation to $\mathbf{C}$ which is holomorphic at $s = 0$. With this, the **analytic torsion**, also called the **determinant of the Laplacian**, is defined by

$$(1.6) \qquad \det{}^* \Delta_{\mathcal{E}} = \exp(-\zeta_{\mathcal{E}}'(0)),$$

where the asterisk reflects the fact that the zero eigenvalues have been omitted in the series (1.5). If $\Delta_{\mathcal{E}}$ has no zero eigenvalues, the asterisk in (1.6) is omitted. In that case, one simply writes $\det \Delta_{\mathcal{E}} = \exp(-\zeta_{\mathcal{E}}'(0))$.

Given a compact Riemann surface X with positive $(1,1)$ form μ, let us define the **structure constant** $c_{\Delta,\mu}(X)$ to be

$$(1.7) \qquad c_{\Delta,\mu}(X) = \log \left(\frac{\det{}^* \Delta_{\mathcal{O},\mu}}{\mathrm{vol}_{\mu}(X)} \right).$$

Basically, we have introduced $c_{\Delta,\mu}(X)$ for notational convenience.

§2. Faltings, Quillen and Theta Metrics.

Let X denote a compact Riemann surface of positive genus g. The **canonical metric** μ_{ca} on X is defined as follows. Let $\omega_1, \ldots, \omega_g$ denote an orthonormal basis of $H^0(X, \mathcal{K})$ relative to the inner product

$$(2.1) \qquad\qquad <\eta, \omega> = \frac{i}{2} \int_X \eta \wedge \bar{\omega}.$$

The canonical metric is the metric associated to the positive $(1,1)$ form

$$\mu_{ca} = \frac{i}{2g} \sum_{j=1}^{g} \omega_j \wedge \bar{\omega}_j.$$

An **Arakelov (canonically admissible)** metric ρ on a line sheaf $\mathcal{L}$ is a metric for which

$$c_1(\rho) = (\deg \mathcal{L}) \, \mu_{ca},$$

where $c_1(\rho)$ is the first Chern form of the metric ρ (see [A], or [La2] chapter 1). For example, if $\mathcal{L} = \mathcal{O}(P)$ and $1\!\!\!/_P$ is a non-zero element of $H^0(X, \mathcal{O}(P))$, then an Arakelov metric is given by

$$(2.2) \qquad\qquad \|1\!\!\!/_P(Q)\|^2 = G(P, Q),$$

where $G(P, Q)$ is the Green's function relative to the canonical metric (see [Fa], [J2], or [La2]). An Arakelov metric on a degree zero line sheaf is, by definition, a metric ρ for which $c_1(\rho) = 0$. In other language, such a metric is called a **flat metric** on $\mathcal{L}$.

We shall assume that all line sheaves $\mathcal{L}$ on X are given Arakelov metrics.

Given a complex vector space V, let $\det V$ denote the top exterior power of V. This is a one complex dimensional vector space. The

determinant line $\det H(\mathcal{E})$ associated to the vector sheaf $\mathcal{E}$ is defined as the one complex dimensional vector space

$$\det H(\mathcal{E}) = \det H^0(X, \mathcal{E}) \otimes (\det H^1(X, \mathcal{E}))^{-1}.$$

In Faltings's work [Fa], it was necessary to define metrics on the determinant lines that are compatible with cohomology operations. Specifically, we mean the following. If

$$\mathcal{L}' \cong \mathcal{L} \otimes \mathcal{O}(P),$$

then the long exact sequence in cohomology induces the following isomorphism of determinant lines.

(2.3) $$\det H(\mathcal{L}') \cong \det H(\mathcal{L}) \otimes \mathcal{L}'[P],$$

where $\mathcal{L}'[P]$ denotes the fibre of $\mathcal{L}'$ over the point P.

In [Fa], Faltings presented a construction of metrics on determinant lines using the Riemann theta function (1.1) and the canonical Green's function (2.2). Faltings's theorem, as stated in [Fa], is the following result.

Proposition 2.1. *For every holomorphic line sheaf $\mathcal{L}$ on X with Arakelov metric, there exists a hermitian metric $H_{F,\mathcal{L}}$ on the determinant line $\det H(\mathcal{L})$ such that the following properties hold.*

(i) An isomorphism of line sheaves induces an isometry of determinant lines.

(ii) If the metric on $\mathcal{L}$ is scaled by a constant $\alpha > 0$, then the metric on $\det H(\mathcal{L})$ is scaled by $\alpha^{\chi(\mathcal{L})}$, where

$$\chi(\mathcal{L}) = \deg \mathcal{L} + 1 - g.$$

(iii) The isomorphism (2.3) is an isometry, when $\mathcal{O}(P)$ is given the Arakelov metric as defined by (2.2).

Furthermore, the metrics on the determinant lines $\det H(\mathcal{L})$ *are unique up to a common scale factor.*

For a proof of Proposition 2.1, see [La2], chapter VI or [Fa], section 3.

The Faltings metrics $H_{F,\mathcal{L}}$ on the determinant lines $\det H(\mathcal{L})$ are normalized as follows. For any $Q \in X$, the complex vector spaces $H^0(X, \mathcal{K})$ and $H^0(X, \mathcal{K} \otimes \mathcal{O}(Q))$ are equal; hence, by Proposition 2.1, the corresponding determinant lines are isometric. Let

$$\phi : H^0(X, \mathcal{K}) \longrightarrow H^0(X, \mathcal{K} \otimes \mathcal{O}(Q))$$

denote such an isomorphism. Let $\mathcal{L} = \mathcal{K} \otimes \mathcal{O}(Q)$. For any basis $\{\eta_j\}_1^g$ of $H^0(X, \mathcal{K})$, the Faltings metric $H_{F,\mathcal{L}}$ is normalized by setting

$$(2.4) \qquad H_{F,\mathcal{L}}(\phi(\eta_1) \wedge \cdots \wedge \phi(\eta_g)) = \det(< \eta_i, \eta_j >),$$

where the inner product in (2.4) is the natural inner product on $H^0(X, \mathcal{K})$ defined by integration (2.1).

The sheaf $\mathcal{O}(-\Theta)$ on $J(X)$ possesses a natural metric arising from the Riemann theta function. This metric is defined as follows. Let $1\!\!\!/\!\!\Theta$ denote the holomorphic section of $\mathcal{O}(-\Theta)$ determined by the Riemann theta function (1.1). The **theta metric** H_Θ on $\mathcal{O}(-\Theta)$ is defined by

$$(2.5) \qquad H_\Theta(1\!\!\!/\!\!\Theta)(z) = ||\theta||^2(z, \Omega).$$

The Abel-Jacobi map $\psi : X \to J(X)$ induces an isomorphism between sections of holomorphic one-forms. Therefore, there exists a basis $\omega_1^J, \ldots, \omega_g^J$ of holomorphic differentials on $J(X)$ such that $\psi^* \omega_1^J, \ldots, \psi^* \omega_g^J$ is an orthonormal basis of $H^0(X, \mathcal{K})$ on X. Define the canonical $(1,1)$ form on $J(X)$ to be

$$\mu_J = \frac{i}{2g} \sum_{j=1}^g \omega_j^J \wedge \bar{\omega}_j^J.$$

By definition, $\psi^* \mu_J = \mu_{ca}$ (see [La2], page 143). In [La1] and [La2], the following theorem is proved.

Proposition 2.2. *Let Θ be the theta divisor on $J(X)$. There exists a metric H on $\mathcal{O}(-\Theta)$ such that*

$$c_1(H) = -g\mu_J.$$

Such a metric is unique up to a constant factor.

The existence of the metric claimed in Proposition 2.2 is proved in Chapter 13 of [La1]. In fact, it is shown that for the theta metric H_Θ we have

$$c_1(H_\Theta) = -g\mu_J.$$

Uniqueness of the metric, up to a constant factor, is proved in Chapter 1 of [La2].

Using the isomorphism (1.3) and the isometry (2.3), Faltings [Fa] shows that the Faltings metric induces a metric on $\mathcal{O}(-\Theta)$ over $J(X)$. Indeed, in the proof of Proposition 2.1, Faltings proves that

$$c_1(H_F) = -g\mu_J.$$

Therefore, for any line sheaf $\mathcal{L}$, the metrics $H_{F,\mathcal{L}}$ and $H_{\Theta,\mathcal{L}}$ differ by a constant that depends solely on the algebraic curve X. Let

$$h_F = \log H_F \quad \text{and} \quad h_\Theta = \log H_\Theta.$$

As stated in [La2], page 144, the **Faltings delta function** $\delta(X)$ can be realized by the following equation.

$$h_{F,\mathcal{L}} = \frac{1}{4}\delta(X) + h_{\Theta,\mathcal{L}}$$

for any determinant line $\det H(\mathcal{L})$ (also see [Fa]). In other words, the Faltings metric and the Theta metric are constant multiples of one another, and this constant defines the Faltings's delta function.

In [Q], Quillen defined a metric on determinant lines using analytic torsion (1.6). The (square of the) **Quillen norm** on $\det H(\mathcal{E})$ is defined by

$$(2.6) \qquad H_{Q,\mathcal{E}} = H_{L^2,\mathcal{E}} \cdot (\det {}^*\mathbf{\Delta}_{\mathcal{E}})^{-1}$$

where

$$(2.7) \qquad H_{L^2}(\Upsilon) = \det(<\eta_i, \eta_j>)(\det(<\omega_i, \omega_j>))^{-1}$$

with Υ, $\{\eta_i\}$ and $\{\omega_i\}$ as defined above. Let us denote the logarithm of the Quillen metric by

$$h_{Q,\mathcal{E}} = \log H_{Q,\mathcal{E}}.$$

The natural question is to understand the relation between the Faltings metric and the Quillen metric. The following result, due to Quillen [Q], answers that question.

Proposition 2.3. *The metric* H_Q *induces a metric on* $\mathcal{O}(-\Theta)$ *over* $J(X)$, *and*

$$c_1(H_Q) = -g\mu_J.$$

Aside from Quillen's paper, proofs of Proposition 2.3 can be found in [ABMNV], [AMV], [F1] and [Sm].

Combining Propositions 2.2 and 2.3, one concludes that the Faltings metric and the theta metric differ by a constant which depends solely on the Riemann surface X. To evaluate this constant it suffices to evaluate the relation for a single line sheaf. Recall that the Faltings metric was normalized by equation (2.4). By comparing (2.4) and (2.7), one obtains that

$$h_F = c_{\mathbf{\Delta},\mu_{\mathrm{Ar}}}(X) + h_Q,$$

where μ_{Ar} denotes the Arakelov metric on X.

§3. Analytic Torsion on Flat Line Sheaves

The contents of Proposition 2.3 can be read as follows. The Quillen metric for degree zero line sheaves induces a metric on the sheaf $\mathcal{O}(-\Theta)$ over $J(X)$. The computation of the first Chern form shows that analytic torsion satisfies as reasonably simple differential equation. The following theorem from [J1] presents an explicit solution of the differential equation.

Theorem 3.1. *Let $\mathcal{L}$ denote a non-trivial degree zero line sheaf on X. Let $P = P_g$ denote a fixed point on X and let $P_1, \ldots, P_{g-1}$ denote generic, distinct points on X. Let $\eta_{\mathcal{L}}^1, \ldots, \eta_{\mathcal{L}}^{g-1}$ denote a basis of $H^0(X, \mathcal{K} \otimes \mathcal{L})$. Let*

$$\mathcal{N} = \mathcal{S} \otimes \mathcal{O}(-P_1 - \cdots - P_{g-1})$$

and $\mathcal{M} = \mathcal{N} \otimes \mathcal{L}$. Then for any metric μ on X, we have the following equation.

$$\log\left(\frac{\det \mathbf{\Delta}_{\mathcal{L},\mu}}{\det(< \eta_{\mathcal{L}}^i, \eta_{\mathcal{L}}^j >)}\right) = 2\log 2\pi + c_{\mathbf{\Delta},\mu}(X) + \log \|\theta\|^2([\mathcal{M}], \Omega)$$

$$- \log\left(\left|\frac{\det((\omega_i)(P_j))}{\det((\eta_{\mathcal{L}}^i)(P_j))} \sum_{j=1}^{g} \frac{\partial \|\theta\|}{\partial z_j}([\mathcal{N}], \Omega)\zeta_j(P_g)\right|^2\right).$$

If we let all points $P_1, \cdots, P_{g-1}$ coalesce to $P = P_g$ on X, Theorem 3.1 yields the following result.

Corollary 3.2. *Let us assume the notation as in Theorem 3.1 except set*

$$\mathcal{N} = \mathcal{S} \otimes \mathcal{O}((1-g)P).$$

Let I denote a g-tuple of integers. Then, in multi-index notation, *we have the following equation.*

$$\log\left(\frac{\det\mathbf{\Delta}_{\mathcal{L},\mu}}{\det(<\eta_{\mathcal{L}}^i,\eta_{\mathcal{L}}^j>)}\right) = 2\log 2\pi + c_{\mathbf{\Delta},\mu}(X) + \log||\theta||^2([\mathcal{M}],\Omega)$$

$$- \log\left(\left|\frac{\mathbf{W}((\omega_i)(P))}{\mathbf{W}((\eta_{\mathcal{L}}^i)(P))}\sum_{|I|=g}\frac{\partial^g||\theta||}{\partial z^I}([\mathcal{N}],\Omega)\zeta^I(P)\right|^2\right),$$

where $\mathbf{W}$ *denotes a Wronskian of the appropiate space of forms.*

Theorem 3.1 completely expresses analytic torsion using theta functions, holomorphic data from X, and the structure constant $c_{\mathbf{\Delta},\mu}(X)$. From Theorem 3.1 it is immediate that if $\mathcal{L}$ and $\mathcal{L}'$ are degree zero line sheaves, then the quotient

$$\frac{\det\mathbf{\Delta}_{\mathcal{L},\mu}}{\det\mathbf{\Delta}_{\mathcal{L}',\mu}}$$

does not depend on the metric μ (see [RS], Theorem 2.1). Theorem 3.1 does not contain any undetermined factors or constants. This is an improvement over similar results stated in [ABMNV], [F1] and [Sm]. Unfortunately, note that in order to evaluate any of these expressions, one must choose a point P on X. Thus, we do not see a way to extend the analytic torsion or the structure constant from a function on the moduli space of algebraic curves to all of Siegel upper half space.

Order the variables which parameterize $J(X)$ by $z_1,\ldots,z_g,\bar{z}_1,\ldots,\bar{z}_g$. In the proof of Theorem 3.1, it is shown that $\det\mathbf{\Delta}_{\mathcal{L},\mu}$ has a second order zero as $[\mathcal{L}]$ approaches the origin in $J(X)$. Relative to the given ordering, it is shown ([J1], section 4) that the Hessian of $\det\mathbf{\Delta}_{\mathcal{L},\mu}$ near the origin is

$$(3.1)\qquad\qquad (2\pi)^2 C_{\mathbf{\Delta},\mu}(X)\begin{pmatrix} 0 & M_X \\ M_X & 0 \end{pmatrix},$$

where $\log C_{\Delta,\mu}(X) = c_{\Delta,\mu}(X)$. Equation (3.1) comes from studying the behavior of the smallest eigenvalue $\lambda_1(\mathcal{L})$ as $[\mathcal{L}]$ approaches zero in $J(X)$ (see section 4 of [J1] or [PS]). Let us now study the asymptotic behavior of the equation in Theorem 3.1 for as $[\mathcal{L}]$ approaches the origin along a complex line $\mathcal{J}$ in $J(X)$.

By the Riemann-Roch theorem ([FK], page 127), $\dim H^0(X, \mathcal{K} \otimes \mathcal{L}) = g - 1$ if $[\mathcal{L}] \neq 0$ and $\dim H^0(X, \mathcal{K}) = g$. In light of (3.1), we are led to the following problem. Let the degree zero line sheaf $\mathcal{L}$ approach the origin in $J(X)$ along a complex line $\mathcal{J}$. The vector space $H^0(X, \mathcal{K} \otimes \mathcal{L})$ will approach a codimension one subspace of $H^0(X, \mathcal{K})$. The problem is to identify this subspace of $H^0(X, \mathcal{K})$ as a function of the direction by which $[\mathcal{L}]$ approaches zero in $J(X)$. Because of the nature of the Abel-Jacobi map, it is natural to expect an answer in terms of the canonical basis $\zeta_1, \ldots, \zeta_g$ of $H^0(X, \mathcal{K})$, as defined in section 1. In [J3], the following lemma is proved.

Lemma 3.3. *Let d denote a non-zero point in $\mathbf{C}^g$, and let $\mathcal{J}$ denote the complex line through the origin in $J(X)$ determined by d. Let t denote a complex number of small modulus so that $\mathcal{J}$ is parameterized near the origin by $\{td \,|\, |t| < \epsilon\}$ for some small epsilon. Let $\mathcal{L}_{td}$ denote a degree zero sheaf for which $[\mathcal{L}_{td}]$ lies on $\mathcal{J}$. Then*

$$\lim_{t \to 0} H^0(X, \mathcal{K} \otimes \mathcal{L}_{td}) = \{\Sigma a_j \zeta_j | \Sigma a_j d_j = 0\}.$$

Lemma 3.3 identifies the codimension one subspace of $H^0(X, \mathcal{K})$ obtained by taking the limit as t approaches zero of $H^0(X, K \otimes \mathcal{L}_{td})$ for fixed nonzero $d \in \mathbf{C}^g$. Note that one can identify the space of codimension one subspaces of $\mathbf{C}^g$ with $\mathbf{P}^{g-1}$. Lemma 3.3 gives an explicit bijection between this $\mathbf{P}^{g-1}$ and the space of directions d from 0 in $J(X)$, which is also a $\mathbf{P}^{g-1}$.

Combining Theorem 3.1, Lemma 3.3 and (3.1), the following result is proved.

Theorem 3.4. *Let d^1, d^2 denote non-zero points in $\mathbf{C}^g$. Let $P_1, \ldots, P_{g-1}$ denote generic points on X and let*

$$\mathcal{L} = \mathcal{S} \otimes \mathcal{O}(-P_1 - \cdots - P_{g-1}).$$

Then the following identity holds.

$$\frac{\sum_1^g \frac{\partial \theta}{\partial z_j}([\mathcal{L}])d_j^1}{\sum_1^g \frac{\partial \theta}{\partial z_j}([\mathcal{L}])d_j^2} = \frac{\det\left[d^1 \,|\,{}^t\zeta(P_1) \ldots |\, {}^t\zeta(P_{g-1})\right]}{\det\left[d^2 \,|\,{}^t\zeta(P_1) \ldots |\, {}^t\zeta(P_{g-1})\right]}.$$

The matrices above have been expressed by indicating each of the g columns.

Theorem 3.4 and the following corollaries are proved in [J3]. In the case when d lies along a coordinate axis in $\mathbf{C}^g$, similar results are presented in [Fr]. In that work, Farkas uses a proof similar to the classical proof of the Riemann vanishing theorem, as given in [FK].

In Theorem 3.4, if we let points $P_1 \ldots P_{g-1}$ coalesce to a point P we obtain the following interesting corollary.

Corollary 3.5. *With the notation as above,*

$$\frac{\sum_1^g \frac{\partial \theta}{\partial z_j}([\mathcal{S} \otimes \mathcal{O}((1-g)P)])d_j^1}{\sum_1^g \frac{\partial \theta}{\partial z_j}([\mathcal{S} \otimes \mathcal{O}((1-g)P)])d_j^2} = \frac{\det[d^1 \,|\, {}^t\zeta(P) \ldots |\, {}^t\zeta^{(g-2)}(P)]}{\det[d^2 \,|\, {}^t\zeta(P) \ldots |\, {}^t\zeta^{(g-2)}(P)]}$$

where $\zeta^{(h)}$ denotes the column of h derivatives of the holomorphic one-forms $\zeta_1 \ldots \zeta_g$.

Corollary 3.5 follows immediately from Theorem 3.4 except for the fact that the left-hand-side involves the *first* derivatives of the theta function. For this, note that if points P_j coalesce to a single point, (3.1) implies that during this process the term

$$(3.2) \qquad \sum_1^g \frac{\partial \theta}{\partial z_j}([\mathcal{S} \otimes \mathcal{O}(-P_1 - \cdots - P_{g-1}) \otimes \mathcal{L}])\zeta_j(P)$$

vanishes to order g. Hence, if $d^1 = \zeta(P)$ and d^2 is fixed, this vanishing in (3.1) implies

$$(3.3) \qquad \sum \frac{\partial \theta}{\partial z_j}([\mathcal{S} \otimes \mathcal{O}((1-g)P)])d_j^2$$

does not vanish identically in P.

Corollary 3.6. *Let $d = (d_j)$ denote a fixed non-zero point in $\mathbf{C}^g$. The zeroes of*

$$\sum_1^g \frac{\partial \theta}{\partial z_j}([\mathcal{S} \otimes \mathcal{O}((1-g)P)])d_j$$

are the Weierstrass points of the codimension one subspace of holomorphic one-forms given by $\{\sum a_j \zeta_j \mid \sum a_j d_j = 0\}$, with corresponding multiplicities.

If we take $d^1 = \zeta(P)$ in Theorem 3.4 and let the points coalesce we obtain

Corollary 3.7. *The zeroes of*

$$\sum_{|I|=g} \frac{\partial^g \theta}{\partial z^I}([\mathcal{S} \otimes \mathcal{O}((1-g)P)])\zeta^I(P)$$

are the Weierstrass points of X, with corresponding multiplicities.

In summary, Theorem 3.1 provides a useful tool for studying Weierstrass points and codimension one Weierstrass points on surfaces of genus $g \geq 2$. On surfaces of genus 1, one can use Theorem 3.1 to reprove know results such as the Jacobi derivative and Kronecker's limit formula. These proofs are given in section 7 of [J1].

§4. **Limits of Weierstrass Points on Degenerating Curves**

By a **degenerating family of Riemann surfaces** we mean a
holomorphic map of the unit disc $\mathbf{D}$ into the stably compactified
moduli space $\overline{\mathcal{M}}_g$ such that the restriction of this map to the punctured disc $\mathbf{D} \setminus \{0\}$ is a holomorphic map into the moduli space $\mathcal{M}_g$.
The fibre over the origin in $\mathbf{D}$ is a noded algebraic curve. For a degenerating family of Riemann surfaces, we define a **holomorphic**
family of principal polarizations to mean the following. The
matrices associated to the hermitian forms vary real analytically
over $\mathbf{D}$, and the limiting matrix corresponding to the origin gives a
principal polarization of the normalization of the noded curve. Rational components of the normalization contribute blocks of zeroes
to the diagonal of the limiting matrix.

Let us give a 'hands-on' construction of a degenerating family
of Riemann surfaces. This will show that a holomorphic family of
principal polarizations is possible. Although we will describe the
pinching to a single node, it is clear that one can easily describe
a family so that the limit curve has many nodes. This material
follows chapter 3 of [F2].

Let X_1 and X_2 denote compact non-singular Riemann surfaces
of positive genera g_1 and g_2, respectively, each with a distinguished
point. Denote the distinguished points on X_i by P_i. Choose a local
coordinate neighborhood U_i of P_i with local uniformizing parame-
ter $z_i : U_i \to \mathbf{D}$ with $\mathbf{D}$ denoting the unit disc in $\mathbf{C}$. Let

$$W_i^t = \{x_i \in X_i \mid x_i \in X_i \setminus U_i \text{ or } |z_i(x_i)| > |t|\}$$

and

$$\mathcal{C}_t = \{(X, Y) \in \mathbf{D} \times \mathbf{D} \mid w_1 \cdot w_2 = t\}$$

be a "complex hyperbola", where t is a complex number with $|t|$
small. We form a Riemann surface of genus $g = g_1 + g_2$ by sewing

together W_1^t and W_2^t with the collar $\mathcal{C}_t$ as follows:

$$x_1 \in W_1^t \cap U_1 \text{ is identified with } (z_1(x_1), \frac{t}{z_1(x_1)}) \in \mathcal{C}_t$$

and

$$x_2 \in W_2^t \cap U_2 \text{ is identified with } (\frac{t}{z_2(x_2)}, z_2(x_2)) \in \mathcal{C}_t.$$

Holomorphic coordinates on $\mathcal{C}_t$ are $x = \frac{1}{2}(w_1 + w_2)$ and $y = \frac{1}{2}(w_1 - w_2)$, so the topological space $W_1^t \cup W_2^t \cup \mathcal{C}_t$ with the above identification and complex structure is a compact non-singular Riemann surface X_t of genus $g = g_1 + g_2$. The fibre over the origin is a reducible uninodal curve X_0 which has two components (of genera g_1 and g_2) meeting at a double point.

Choose a canonical homology basis of X_i such that none of the representative curves pass through the neighborhood U_i. We shall take the union of the bases of $H_1(X_1, \mathbf{Z})$ and $H_1(X_2, \mathbf{Z})$ to form a canonical basis of $H_1(X_t, \mathbf{Z})$. Let $\zeta_1(z; X_1), \ldots, \zeta_{g_1}(z; X_1)$ (resp. $\zeta_1(z; X_2), \ldots, \zeta_{g_2}(z; X_2)$) denote a canonical basis of $H^0(X_1, \mathcal{K})$ (resp. $H^0(X_2, \mathcal{K})$). Let $\omega_k^j(z_j; P_j)$ denote a meromorphic one-form on X_j with zero A-periods and local expansion

$$\omega_k^j(z_j; P_j) = \frac{dz_j}{z_j^k} + \text{ terms regular in } z_j$$

in the local coordinate z_j. With the above choices, the differential $\omega_k^j(z_j; P_j)$ is uniquely determined.

With this setup, Fay presents first order asymptotics of the dual basis of holomorphic abelian differentials $\zeta_i(\cdot, X_t)$ on X_t. In [J2], this is extended to give full asymptotics for $\zeta_i(z, X_t)$ as a power series in t uniformly for $z \in X_t$ away from $\mathcal{C}_t$ (see also [Y]). The result is the following.

Proposition 4.1. *Let X_t denote a degenerating family of compact Riemann surfaces such that X_0 is reducible and uninodal. Let $\mathbf{H}_{X_t}$ denote a holomorphic family of principal polarizations with associated matrices M_{X_t} and period matrices Ω_t. The holomorphic differentials $\zeta_i(\cdot, X_t)$ have the following expansion away from the developing node. If $j \leq g_1$, then*

$$\zeta_j(z, X_t) = \left\{ \begin{array}{ll} \zeta_j(z, X_1) & + \sum_{n=1}^{\infty} \eta_{j,n}^1(z; P_1) t^n \ ; \ z \in X_1 \setminus U_1 \\ 0 & + \sum_{n=1}^{\infty} \eta_{j,n}^2(z; P_1) t^n \ ; \ z \in X_1 \setminus U_1 \end{array} \right\}$$

where the differential $\eta_{j,n}^1(z; P_1)$ (resp. $\eta_{j,n}^2(z; P_1)$) is a linear combination of the meromorphic differentials $\omega_2^1(z; P_1), \ldots, \omega_{n+1}^1(z; P_1)$ (resp. $\omega_2^2(z; P_1), \ldots, \omega_{n+1}^2(z; P_1)$) with coefficients that depend solely on the first $n-1$ derivatives of $\zeta_j(z; X_1)$ evaluated at P_1 in the local coordinate z_1. Similar results hold for $j > g_1$.

The proof of Proposition 4.1 is given in section 3 of [J2] (see also [Y]). Our proof builds on the results in chapter 3 of [F2]. We shall now use Proposition 4.1 to study the limits of Weierstrass points for a degenerating family of Riemann surfaces whose limit is a uninodal reducible curve with two components of genera g_1 and g_2.

Theorem 4.2. *Let $\eta_1^1, \ldots, \eta_g^1$ denote a basis of $H^0(X_1, \mathcal{K} \otimes \mathcal{O}((g_2+1)P_1))$. Then as the parameter t approaches zero,*

$$\mathrm{div}_{(X_1 \setminus U_1)^g}[\det(\zeta_j(R_i; X_t))] \to \mathrm{div}_{(X_1 \setminus U_1)^g}[\det(\eta_j^1(R_i))].$$

Let $\eta_1^2, \ldots, \eta_g^2$ denote a basis of $H^0(X_2, \mathcal{K} \otimes \mathcal{O}((g_1+1)P_2))$. Then as the parameter t approaches zero,

$$\mathrm{div}_{(X_2 \setminus U_2)^g}[\det(\zeta_j(R_i; X_t))] \to \mathrm{div}_{(X_2 \setminus U_2)^g}[\det(\eta_j^2(R_i))].$$

The proof of Theorem 4.2 follows directly from Proposition 4.1.

In section 1 we defined a Weierstrass point on a compact Riemann surface X of genus g. Equivalently, one can define a Weierstrass point to be a zero of the Wronskian $\mathbf{W}(\zeta_j(R))$ where $\{\zeta_j\}$ denotes a basis of $H^0(X, \mathcal{K})$ (see [FK], page 84). The multiplicity of the zero is the weight of the Weierstrass point. Since $\mathbf{W}(\zeta_j(R))$ is a holomorphic form of weight $\frac{1}{2}g(g+1)$, the number of zeroes, counting multiplicity, equals $g(g^2 - 1)$. Theorem 4.2 allows us to study the asymptotics of the Weierstrass points on a degenerating family of Riemann surfaces X_t such that X_0 is uninodal and reducible. As a corollary of Theorem 4.2, we have the following result.

Corollary 4.3. *With notation as in Theorem 4.2,*

$$\text{div}_{(X_1 \setminus U_1)}[\mathbf{W}(\zeta_j(R; X_t))] \to \text{div}_{(X_1 \setminus U_1)}[\mathbf{W}(\eta_j^1(R))],$$

and

$$\text{div}_{(X_2 \setminus U_2)}[\mathbf{W}(\zeta_j(R; X_t))] \to \text{div}_{(X_2 \setminus U_2)}[\mathbf{W}(\eta_j^2(R))].$$

From Corollary 4.3, one can obtain finer information concerning the limits of Weierstrass points. The Wronskian $\mathbf{W}(\eta_j^1(R))$ is a form of weight $\frac{1}{2}g(g+1)$ on X_1. Hence its degree is

$$\frac{1}{2}g(g+1) \cdot (2g_1 - 2) = g(g+1)(g_1 - 1).$$

The Wronskian has a pole at P_1 which, for generic P_1, is of order

$$(g_1 + 2) + \cdots + (g_1 + 2g_2) = g_2(g+1).$$

Therefore, the Wronskian $\mathbf{W}(\eta_j^1(R))$ has Z zeroes on X_1, where

$$\begin{aligned} Z &= g(g+1)(g_1 - 1) + g_2(g+1) \\ &= g_1(g^2 - 1). \end{aligned}$$

These zeroes are the limits of the Weierstrass points from X_t. Note that Proposition 4.1 also implies that the limits of the Weierstrass points do not depend on t. Similarly, if P_2 is a pole of $\mathbf{W}(\eta_j^2(R))$ of order $g_1(g+1)$, then precisely $g_2(g^2-1)$ points on X_2 are limits of Weierstrass points from X_t.

By the Riemann-Roch Theorem, the points P on X_1 (resp. X_2) that are limits of Weierstrass points from X_t are the points for which $gP - (g_2+1)P_1$ (resp. $gP - (g_1+1)P_2$) is linearly equivalent to an effective divisor of degree $g_1 - 1$ (resp. $g_2 - 1$). In the special case when $g_1 = 1$, we can take P_1 to the be origin of the elliptic curve X_1. The above results state that the limits of the Weierstrass points of X_t on X_1 are precisely the g-torsion points on X_1.

As in the above discussion, one can describe a degenerating family of curves whose limiting fibre is an irreducible uninodal curve. The construction is as follows. Let X denote a compact Riemann surface of positive genus g with distinguished points P_1 and P_2. Choose local coordinates $z_i : U_i \to D$ in disjoint neighborhoods of U_i of P_i. Define $\mathcal{C}_t$ as above and

$$W_t = \{P \in X \mid P \in X \setminus (U_1 \cup U_2) \text{ or } P \in U_i \text{ with } |z_i(P)| > |t|\}$$

for any non-zero $t \in \mathbf{D}$. Form a Riemann surface of genus $g+1$ by sewing together W_t and U_t with the following identification:

$$P \in W_t \cap U_1 \text{ is identified with } (z_1(P), \frac{t}{z_1(P)}) \in \mathcal{C}_t$$

and

$$P \in W_t \cap U_2 \text{ is identified with } (\frac{t}{z_2(P)}, z_2(P)) \in \mathcal{C}_t.$$

Holomorphic coordinates on $\mathcal{C}_t$ are $x = \frac{1}{2}(w_1 + w_2)$ and $y = \frac{1}{2}(w_1 - w_2)$. At $t = 0$, the surface X_0 is a singular curve with one component and one node.

A canonical homology basis for X_t is constructed as follows. Choose a basis of $H_1(X, \mathbf{Z})$ with curves $A_1, \ldots, B_g$ lying in $X \setminus (U_1 \cup U_2)$. Choose A_{g+1} to be the curve ∂U_1. The difficult part is to describe B_{g+1}. Fix a path γ from $z_1^{-1}(\frac{1}{2})$ to $z_2^{-1}(\frac{1}{2})$ lying entirely in $X \setminus (z_1^{-1}(D_{\frac{1}{2}}) \cup z_2^{-1}(D_{\frac{1}{2}}))$ but not intersecting any of the curves $A_1, \ldots, B_g$. Let $\gamma_{i,t}$ be a path in U_i from $z_i^{-1}(\frac{1}{2})$ to $z_i^{-1}(\sqrt{t})$ lying in $|t| < |z_i| < 1$. With properly chosen orientations, set $B_{g+1} = \gamma \cup \gamma_{1,t} \cup \gamma_{2,t}$. When $t = 0$, B_{g+1} is a path passing through the node. Thus we have described a well-defined basis of $H_1(X_t, \mathbf{Z})$ for all t in $\mathbf{D}$ minus some path from the origin to the boundary.

Fay's results in [F2] yield a complete asymptotic expansion for the dual basis $\zeta_i(x, X_t)$ of holomorphic abelian differentials on X_t.

Proposition 4.4. *Let σ denote a path in $\mathbf{D}$ from the origin to the boundary. For all t in $\mathbf{D} \setminus \sigma$, let X_t denote a degenerating family of compact Riemann surfaces such that X_0 is irreducible and uninodal. Let $\mathbf{H}_{X_t}$ denote a holomorphic family of principal polarizations with associated matrices M_{X_t} and period matrices Ω_t. The holomorphic differentials $\zeta_i(\cdot, X_t)$ have the following expansion: If $j \leq g$ then*

$$\zeta_j(z; X_t) = \zeta_j(z; X_0) + o(1).$$

Let $D = P_1 + P_2$ and let ω_D denote the meromorphic one form on X whose polar divisor is D and has residue -1 at P_1 and $+1$ at P_2. Then for z on $X_t \setminus (U_1 \cup U_2)$,

$$\zeta_{g+1}(z; X_t) = \omega_D(z) + o(1).$$

Proposition 4.4 is proved in [F2]. Analogous to Theorem 4.2, we have the following result.

Theorem 4.5. *Let* $\eta_1^1, \ldots, \eta_{g+1}^1$ *denote a basis of* $H^0(X, \mathcal{K} \otimes \mathcal{O}(P_1 + P_2))$. *Then as the parameter* t *approaches zero,*

$$\mathrm{div}_{X \backslash (U_1 \cup U_2)^{g+1}} [\det (\zeta_j (R_i; X_t))]$$
$$\rightarrow \mathrm{div}_{X \backslash (U_1 \cup U_2)^{g+1}} [\det (\eta_j^1 (R_i))].$$

The proof of Theorem 4.5 follows directly from Proposition 4.4. By letting the points $R_1, \ldots, R_{g+1}$ coalesce, we obtain the following result.

Corollary 4.6. *With notation as in Theorem 4.5,*

$$\mathrm{div}_{X \backslash (U_1 \cup U_2)} [\mathbf{W} (\zeta_j (R; X_t))] \rightarrow \mathrm{div}_{X \backslash (U_1 \cup U_2)} [\mathbf{W} (\eta_j^1 (R))].$$

Since X is a genus $g + 1$ uninodal irreducible curve, the Wronskian $\mathbf{W}(\eta_j^1 (R))$ is a form of weight $\frac{1}{2}(g + 1)(g + 2)$. Hence, its degree is

$$\frac{1}{2}(g + 1)(g + 2) \cdot (2g - 2) = (g + 2)(g^2 - 1).$$

The Wronskian $\mathbf{W}(\eta_j^1 (R))$ has poles at P_1 and P_2 of order $g + 1$. Therefore, $\mathbf{W}(\eta_j^1 (R))$ has Z zeroes on X, where

$$Z = (g + 2)(g^2 - 1) + (2g + 2)$$
$$= g(g + 1)^2.$$

These zeroes are the limits of the Weierstrass points from X_t. By the Riemann-Roch theorem, these are precisely the points P on X such that $(g + 1)P - D$ is linearly equivalent to an effective divisor of degree $g - 1$.

The results presented above are discussed in [Dz], pages 60-66 and [EH], pages 498 and 507. Propositions 4.1 and 4.4 can be generalized to a degenerating family of Riemann surfaces whose limit surfaces is a noded curve with an arbitrary number of nodes. These results can be applied to study limits of Weierstrass points for a general degenerating family of Riemann surfaces. One could also study limits of codimension one Weierstrass points which could include using judicious choices of the subspaces of holomorphic one forms, possibly allowing for dependence on the t-parameters. Results in these directions will be presented elsewhere.

BIBLIOGRAPHY

[ABMNV] ALVAREZ-GAUME, L., BOST J.-B., MOORE, G., NELSON, P., VAFA, C. : Bosonization on higher genus Riemann surfaces. Commun. Math. Phys. 112, 503-552 (1987).

[AMV] ALVAREZ-GAUME, L., MOORE, G., VAFA, C.: VAFA, C. : Theta functions, modular invariance, and strings. Commun. Math. Phys. 106, 1-40 (1986).

[A] ARAKELOV, S.J.: An intersection theory for divisors on an arithmetic surface. Izv. Akad. Nauk. 35, 1269-1293 (1971).

[DM] DELIGNE, P., and MUMFORD, D.: The Irreducibility of the Space of Curves of Given Genus. Publ. I.H.E.S. 36, 75-110 (1969).

[Dz] DIAZ, S.: Exceptional Weierstrass points and the divisor on moduli space that they define. Memoirs of the AMS 327, Providence: AMS 1985.

[EH] EISENBUD, D., AND HARRIS, J.: Existence, decomposition and limits of certain Weierstrass points. Invent. Math. 87, 495-515 (1987).

[Fa] FALTINGS, G.: Calculus on Arithmetic surfaces. Ann. of Math. 119, 387-424 (1984).

[Fr] FARKAS, H.: Identities on compact Riemann surfaces. Preprint 1989.

[FK] FARKAS, H., and KRA, I.: Riemann Surfaces, volume 71

of Graduate Texts in Mathematics, New York: Springer-Verlag, 1980.

[F1] FAY, J.: Perturbation of Analytic Torsion on Riemann Surfaces. Preprint (1989).

[F2] FAY, J.: Theta functions on Riemann surfaces. LNM 352, New York: Springer-Verlag 1973.

[F3] FAY, J.: On the even-order vanishing of Jacobian theta functions. Duke Mathematical Journal 51, 109-132, (1984).

[GH] GRIFFITHS, P., and HARRIS, J.: Principles of Algebraic Geometry. New York: John Wiley and Sons (1978).

[Ha] HARTSHORNE, R.: Algebraic Geometry, volume 52 of Graduate Texts in Mathematics, New York: Springer-Verlag, 1977.

[J1] JORGENSON, J.: Analytic Torsion for line bundles on Riemann surfaces. Duke Mathematical Journal 62, 527-549 (1991).

[J2] JORGENSON, J.: Asymptotic behavior of Faltings's delta function. Duke Mathematical Journal 61, 221-254 (1990).

[J3] JORGENSON, J.: On the directional derivatives of the theta function along its divisor. To appear in Israel Journal of Math.

[La1] LANG, S.: Fundamentals of Diophantine Geometry, New York: Springer-Verlag (1983).

[La2] LANG, S.: Introduction to Arakelov Theory. New York: Springer-Verlag (1988).

[La3] LANG, S.: Introduction to Complex Hyperbolic Spaces. New York: Springer-Verlag (1987).

[La4] LANG, S.: Introduction to Algebraic and Abelian functions, second edition, volume 89 of Graduate Texts in Mathematics, New York: Springer-Verlag, 1982.

[M1] MUMFORD, D.: Tata Lectures on Theta I, volume 28 of

Progress in Mathematics, Boston: Birkhauser (1983).

[M2] MUMFORD, D.: Tata Lectures on Theta II, volume 43 of Progress in Mathematics, Boston: Birkhauser (1984).

[PS] PHILLIPS, R., AND SARNAK, P.: Geodesics in Homology Classes. Duke Mathematical Journal 55, 287-297 (1987).

[Q] QUILLEN, D.: Determinants of Cauchy-Riemann operators over a Riemann surface. Func. Anal. Appl. 19, 31-34 (1986).

[RS] RAY, D., and SINGER, I.: Analytic torsion for complex manifolds. Ann. Math. 98, 154-177 (1973).

[Se] SEELEY, R.: Complex powers of an elliptic operator. Proc. Symp. Pure Math. 10, 288-307 (1967).

[Sm] SMIT, D.-J.: String theory and algebraic geometry on moduli spaces. Commun. Math. Phys. 114, 645-685 (1988).

[W] WENTWORTH, R. A.: The asymptotic behavior of the Arakelov Green's function and Faltings's invariant. Commun. Math. Phys. 137, 427-459 (1991).

[Y] YAMADA, A.: Precise variational formulas for Abelian differentials. Kodai Mathematical Journal 3, 114-143 (1980).

Department of Mathematics
Yale University
Box 2155 Yale Station
New Haven, CT 06520

Contemporary Mathematics
Volume **136**, 1992

A Problem of Narasimhan

GEORGE R. KEMPF

This is development of some ideas that I learned from M. S. Narasimhan a few years ago. The basic problem is to compute the Hermitian-Einstein metric on the Picard bundles of a compact Riemann surface.

Part of the discussion is valid over an algebraic closed field k of arbitrary characteristic. When appropriate I will specialize k to $\mathbf{C}$.

§1. The Picard bundles of algebraic curves

Let C be a complete smooth algebraic curve over k of genus g. Let J be the Jacobian of C with theta divisor θ. Then

$$\mathcal{P} = \mathcal{O}_{J \times J}\left((\pi_1 + \pi_2)^{-1}\theta - \pi_1^{-1}\theta - \pi_2^{-1}\theta\right)$$

is a Poincaré sheaf on $J \times J$ where we identify J with its dual abelian variety. As usual $\mathcal{Q} = \mathcal{P}|_{C \times J}$ is a Poincaré sheaf on $C \times J$ where we regard C as contained in J by $\int_o^* : C \to J$ where o is arbitrary point of C.

Let $\mathcal{L}$ be an invertible sheaf on C of degree d. Then $\pi_C^* \mathcal{L} \otimes \mathcal{Q}$ is a universal family of invertible sheaves of C of degree d parameterized by J. Assume that $d > 2g - 2$. Then $\pi_{J_*}(\pi_C^* \mathcal{L} \otimes \mathcal{Q}) \equiv \mathcal{W}_d$ is a locally free $\mathcal{O}_J$-module of rank $d - g + 1$ and $R^i \pi_{J_*}(\pi_C^* \mathcal{L} \otimes \mathcal{Q}) = 0$ if $i > 0$. Here $\mathcal{W}_d$ is called the d-th Picard bundle on J.

A new result which has been proven by Ein and Lazarsfeld [2] after partial results in [7] and [5].

Theorem 1. *The Picard bundles $\mathcal{W}_d$ are stable vector bundles on J with respect to the polarization θ.*

Now if $\mathbf{C} = k$ we can do the following. We can give J the Riemaniann metric; *i.e.* the Kähler metric whose Kähler form ω_J is the invariant differential 1-1 form in the class $c_1(\mathcal{O}_J(\theta))$. Recall that Riemann computed ω_J explicitly.

Partially supported by the National Science Foundation Grant DMS-8800583.

1991 *Mathematics Subject Classification.* Primary 14K07, 14H40, 30F99.

This paper is in final form and no version of it will be submitted for publication elsewhere.

A Hermitian metric h on a complex vector bundle $\mathcal{W}$ over a Kähler manifold (J,ω) is Hermitian-Einstein if the Kähler trace of its curvature is a fixed multiple of ω times the identity endomorphism of $\mathcal{W}$. Recall by the Donaldson-Ühlenbeck-Yau Theorem we have

Corollary 2. *There exists (unique up to constant multiple) Hermitian-Einstein metric on $\mathcal{W}_d$ over (J,ω_J).*

The basic problem of Narasimhan is to determine this metric in an independent way which we will discuss next. Naively the Corollary gives a (up to constant) Hermitian metric on $\Gamma(C,\mathcal{M})$ where $\mathcal{M}$ is an invertible sheaf of degree d on C. One may ask if it is possible to determine this metric for one fixed $\mathcal{M}$ without recourse to doing global differential equations on the space of all $\mathcal{M}$.

§2. One possible solution to the problem

Narasimhan has suggested that the Hermitian-Einstein metric on $\mathcal{W}_d$ is an appropriate multiple of the standard type L^2-metric on $\mathcal{W}_d$. I will consider this idea in detail.

C has a Kähler metric $\omega_C = \int^* \omega_J$. We will give $\pi_C^* \mathcal{L} \otimes \mathcal{Q}$ a product Hermitian metric coming from metric on $\mathcal{L}$ with curvature a scalar multiple of ω_C and a metric on $\mathcal{P}$ with invariant curvature. Then if $\mathcal{L}_j = \pi_C^* \mathcal{L} \otimes \mathcal{Q}|_j$ from any j in J, the L^2-metric on $\mathcal{W}_d|_j \cong \Gamma(C,\mathcal{L}_j)$ is just

$$\int_C <f,g> \omega_C = <f,g> .$$

This metric depends in a C^∞-manner on j. Thus it defines a Hermitian metric on $\mathcal{W}_d$.

Question. Is the Hermitian-Einstein metric on $\mathcal{W}_d$ have the form $\mu(j) <,>$ *where $\mu(j)$ is a positive C^∞-function on J?*

We can easily determine μ if the answer is positive. Recall that $\lambda = \det \mathcal{W}_d$ has the induced L_2-metric. There is a Quillen multiplier $\alpha(j)s.t.$ curvature of $\alpha \cdot <>_{L'}$ is given by the formula of Bismut-Gillet-Soulé [1]. By the formula this curvature is a multiple of ω_J, which should be the trace of the curvature of the Hermitian-Einstein metric on $\mathcal{W}_d$. We can conclude that $\mu(j)^{rank\ \mathcal{W}_d} = \alpha(j)$.

Now that we have stated the problem; *i.e.*, does the L_2-metric satisfy a certain kind of differential equation, we will show that the question has a positive answer when $g = 1$; *i.e.* C is an abelian curve. The method, in fact, generalizes to an abelian variety of arbitrary dimension.

§3. The Picard bundles on abelian varieties

Let k be general. Let X be an abelian variety and $X^\wedge$ be the dual abelian variety. The $\mathcal{P}$ be a Poincaré sheaf on $X \times X^\wedge$. If $\mathcal{L}$ is an ample invertible sheaf on X, then we have the locally free sheaf $\mathcal{W}(\mathcal{L}) = \pi_{X^\wedge *}(\pi_X^* \mathcal{L} \otimes \mathcal{P})$ on $X^\wedge$ which is a Picard bundle.

These Picard bundles are easier than those of curves of higher genus.

Theorem 3. $\mathcal{W}(\mathcal{L})$ *is stable with respect to any polarization of* $X^\wedge$.

Proof. Let $\psi_\mathcal{L} : X \to X^\wedge$ be the usual isogony send x to $(T_x)^* \mathcal{L} \otimes \mathcal{L}^{\otimes -1}$. Then $(1_X, \psi_\mathcal{L})^* \rho \approx (\pi_1 + \pi_2)^* \mathcal{L} \otimes \pi_1^* \mathcal{L}^{\otimes -1} \otimes \pi_2^* \mathcal{L}^{\otimes -1}$. It follows that $\psi_\mathcal{L}^* \mathcal{W}(\mathcal{L}) \approx \pi_{2*}(\pi_1 + \pi_2)^* \mathcal{L} \otimes \mathcal{L}^{\otimes -1}$. Using the π_2 automorphism $(x, y) \to (x+y, y)$ we see that $\psi_\mathcal{L}^* \mathcal{W}(\mathcal{L}) \approx \Gamma(X, \mathcal{L}) \otimes_k \mathcal{L}^{\otimes -1}$ (see [4]). $\mathcal{W}(\mathcal{L})$ is gotten from $\Gamma(X, \mathcal{L}) \otimes_k \mathcal{L}^{\otimes -1}$ by taking the quotient under the $K(\mathcal{L}) = \ker \psi_\mathcal{L}$-action induced by the $H(\mathcal{L})$ action on $\Gamma(X, \mathcal{L}) \otimes$ the contragradient action on $\mathcal{L}^{\otimes -1}$ where $1 \to \mathbf{C}_m \to H(\mathcal{L}) \to K(\mathcal{L}) \to 0$ is Mumford's theta group.

Now let $0 \subsetneq \mathcal{F} \subsetneq \mathcal{W}(\mathcal{L})$ be a coherent sheaf such that $\mathcal{W}/\mathcal{F}$ is torsion-free. Then $\mathcal{F}' = \psi_\mathcal{L}^* \mathcal{W}(\mathcal{L})$ is a subsheaf of $\Gamma(X, \mathcal{L}) \otimes_k \mathcal{L}^{\otimes -1}$ with the same torsion-free property. By linear algebra $\mathcal{F}' \hookrightarrow \oplus^i \mathcal{L}^{\otimes -1}$ where i is the rank of $\mathcal{F}$. The shape of $\mathcal{F} = \frac{\deg \mathcal{F}}{\operatorname{rank} \mathcal{F}} /$ slope of $\mathcal{W}(\mathcal{L}) = \frac{\deg \mathcal{W}(\mathcal{F})}{\operatorname{rank}} = M$ is invariant under isogony. Thus $M \leq - \deg \mathcal{L} / - \deg \mathcal{L} \leq 1$. Therefore $\mathcal{W}(\mathcal{L})$ is semi-stable. To prove stability we have to eliminate the case $M = 1$. Then $\mathcal{F}' \approx \oplus^i \mathcal{L}^{\otimes -1}$ and $\mathcal{F}' = \mathcal{V} \otimes_k \mathcal{L}^{\otimes -1}$ where $\mathcal{V}$ is $H(\mathcal{L})$-invariant subspace of $\Gamma(X, \mathcal{L})$ of dimension $0 \neq i \neq \dim \Gamma(X, \mathcal{L})$. As $M(X, \mathcal{L})$ is an irreducible representation of $H(\mathcal{L})$ this is impossible.

$$Q.E.D.$$

We note that

Theorem 4. *The family* $\{\mathcal{W}(\mathcal{L})\}$ *where* $\mathcal{L}$ *runs through a complete family of algebraically equivalent* $\mathcal{L}$ *is a versal deformation of the Picard bundle.*

Proof. The mapping $\mathcal{L}$ to $\mathcal{W}(\mathcal{L})$ is injective. This follows easily from Mukai's [6] definition of Fourier transformation; *i.e.*, $\mathcal{W}(\mathcal{L})$ is the Fourier transformation of $\mathcal{L}$. So $\mathcal{L}$ is the Fourier transform of $\mathcal{W}(\mathcal{L})$. It remains to check that the tangent space has dimension = dimension of X. This follows from

Lemma 5. $\oplus H^i(X^\wedge, \mathcal{H}om(\mathcal{W}(\mathcal{L}), (\mathcal{W}(\mathcal{L}))$ *is an exterior algebra* $\simeq \Lambda(H^1(X, O_X))$.

Proof. There is an invertible sheaf $\mathcal{M}$ on an abelian variety, such that

$$H^i(Y, \mathcal{M}) \approx H^i((X^\wedge, \mathcal{H}om(\mathcal{W}(\mathcal{L}), \mathcal{W}(\mathcal{L})) \text{ for all } i.$$

Explicitly take $Y = X \times X \times X\hat{\ }$ and $\mathcal{M} = \pi_1^*\mathcal{L} \otimes \pi_{13}^*\mathcal{P} \otimes \pi_2^*\mathcal{L}^{\otimes -1} \otimes \pi_{23}^*\mathcal{P}^{\otimes -1}$. Then by the Künneth formula and Serre duality

$$R_{\pi_X}^g \mathcal{M} \approx \mathcal{H}om(\mathcal{W}(\mathcal{L}), (\mathcal{W}(\mathcal{L}))$$

is the only non-zero higher direct image of $\mathcal{M}$. Thus by the Leray spectral sequence we have the desired isomorphism. To compute the cohomology of $\mathcal{M}$ is an easy exercise using [4].

$$Q.E.D.$$

Let $k = \mathbf{C}$. Then $\Gamma(X, \mathcal{L})$ by a unique up to scalar Hermitian metric which is fixed by $H(\mathcal{L})$ up to scalar. $\mathcal{L}^{\otimes -1}$ has a metric with invariant curvature which is similarly invariant. Thus

$$\Gamma(X, \mathcal{L}) \otimes \mathcal{L}^{\otimes -1}$$

has a Hermitian-Einstein metric for any invariant Kähler structure on X which is $K(\mathcal{L})$-invariant. This descends to the Hermitian-Einstein metric on $\mathcal{W}(\mathcal{L})$. Now we note that this comes from integration on the fiber on $\pi_2 X \times X \to X$ on the sheaf $\pi_1^*\mathcal{L} \otimes \mathcal{L}^{\otimes -1}$ which is given the metric with invariant curvature and the volume form on the fiber is Haar measure. We need only note that the L^2-metric on $\Gamma(X, \mathcal{L})$ is clearly $H(\mathcal{L})$-covariant. Therefore

Corollary 6. *The Hermitian-Einstein metric on $\mathcal{W}(\mathcal{L})$ comes from the L^2-metric.*

REFERENCES

[1] J.-M. Bismut, H. Gillet and C. Soulé, Analytic Torsion and holomorphic determinantal bundles, to appear.

[2] L. Ein and R. Lazarsfeld, Stability and restrictions of Picard bundles, with an application to normal bundles of elliptic curves, to appear.

[3] G. Kempf, Appendix (about cohomology of degenerate line bundles on abelian varieties) to D. Mumford's article, in Questions and Algebraic varieties, Centro. Int. Math. Estivo. Roma, 1970.

[4] _____, Toward the inversion of abelian integrals II, AJM 101(1979) pp. 184-202.

[5] _____, Rank g Picard bundles are stable, AJM to appear.

[6] S. Mukai, Duality between $D(X)$ and $D(\hat{X})$ with its application to Picard sheaves, Nag. Math. J. 81(1981), 153-175.

[7] H. Umemura, On a property of symmetric products of a curve of genus 2, Proc. dut. Symp. on Alg. Geom., Kyoto 1977, pp. 709-721.

Contemporary Mathematics
Volume **136**, 1992

On the Reduction of Abelian Integrals and a Problem of H. Hopf

HENRIK H. MARTENS

ABSTRACT. The purpose of this paper is to draw attention to an apparently little–known constructive aspect of Poincaré's work on the reduction of Abelian integrals, and to suggest its relevance for the homology classification of branched coverings.

Introduction

Let X and Y be closed oriented surfaces of positive genera g and p and let $f : X \to Y$ be a continuous map. Hopf [5] showed that any abstractly given homomorphism of the homotopy groups is induced by a surface map, and asked for a characterization of the abstractly given homomorphisms of the first homology groups which can be induced by surface maps.

The problem is closely related to that of characterizing those homomorphisms of Jacobian varieties which are induced by maps of the associated closed Riemann surfaces. If $p = 1$, any such homomorphism is induced by the surface map obtained by restriction to the imbedded image, but for $p > 1$ the situation is more complicated. If X and Y are closed Riemann surfaces of genera $g > p > 1$, a homomorphism $h : J(X) \to J(Y)$ need not be induced by a map of the surfaces (take one that is, and compose with multiplication by an integer in $J(Y)$, for instance.)

More generally, there is the problem of *reduction* which occupied several analysts in the latter half of the last century. Having observed that an abelian

1991 *Mathematics Subject Classification*. Primary 30F10, 14E20; Secondary 57M12.

Work on this paper was carried out with a grant from NAVF while the author enjoyed the hospitality and support of the Institut Mittag-Leffler, the Mathematical Institute of the University of Copenhagen, and the Centre de Recerca Matemàtica, Barcelona.

Some of the problems considered here were the topic of (unpublished) joint work with Hershel Farkas.

This paper is in final form and no version of it will be submitted for publication elsewhere .

integral of a given genus sometimes could be reduced to one of lower genus, in particular to an elliptic integral, one realized that the "lower genus" integral could fail to come from a surface, even if the one of higher genus did. In modern terms, a homomorphism $J(X) \to A$, of a Jacobian variety onto an abelian variety may occur without the abelian variety being a Jacobian one. The problem of deciding when it is, and when the homomorphism is induced by a map of the associated surfaces, could be regarded as a Schottky problem for maps.

Since knowledge of the induced homology homomorphism is equivalent to that of the induced homomorphism of Jacobian varieties and of the map itself (see [8]), a solution of Hopf's problem would be a natural first step. It appears that Poincaré's work on the reduction of abelian integrals may contain a clue to this, – a fact that may have been overlooked in the past. Our objective here is to discuss some relevant aspects of this work.

1. Reducible Abelian Integrals and Theta Functions

In a paper of 1935 A. A. Albert [1] gave a proof of the fact that if a Riemann matrix is on the form

$$\begin{pmatrix} \omega_1 & 0 \\ \omega_3 & \omega_2 \end{pmatrix},$$

then it is isogenous (*isomorphic* in Albert's terminology) to the matrix

$$\begin{pmatrix} \omega_1 & 0 \\ 0 & \omega_2 \end{pmatrix},$$

and ω_1 and ω_2 are Riemann matrices. He attributed the result to Poincaré, and it has since been referred to as the *Poincaré Complete Reducibility Theorem*. Albert's goal was the classification of multiplier algebras of Riemann matrices, and the theorem plays a key role in his decisive work on this problem.

Poincaré, however, was concerned with an entirely different problem. In a manuscript of 1874, (which later appeared in the Acta Mathematica, [6]) S. Kowalevsky had quoted two results of Weierstrass on the reduction of abelian integrals to elliptic integrals, and the expression of the associated theta functions in terms of products of theta functions of fewer variables, without proof. Poincaré's goal was to provide a proof of the theorems, and of their generalisation to arbitrary cases of reduction. While the *existence* of a representation of theta functions of g variables in terms of theta functions of fewer variables in the presence of reduction may be deduced from Albert's proof, Poincaré's papers provide algorithms for the explicit computation of the representation. This aspect of his work does not appear to have been adequately reported in the literature, possibly because of the sketchy nature of his papers (see [10], [11] and [9], p. 53.)

Let Z be a $g \times g$ matrix of complex numbers satisfying the usual *Riemann relations:*

$$^t Z = Z$$

$$\Im(Z) > 0,$$

i.e. Z is symmetric and the matrix of imaginary parts of its entries is positive definite. Consider its associated theta function

$$\theta(z; Z) = \sum_m \exp[\pi i\,{}^t m(Zm + 2z)],$$

where the summation is over all g-vectors m of integers, $z \in C^g$, and m and z are to be thought of as column vectors.

If the matrix Z splits into a direct sum

$$Z = \begin{pmatrix} Z_1 & 0 \\ 0 & Z_2 \end{pmatrix}$$

of a $p \times p$ and a $q \times q$ matrix, then, with $z = \begin{pmatrix} z_1 \\ z_2 \end{pmatrix}$ the theta function will split into a product

$$\theta(z; Z) = \theta(z_1; Z_1)\theta(z_2; Z_2),$$

by an obvious rearrangement of the defining series.

More generally, reduction to theta functions of lower dimension is possible when Z is on the form

$$Z = \begin{pmatrix} Z_1 & Q \\ {}^t Q & Z_2 \end{pmatrix},$$

where Q is a $p \times q$ matrix of rationals, as shown by the following argument:

We have

$${}^t m(Zm + 2z) = {}^t m_1(Z_1 m_1 + 2z_1 + 2Qm_2) + {}^t m_2(Z_2 m_2 + 2z_2).$$

Multiplying by πi, exponentiating, and summing over m_1, we get

$$\theta(z; Z) = \sum_{m_2} \theta(z_1 + Qm_2; Z_1)\exp[\pi i\,{}^t m_2(Z_2 m_2 + 2z_2)].$$

Now, let $D = \mathrm{diag}(d^1, d^2, \ldots, d^q)$ be a diagonal matrix of positive integers such that QD is a matrix of integers. Write

$$m_2 = \bar{m}_2 + Dk_2,$$

where $0 \leq \bar{m}_2^j < d^j$, and observe that

$$\theta(z_1 + Qm_2; Z_1) = \theta(z_1 + Q\bar{m}_2; Z_1).$$

Fixing $\bar{m}_2$ we may therefore sum over k_2 to get the finite sum representation

$$(*) \qquad \theta(z; Z) = \sum_{\bar{m}_2} \theta(z_1 + Q\bar{m}_2; Z_1) \times \theta \begin{bmatrix} D^{-1}\bar{m}_2 \\ 0 \end{bmatrix} (Dz_2; DZ_2D),$$

where

$$\theta \begin{bmatrix} D^{-1}\bar{m}_2 \\ 0 \end{bmatrix} (Dz_2; DZ_2D) =$$

$$\sum_{k_2} \exp[\pi i\,{}^t(k_2 + D^{-1}\bar{m}_2)(DZ_2D(k_2 + D^{-1}\bar{m}_2) + 2Dz_2)] =$$

$$\sum_{k_2} \exp[\pi i^t m_2 (Z_2 m_2 + 2z_2)]$$

is the standard *theta function with characteristics* defined generally, in dimension
g, by

$$\theta \begin{bmatrix} r \\ s \end{bmatrix} (z; Z) = \sum_m \exp[\pi i^t (m+r)(Z(m+r) + 2(z+s))],$$

where r and s are real g–vectors (see, e.g., Conforto [4].)

We say, in general, that a $g \times 2g$ period matrix $(E \quad Z)$ *admits reduction* if
it satisfies an equation

$$H \times (E \quad Z) = \Pi \times M,$$

where Π is a $p \times 2p$ matrix of complex numbers, H is a maximal rank $p \times g$
matrix of complex numbers, and M is a maximal rank $2p \times 2g$ matrix of integers,
$1 \le p < g$. It is easily seen that a period matrix of the form considered above
necessarily admits reduction.

Poincaré showed, conversely, that if a period matrix $(E \quad Z)$ admits reduction, then it is *symplectically equivalent* to a matrix on the form indicated.

THEOREM (WEIERSTRASS – POINCARÉ). *Let $(E \quad Z)$ be a $g \times 2g$ matrix
satisfying the Riemann conditions and admitting reduction*

$$H \times (E \quad Z) = \Pi \times M$$

where Π is an $p \times 2p$ matrix of complex numbers, $1 \le p < g$.

*Then there is a $g \times g$ non-singular matrix A of complex numbers, and a $2g \times 2g$
symplectic unimodular matrix T such that*

$$(E \quad Z) \times T = A \times \begin{pmatrix} E_1 & 0 & Z_1 & Q \\ 0 & E_2 & {}^t Q & Z_2 \end{pmatrix},$$

*where Z_1 and Z_2 are $p \times p$ and $(g-p) \times (g-p)$ matrices satisfying the Riemann
relations, and Q is an $p \times (g-p)$ matrix of rational numbers whose non–zero
entries, if any, are confined to an initial string along the main diagonal, q_{jj}.*

Poincaré's proof is based on the existence of a normal form for M, which is
of independent interest, especially in the context of surface maps:

NORMAL FORM LEMMA. *Let $1 \le p < g$, and let M be a $2p \times 2g$ matrix of
maximal rank with integer entries such that $M J^t M$ is non–singular. Then*

$$M = SNT$$

*where S is a $2p \times 2p$ non–singular matrix of integers, T is a $2g \times 2g$ symplectic
unimodular matrix, and N is a $2p \times 2g$ matrix of integers on block form*

$$N = \begin{pmatrix} E_1 & 0 & 0 & 0 \\ 0 & X & \Delta & 0 \end{pmatrix},$$

*where E_1 is a $p \times p$ identity matrix, Δ is a diagonal matrix of integers each of
which is a multiple of the following, and X is a $p \times (g-p)$ matrix where $x_{jj} = 1$
for all $j \le r$ for some r with $0 \le r < g-p$, and the remaining entries are zero.*

The non-zero entries of X correspond to the entries of Δ which are different from 1. Moreover, the greatest common divisor of the $2m \times 2m$ subdeterminants of N is 1.

Poincaré's proof of this lemma is rather sketchy, but a reasonably complete treatment will appear shortly in *Publicacions Matemàtiques*. Given the lemma, the proof of the theorem is straightforward. If a matrix M occurs in a reduction equation

$$H \times (E \quad Z) = \Pi \times M,$$

then MJ^tM must be non–singular.

To see this, observe that by an easy calculation the Riemann conditions for Z are equivalent to the equations

$$(E \quad Z)J^t(E \quad Z) = 0$$

and

$$i(E \quad Z)J^t(E \quad \overline{Z}) > 0.$$

It follows that $\Pi MJ^tM^t\Pi = 0$ and that $i\Pi MJ^tM^t\overline{\Pi}$ is positive definite, since H is of maximal rank. Since

$$\begin{pmatrix} \Pi \\ \overline{\Pi} \end{pmatrix}(MT^tM)^t\begin{pmatrix} \Pi \\ \overline{\Pi} \end{pmatrix} = \begin{pmatrix} 0 & \Pi MJ^tM^t\overline{\Pi} \\ \overline{\Pi}MJ^tM^t\Pi) & 0 \end{pmatrix},$$

it follows that the factors on the left are nonsingular.

Now, from the reduction equation we get, by the lemma,

$$H \times (E \quad Z) = \Pi \times S \times \begin{pmatrix} E_1 & 0 & 0 & 0 \\ 0 & X & \Delta & 0 \end{pmatrix} \times T.$$

Writing $\Pi \times S$ as $(\Pi_1 \quad \Pi_2)$, and multiplying out, we have

$$H \times (E \quad Z) = (\Pi_1 \quad \Pi_2X \quad \Pi_2\Delta \quad 0) \times T$$
$$= (\Pi_2\Delta \quad 0 \quad -\Pi_1 \quad -\Pi_2X) \times JT.$$

Since the matrix JT is symplectic unimodular,

$$(E \quad Z)(JT)^{-1} = G \times (E \quad Z'),$$

where Z' is a $g \times g$ matrix satisfying the Riemann relations and G is a non–singular $g \times g$ matrix. (Write $(E \quad Z)(JT)^{-1} = (\Omega_1 \quad \Omega_2) = \Omega$, show that $i\Omega J^t\overline{\Omega}$ is positive definite and conclude that Ω_1 and Ω_2 must be non–singular.) Then

$$HG(E \quad Z') = (\Pi_2\Delta \quad 0 \quad -\Pi_1 \quad -\Pi_2X)$$

whence it follows that

$$HG = (\Pi_2\Delta \quad 0).$$

Since H was assumed of maximal rank, $\Pi_2\Delta$ must be non–singular. Then

$$(E \quad Z') = \begin{pmatrix} E_1 & 0 & -\Delta^{-1}\Pi_2^{-1}\Pi_1 & -\Delta^{-1}X \\ 0 & E_2 & -^tX\Delta^{-1} & Z_2 \end{pmatrix},$$

with some $(g - p) \times (g - p)$ matrix Z_2. Since this is on the postulated form, the proof of the theorem is completed.

Note that the matrix Z' may be computed from Z once the symplectic matrix T of the lemma has been determined. The proof of the lemma gives an algorithm to determine T, given M. Thus the proof of the theorem is constructive.

Bounds on the number of terms in the representation $(*)$ may be obtained from the entries of the matrix Δ and the non–zero entries of X. When the reduction arises as a consequence of mappings of Riemann surfaces, the entries of Δ will be divisors of the degree of the map.

2. Hopf's Problem

The Normal Form Lemma of Poincaré does not immediately represent a normal form for induced homology maps, since T is not symplectic. But much relevant information can nevertheless be read out of it. If all entries of the matrix X are zero, for instance, the matrix Z' will split into a direct sum. This cannot happen when $(\,E \quad Z\,)$ is the canonical period matrix of a closed Riemann surface, and thus excludes certain matrices M from representing induced homology maps of surfaces.

In the paper already cited Hopf established the following necessary condition:

CONDITION OF HOPF. *If the induced homology homomorphism of the map is represented with respect to canonical homology bases by a $2p \times 2g$ matrix of integers, M, then $g - 1 \le d(p - 1)$, and*

$$M J^t M = d J',$$

where d is the degree of the map, $J = \begin{pmatrix} 0 & E \\ -E & 0 \end{pmatrix}$ is the canonical intersection matrix of the basis on X, and J' is the analogous matrix for Y.

Consider now matrices M satisfying the necessary conditions of Hopf. We shall say that such a matrix *represents a covering (resp. map)* if it represents the induced homology homomorphism, with respect to canonical bases, of a (possibly branched) covering (resp. continuous map) of closed oriented surfaces. We shall also say that two matrices are *symplectically equivalent* if one can be obtained from the other by multiplying on the right and the left with symplectic unimodular matrices (corresponding to a change of canonical bases.) We assume, without loss of generality, that d is positive.

For $p = 2$, $g = 3$ and $d = 2$, (representing the case of a double cover of a genus 2 surface by one of genus 3,) the only possibility for N is

$$\begin{pmatrix} 1 & 0 & 0 & 0 & 0 & 0 \\ 0 & 1 & 0 & 0 & 0 & 0 \\ & & & & & \\ 0 & 0 & 1 & 2 & 0 & 0 \\ 0 & 0 & 0 & 0 & 1 & 0 \end{pmatrix}.$$

For a double covering of a genus 2 surface by one of genus 4, there are two possibilities:

$$\begin{pmatrix} 1 & 0 & 0 & 0 & & 0 & 0 & 0 & 0 \\ 0 & 1 & 0 & 0 & & 0 & 0 & 0 & 0 \\ & & & & & & & & \\ 0 & 0 & 1 & 0 & & 2 & 0 & 0 & 0 \\ 0 & 0 & 0 & 0 & & 0 & 1 & 0 & 0 \end{pmatrix},$$

and

$$\begin{pmatrix} 1 & 0 & 0 & 0 & & 0 & 0 & 0 & 0 \\ 0 & 1 & 0 & 0 & & 0 & 0 & 0 & 0 \\ & & & & & & & & \\ 0 & 0 & 1 & 0 & & 2 & 0 & 0 & 0 \\ 0 & 0 & 0 & 1 & & 0 & 2 & 0 & 0 \end{pmatrix}.$$

In the second case, S must be symplectic and N is symplectically equivalent to M. This case can be realized by covering the surface with two copies of itself cross-connected over a slit.

For double coverings of a genus 2 surfaces by surfaces of higher genera, there are again only two possibilities for N : those obtained from the above by adding zero columns to the two 4×4 blocks. If the genus 4 cases can be realized by coverings, then the higher genus cases can be realized by coverings obtained by introducing slits and cross–connections to raise the genus appropriately.

When M and N are not symplectically equivalent, a further study is necessary. I have not been able to get a good normal form for maps (with respect to canonical bases) in general, but in the case of maps of prime degree, one can prove the following:

NORMAL FORM LEMMA (PRIME COVERS). *If d is prime, and M is a $2m \times 2n$ matrix of integers satisfying the equation $MJ^tM = dJ'$, then*

$$M = T_1 \begin{pmatrix} E & 0 & 0 & 0 \\ 0 & X & dE & 0 \end{pmatrix} T_2,$$

where the T_j are symplectic unimodular, and the X of the middle factor is the same as that of the Poincaré normal form for M.

From this we conclude that the only possible normalized matrix representation for a double cover of a genus 2 surface by one of genus 3 must be

$$\begin{pmatrix} 1 & 0 & 0 & & 0 & 0 & 0 \\ 0 & 1 & 0 & & 0 & 0 & 0 \\ & & & & & & \\ 0 & 0 & 1 & & 2 & 0 & 0 \\ 0 & 0 & 0 & & 0 & 2 & 0 \end{pmatrix},$$

which represents the covering obtained by taking a genus 3 surface represented by a torus with two symmetrically placed handles and forming the quotient under the obvious involution.

The first possibility for a covering by a genus 4 surface is obtained by adding zero columns to this matrix. Such a covering must have two branch points corresponding to a cross–connection over slits. Removing this will eitner disconnect the surface (as in the case already discussed) or reduce the genus. Introducing slits in the genus 3 covering above and cross–connecting, however, produces a matrix of the second kind. Thus it appears that this possibility must be ruled out.

More generally, any covering of a genus 2 surface of prime degree d has a representation with respect to canonical bases on one of the forms

$$\begin{pmatrix} 1 & 0 & 0 & 0 & 0 & \cdots & 0 & 0 & 0 & 0 & 0 & 0 & \cdots & 0 \\ 0 & 1 & 0 & 0 & 0 & \cdots & 0 & 0 & 0 & 0 & 0 & 0 & \cdots & 0 \\ \\ 0 & 0 & 1 & 0 & 0 & \cdots & 0 & d & 0 & 0 & 0 & 0 & \cdots & 0 \\ 0 & 0 & 0 & 0 & 0 & \cdots & 0 & 0 & d & 0 & 0 & 0 & \cdots & 0 \end{pmatrix},$$

or

$$\begin{pmatrix} 1 & 0 & 0 & 0 & 0 & \cdots & 0 & 0 & 0 & 0 & 0 & 0 & \cdots & 0 \\ 0 & 1 & 0 & 0 & 0 & \cdots & 0 & 0 & 0 & 0 & 0 & 0 & \cdots & 0 \\ \\ 0 & 0 & 1 & 0 & 0 & \cdots & 0 & d & 0 & 0 & 0 & 0 & \cdots & 0 \\ 0 & 0 & 0 & 1 & 0 & \cdots & 0 & 0 & d & 0 & 0 & 0 & \cdots & 0 \end{pmatrix}.$$

The unbranched case of the first can be realized as the quotient map of a torus with d handles symmetrically placed to give a rotational symmetry of period d. The second can be realized as follows: cut open a handle of a genus 2 surface, take a second copy of the surface and cut the corresponding handle. Reassemble the handles cross–wise to get a surface of genus 3, and map onto a genus 2 surface by the obvious identification. Continue the process, cutting open previously unused handles and attaching new copies until the desired degree is attained. Branched versions of the second case may be produced by first constructing a $(d-1)$-fold covering in the manner indicated and then attaching the last copy by a slit and cross–connection. Higher genus covers with the same matrix may then be produced by introducing additional slits and cross-connections.

SUMMARY. *It thus appears that any $4 \times 2g$ matrix of integers satisfying the necessary conditions of Hopf with d a prime will represent a map, but not necessarily a covering, provided $X \neq 0$ in the Poincaré normal form. There are two possible normal forms for such matrices, distinguished by the number of non–zero columns of X in the Poincaré normal form. These can be determined by computing the greatest common divisor of the 4×4 subdeterminants of the matrix. Matrices with normal foms of the second kind can always represent coverings, while those with normal forms of the first kind have only been shown (here) to represent coverings in the unbranched case.*

The general case of a covering of prime degree may be discussed along the same lines. For a matrix satisfying the Hopf conditions there will be $p+1$

possibilities for X, (including the forbidden $X = 0$.) They may be determined by computing the greatest common divisor of the $2p \times 2p$ subdeterminants of the matrix. Constructions like the ones used above show easily that the case when X has p non–zero columns always represents a covering, while the case where X has only $p - 1$ non–zero columns represents a covering in the unbranched case (and hence always represents a map.)

These inconclusive musings of a topological dilettante should suggest that something may be gained from a closer study of the Poincaré normal form.

3. Exercise

Show that there exist closed Riemann surfaces X and X' of positive genera whose period matrices satisfy a reduction equation

$$H \times (E \quad Z) = (E' \quad Z') \times M,$$

where M is a matrix representing a covering (for some pair of surfaces), but where the corresponding map $J(X) \to J(X')$ is not induced by a holomorphic map $X \to X'$.

4. An Historical Remark

As Igusa has pointed out ([6], p. 165,) Poincaré regarded his observation that the completely reducible cases are dense in the Siegel upper half space as a key result in the theory of abelian functions. Altogether he devoted half a dozen papers to the problem of reduction, and it is one of the six topics that he selected for discussion in a series of lectures in Göttingen in April, 1909. (The others were: Fredholm Equations, (two) Applications of Integral Equations in Physics, Mécanique Nouvelle, and Transfinite Numbers.) Weierstrass assigned research problems from the area to his students Koenigsberger and Kowalevsky, other contributions came from P. Appell and E. Picard (another half dozen papers.) In view of this it is hard to decide what significance, if any, should be attached to the repeated assertions of R. Cooke [3] that the topic "was not one of the central parts of the theory."

It is amazing that Poincaré's results have not found their way into the expository literature. The only result cited in the German Encyclopädie, and in Krazer's book on theta functions, seems to be the special case of the Normal Form Lemma when $p = 1$. Baker's books have chapters on reduction which do not mention the normal form, although Poincaré's papers are mentioned. In his paper [1] Albert states that all previous proofs are not very simple. This refers in the first place to work of Scorza on the complete reducibility theorem, but could be interpreted as a remark also on Poincaré's work. It is doubtful, however, that Albert could have read Poincaré's papers without noticing that the idea of his simplified proof already appears there. It is sometimes stated that Poincaré only considered period matrices of Riemann surfaces. In fact he treats the case of principally polarized Riemann matrices and their theta functions. In his treatment of what is now called the complete reduction theorem he explicitly deals with higher order transformations, and not with isogenies (which do not yield reduction of theta functions.) In this context Albert's exposition misses an essential point.

Interest in the topic of reduction has recently resurfaced in connection with the applications of theta functions to the solution of non–linear partial differential equations, see [2].

REFERENCES

1. Albert, Adrian A., *A Note on the Poincaré Theorem on Impure Riemann Matrices*, Ann. of Math. **36** (1935), 151–156.
2. Belokolos, E. D., et al., *Algebraic–geometric principles of superposition of finite–zone solutions of integrable non–linear equations*, Russian Math. Surveys **41** (1986), 1–49.
3. Cooke, R., *The Mathematics of Sonya Kovalevskaya*, Springer–Verlag, New York, 1984.
4. Conforto, F., *Abelsche Funktionen und algebraische Geometrie*, Springer–Verlag, Berlin, 1956.
5. Hopf, H., *Beiträge zur Klassifisierung der Flächenabbildungen*, J. Reine Angew. Math. **165** (1931), 225-236.
6. Igusa, J.-I., *Problems on Abelian functions at the time of Poincaré and some at present*, Bull. Amer. Math. Soc. **6** (1982), 161-174.
7. Kowalevsky, S., *Über die Reduction einer bestimmten Klasse Abel'scher Integrale 3ten Ranges auf elliptische Integrale*, Acta Math. **4** (1884), 393–414.
8. Martens, H. H., *Observations on morphisms of closed Riemann surfaces*, Bull. London Math. Soc. **10** (1978), 209–212.
9. Minkowski, H., *Briefe an David Hilbert*, Springer–Verlag, Berlin, 1973.
10. Poincaré, H., *Sur les fonctions abéliennes*, Amer. J. Math. **8** (1886), 239–342.
11. ———, *Sur la réduction des intégrales abéliennes*, Bull. Soc. Math. France **12** (1884), 124–143.

DIVISION OF MATHEMATICAL SCIENCES, THE NORWEGIAN INSTITUTE OF TECHNOLOGY, N–7034 TRONDHEIM, NORWAY

E-mail: henrik@imf.unit.no

Contemporary Mathematics
Volume **136**, 1992

NORMALIZATION OF
THE KRICHEVER DATA

MOTOHICO MULASE*

Institute of Theoretical Dynamics
University of California
Davis, CA 95616, U. S. A.
and
Max-Planck-Institut für Mathematik
Gottfried-Claren-Strasse 26
D-5300 Bonn 3, Germany

1. Introduction.

The purpose of this note is to give a canonical normalization of the Krichever data consisting of algebraic curves and torsion free sheaves on them. Generalizing the original Krichever data of Segal-Wilson [SW] in order to deal with the higher rank cases, the notion of *quintets* was introduced in [M1]. A quintet $(C, p, \pi, \mathcal{F}, \phi)$ is a set of geometric data consisting of a curve C, a point $p \in C$, a locally defined r-sheeted covering π of C ramified at p, a torsion free rank r sheaf $\mathcal{F}$, and a local trivialization ϕ of $\mathcal{F}$ near p. One can define a category $\mathcal{Q}$ of these quintets. The cohomology functors give rise to a fully-faithful contravariant functor χ between $\mathcal{Q}$ and a category $\mathcal{S}$ of some algebraic data [M1]. An object of $\mathcal{S}$ is a pair (A, W) consisting of a subring A of the formal Laurent series ring $k((z))$ and a vector subspace $W \subset k((z))$, which is commensurable with $k[z^{-1}]$ in $k((z))$, and satisfying that

$$A \cdot W \subset W .$$

The case that the sheaf $\mathcal{F}$ has rank one was studied extensively in [SW]. The higher rank cases were studied in [M1] in which a complete geometric classification of all commutative rings of ordinary differential operators was established.

Segal and Wilson [SW] showed that a quintet with a nonsingular curve C and a line bundle $\mathcal{F}$ is in one-to-one correspondence with W, which can be identified

*Research supported in part by the NSF Grant DMS 91–03239.
1991 *Mathematics Subject Classification.* Primary 14D15, 14H60, 58B99.
Research supported in part by NSF Grant DMS 91-03239.
This paper is in final form and no version of it will be submitted for publication elsewhere.

with a point of certain infinite dimensional Grassmannian. In this case, the algebra A can be recovered simply by

$$A = A_W = \{v \in k((z)) \mid vW \subset W\},$$

which we call the maximal stabilizer of W. The algebraic nature of the above statement will be clarified in Lemma 3.1. We will show that if the pair (A, W) satisfies that

(1) W is a rank-one A-module, and
(2) A is a normal ring,

then A is a maximal stabilizer of W. In general, however, maximality does not imply normality. Thus we are led to study normalization of the pair (A, W).

Our result of this paper is that for every quintet $(C, p, \pi, \mathcal{F}, \phi)$, there is a unique quintet $(C', p', \pi', \mathcal{F}', \phi')$ and a canonical morphism

$$(C', p', \pi', \mathcal{F}', \phi') \longrightarrow (C, p, \pi, \mathcal{F}, \phi)$$

such that C' is a normalization of C and $\mathcal{F}'$ is a locally free sheaf on C' having the same rank of $\mathcal{F}$. We construct this normalization by a simple algebraic procedure on the pair (A, W).

A natural supersymmetric generalization of the theory has been obtained in [M2] and [MR]. In particular, a characterization of the Jacobians of algebraic super curves of dimension 1|1 has been established in [M2].

2. The Krichever functor.

Throughout this paper, we work with a field k of an arbitrary characteristic. Let $V = k((z))$ be the set of all formal Laurent series in one variable z. This is the field of fractions of the ring $k[[z]]$ of formal power series. We denote by

$$V^{(\nu)} = k[[z]] \cdot z^{-\nu},$$

which is the set of all formal Laurent series whose pole order at $z = 0$ is less than or equal to $\nu \in \bar{\lambda}$. We say $v \in V$ has *order* ν if $v \in V^{(\nu)} \setminus V^{(\nu-1)}$. For every vector subspace W in V, let γ_W denote the natural map of W into $V/V^{(-1)}$ defined by

$$
\begin{array}{ccc}
V & \xrightarrow{\text{identity}} & V \\
{\scriptstyle \text{inclusion}}\uparrow & & \downarrow{\scriptstyle \text{projection}} \\
W & \xrightarrow{\ \gamma_W\ } & V/V^{(-1)} .
\end{array}
$$

When γ_W is Fredholm, we define the Fredholm index by Index $\gamma_W = \dim_k \mathrm{Ker}\, \gamma_W - \dim_k \mathrm{Coker}\, \gamma_W$.

Definition 2.1. *We call the following set the Grassmannian of index μ:*

$$G(\mu) = \Big\{ \text{vector subspace } W \,\big|\, \gamma_W \text{ is Fredholm of index } \mu \Big\}.$$

Note that $G(\mu)$ has a structure of the pro-algebraic variety in the sense of Grothendieck.

Definition 2.2. *Let r be a positive integer and μ an arbitrary integer. A pair (A, W) is said to be a Schur pair of rank r and index μ if the following conditions are satisfied:*

(1) *W is a point of the Grassmannian $G(\mu)$ of index μ.*
(2) *$A \subset V$ is a k-subalgebra of V such that $k \subset A$, $k \neq A$, $AW \subset W$ and*

$$r = \operatorname{rank} A = \text{G.C.D.} \{\operatorname{ord} a \,\big|\, a \in A\}.$$

We denote by $\mathcal{S}_r(\mu)$ the set of all Schur pairs of rank r and index μ.

Schur [S] showed that every commutative ring of ordinary differential operators can be embedded in the ring of pseudo-differential operators with constant coefficients. Our Schur pair is nothing but an algebraic abstraction of his situation. See [M1] for detail.

Remark 2.3. Let

$$A_W = \{ v \in V \,\big|\, vW \subset W \}.$$

If $k \neq A_W$, then (A_W, W) gives a Schur pair, which we call a *maximal Schur pair*. However, we have always $A_W = k$ for a generic W. In this case, W does not have any interesting geometric information.

Definition 2.4. *We define the category of Schur pairs $\mathcal{S}$ as follows:*

(1) *The set of objects is defined by*

$$Ob(\mathcal{S}) = \bigcup_{\mu \in \bar{\wedge}} \bigcup_{r \in \blacktriangle} \mathcal{S}_r(\mu).$$

(2) *The set of morphisms $Mor\big((A_2, W_2), (A_1, W_1)\big)$ consists of*

$$(\alpha, \iota) : (A_2, W_2) \longrightarrow (A_1, W_1),$$

where $\alpha : A_2 \hookrightarrow A_1$ and $\iota : W_2 \hookrightarrow W_1$ are injective homomorphisms.

Next, let us define a category of geometric data consisting of algebraic curves and torsion free sheaves on them, and construct a contravariant functor from this category to the category of Schur pairs.

Definition 2.5. *Let r be a positive integer and μ an arbitrary integer. We call $(C, p, \pi, \mathcal{F}, \phi)$ a quintet of rank r and index μ if it consists of the following geometric data:*

(1) *C is a reduced irreducible complete algebraic curve defined over k.*

(2) *$p \in C$ is a smooth k-rational point.*

(3) *$\pi : U_0 \to U_p$ is an r-sheeted covering of U_p ramified at p, where U_0 is the formal completion of the affine line $\beth^1_k = \beth^1$ at the origin $0 \in \beth^1$ and U_p is the formal completion of the curve C at p. Once for all, we choose a coordinate z on $\beth^1$ and fix it throughout this paper so that we have $U_0 = \operatorname{Spec} k[[z]]$.*

(4) *$\mathcal{F}$ is a torsion free sheaf of $\mathcal{O}_C$-modules on C of rank r satisfying*

$$\dim_k H^0(C, \mathcal{F}) - \dim_k H^1(C, \mathcal{F}) = \mu \ .$$

(5) *$\phi : \mathcal{F}_{U_p} \xrightarrow{\sim} \pi_* \mathcal{O}_{U_0}(-1)$ is an $\mathcal{O}_{U_p}$-module isomorphism between the formal completion $\mathcal{F}_{U_p}$ of $\mathcal{F}$ at $p \in C$, which is a free $\mathcal{O}_{U_p}$-module of rank r, and the direct image sheaf $\pi_* \mathcal{O}_{U_0}(-1)$.*

Two quintets $(C, p, \pi, \mathcal{F}, \phi)$ and $(C, p, \pi, \mathcal{F}, c\phi)$ are identified if $c \in k^\times$. We also identify $(C, p, \pi_1, \mathcal{F}, \phi_1)$ with $(C, p, \pi_2, \mathcal{F}, \phi_2)$ when the following diagram commutes:

$$
\begin{array}{ccc}
H^0(U_p, \mathcal{F}_{U_p}) & \xrightarrow{\ \phi_1\ } & H^0(U_p, \pi_{1*}\mathcal{O}_{U_0}(-1)) \\
{\scriptstyle \phi_2}\big\downarrow & & \big\downarrow{\scriptstyle \wr} \\
H^0(U_p, \pi_{2*}\mathcal{O}_{U_0}(-1)) & \xrightarrow{\ \sim\ } & H^0(U_0, \mathcal{O}_{U_0}(-1)).
\end{array}
$$

The set of all quintets of rank r and index μ is denoted by $\mathcal{Q}_r(\mu)$.

Remark 2.6. When $r = 1$, π is an isomorphism $\pi : U_0 \xrightarrow{\sim} U_p$. Since we have chosen a coordinate z on U_0, π gives a local coordinate $y = \pi(z)$ on U_p. Thus our quintet $(C, p, \pi, \mathcal{F}, \phi)$ becomes $(C, p, y, \mathcal{F}, \phi)$ with a local parameter y around p and a local trivialization ϕ of $\mathcal{F}$ near the point p. This is the original Krichever data introduced by Segal-Wilson [SW].

Definition 2.7. *We define a category $\mathcal{Q}$ of quintets as follows:*

(1) *The set of objects is defined by*

$$Ob(\mathcal{Q}) = \bigcup_{\mu \in \overline{\Lambda}} \bigcup_{r \in \blacktriangle} \mathcal{Q}_r(\mu).$$

(2) *A morphism*

$$(\beta, \psi) : (C_1, p_1, \pi_1, \mathcal{F}_1, \phi_1) \longrightarrow (C_2, p_2, \pi_2, \mathcal{F}_2, \phi_2)$$

consists of a morphism $\beta : C_1 \to C_2$ of curves and a homomorphism $\psi : \mathcal{F}_2 \to \beta_* \mathcal{F}_1$ of sheaves on C_2 such that

$$\beta(p_1) = p_2,$$

$$U_0 \rightrightarrows U_0$$

$$\pi_1 \downarrow \qquad \qquad \downarrow \pi_2$$

$$U_{p_1} \xrightarrow{\ \widehat{\beta}\ } U_{p_2} \,,$$

i.e. $\pi_2 = \widehat{\beta} \circ \pi_1$, where $\widehat{\beta}$ is the morphism of formal schemes determined by β, and

$$\mathcal{F}_{2U_{p_2}} \xrightarrow{\ \widehat{\psi}\ } \widehat{\beta}_* \mathcal{F}_{1U_{p_1}}$$

$$\phi_2 \downarrow \qquad \qquad \qquad \downarrow \widehat{\beta}_*(\phi_1)$$

$$\pi_{2*}\mathcal{O}_{U_0}(-1) \; =\!=\!= \; \widehat{\beta}_* \pi_{1*}\mathcal{O}_{U_0}(-1) \,,$$

where $\widehat{\psi}$ is the homomorphism of sheaves on U_{p_2} defined by ψ.

For a quintet $(C, p, \pi, \mathcal{F}, \phi)$ of rank r and index μ, we define

$$\begin{cases} A = \pi^* \big(H^0(C \setminus \{p\}, \mathcal{O}_C) \big) \\ W = \phi \big(H^0(C \setminus \{p\}, \mathcal{F}) \big) \,. \end{cases}$$

The identification $U_0 = \operatorname{Spec} k[[z]]$ makes both A and W subsets of $k((z))$. Moreover, we can show that (A, W) is a Scur pair of rank r and index μ. Furthermore, a morphism

$$(\beta, \psi) : (C_1, p_1, \pi_1, \mathcal{F}_1, \phi_1) \longrightarrow (C_2, p_2, \pi_2, \mathcal{F}_2, \phi_2)$$

gives rise to a morphism

$$(\alpha, \iota) : (A_2, W_2) \longrightarrow (A_1, W_1) \,,$$

where

$$\begin{cases} \alpha : \pi_2^* \big(H^0(C_2 \setminus \{p_2\}, \mathcal{O}_{C_2}) \big) \longrightarrow \pi_1^* \big(H^0(C_1 \setminus \{p_1\}, \mathcal{O}_{C_1}) \big) \\ \iota : \phi_2 \big(H^0(C_2 \setminus \{p_2\}, \mathcal{F}_2) \big) \longrightarrow \phi_1 \big(H^0(C_1 \setminus \{p_1\}, \mathcal{F}_1) \big) \end{cases}$$

are defined by the natural cohomology homomorphisms, and (A_i, W_i) is the Schur pair corresponding to the quintet $(C_i, p_i, \pi_i, \mathcal{F}_i, \phi_i)$ for $i = 1, 2$. It was established in [M1] that the above correspondence gives a fully-faithful contravariant functor

$$\chi : \mathcal{Q} \longrightarrow \mathcal{S}.$$

We call this anti-equivalence functor the *Krichever functor*.

3. Normalization of the Schur pairs and the quintets.

In this section, we study the Schur pairs and the quintets of rank one, and define the Krichever map. The injectivity of this map is proved by using a property of normal rings. We show that every rank one nonsingular quintet corresponds to a maximal Schur pair. We then study the *normalization* of the Schur pairs, and show that it corresponds to the geometric normalization of the algebraic curves.

Let us start with

Lemma 3.1. *Let (A, W) be a Schur pair. If A is a normal ring and rank $A = 1$, then A is maximal:*

$$A = A_W = \{v \in V_0 \mid v \cdot W \subset W\} \, .$$

Proof. Since $A \subset A_W$, the rank of A_W is also one. In particular, we have

$$(3.2) \qquad\qquad \dim_k A_W / A < +\infty \, .$$

Let $a \in A_W \setminus A$ and consider the set $A \cdot a \subset A_W$. If $A \cdot a \cap A = 0$, then $A \oplus A \cdot a \subset A_W$ and it contradicts (3.2). Therefore, there are elements a_0 and a_1 in A such that $a = \frac{a_0}{a_1}$. Hence A_W is contained in the field $K(A)$ of fractions of A.

The condition (3.2) also shows that A_W is integral over A. But since A is integrally closed in $K(A)$, we can conclude that $A = A_W$. This completes the proof.

Theorem 3.3. *Let $\mathcal{M}_1(\mu)_{ns}$ be the set of isomorphism classes of quintets of rank one such that the algebraic curve C in the quintet is nonsingular. Then the Krichever functor χ gives an injective map*

$$\chi_1 : \mathcal{M}_1(\mu)_{ns} \longrightarrow G(\mu) \, .$$

Proof. A quintet in $\mathcal{M}_1(\mu)_{ns}$ corresponds to a Schur pair (A, W) of rank one. Since the isomorphism relation among quintets gives the equality of the Schur pairs, the isomorphism class of the quintets determines a unique Schur pair. The smoothness assumption of C implies that the affine coordinate ring A is normal, hence by Lemma 3.1, maximal. But since the maximal Schur pair (A_W, W) is in one-to-one correspondence with the point $W \in G(\mu)$ canonically, the image point W of χ_1 determines the isomorphism class of the quintets. This completes the proof.

Lemma 3.1 tells us that every rank one normal ring A is maximal. It is natural to ask if its converse is true: is the maximal stabilizer A_W of a point W of the Grassmannian a normal ring?

Let us consider an example:

$$\begin{cases} W = k[z^{-2}, z^{-3}] \in G(-1) \\ A = A_W = k[z^{-2}, z^{-3}] \, . \end{cases}$$

Certainly A is maximal and of rank one, but it is not a normal ring. This example leads us to the following:

Theorem 3.4. *For an arbitrary Schur pair (A, W) of rank r, let us denote by A' the integral closure of the ring A in the field $K(A)$ of fractions of A, and $W' = A' \cdot W$. Then*

(1) *(A', W') is also a Schur pair of rank r, which we call the* normalization *of (A, W).*

(2) *Let $(C, p, \pi, \mathcal{F}, \phi)$ and $(C', p', \pi', \mathcal{F}', \phi')$ be the quintets corresponding to the Schur pairs (A, W) and (A', W'), respectively. Then the morphism*

$$(\beta, \psi) : (C', p', \pi', \mathcal{F}', \phi') \longrightarrow (C, p, \pi, \mathcal{F}, \phi)$$

corresponding to $A \hookrightarrow A'$ and $W \hookrightarrow W'$ consists of a normalization

$$\beta : C' \longrightarrow C$$

of the curve C such that $p' = \beta^{-1}(p)$ and a sheaf homomorphism

$$\psi : \mathcal{F} \longrightarrow \beta_* \mathcal{F}' .$$

Here the curve C' is nonsingular and $\mathcal{F}'$ is a locally free sheaf of $\mathcal{O}_{C'}$-modules of rank r.

Proof. It is obvious that $A \subset A'$, $W \subset W'$ and

$$A' \cdot W' = A' \cdot A' \cdot W = A' \cdot W = W' .$$

Since the order of every element of $K(A)$ is a multiple of r, the rank of A' is also r. In order to show the well-definedness of (A', W') as a Schur pair, it suffices to establish the following:

Lemma 3.5. *Let (A, W) be a Schur pair of rank r, and A' the integral closure of A in $K(A)$. Then we have*

$$\begin{cases} A' \cap k[[z]] = k \\ A' \cap k[[z]] \cdot z = 0 . \end{cases}$$

Proof. Suppose that A' has an element y of negative order, say $-\ell r$. Consider the polynomial ring $A[y] \subset A'$. Since y is integral over A, there are m elements $f_1, f_2, \cdots, f_m \in A[y]$ which generate $A[y]$ over A. We can express the element y as $y = \frac{b}{a}$ with $a, b \in A$ and $a \neq 0$ because $A' \subset K(A)$. Thus there is a large positive integer N such that

$$a^N \cdot f_j \in A, \qquad j = 1, 2, \cdots, m.$$

Hence we have $a^N \cdot A[y] \subset A$. It means that

$$a^N \cdot y^n \in A$$

for every $n \geq 0$. So by taking n sufficiently large, we obtain

$$a^N \cdot y^n \in A \cap k[[z]] \cdot z .$$

But since we know $A \cap k[[z]] \cdot z = 0$ from [M1, Section 3], we conclude that $y = 0$.

Therefore, every element of A' has a nonnegative order. Since $k \subset A'$, the only possible order zero elements are nonzero constants. This completes the proof of the lemma.

By this lemma, we see that $W' = A' \cdot W$ satisfies the Fredholm condition. Therefore, (A', W') is a Schur pair.

Of course the normalization of a Schur pair corresponds to the normalization of the algebraic curve C. Since $\mathcal{F}'$ is a torsion free sheaf on the normalization C', it is locally free. This completes the proof of the theorem.

The normalization (A', W') is a maximal Schur pair if the rank r is equal to one. In general, however, A_W is not necessarily included in the field $K(A)$. This gives the difference between the normal Schur pairs and the maximal ones. It should be an interesting project to study geometry of maximal Schur pairs of higher ranks from the point of view of two-dimensional quantum gravity [Sc].

References

[K]   I. M. Krichever: Methods of algebraic geometry in the theory of nonlinear equations, Russ. Math. Surv. **32** (1977) 185–214.

[M1]   M. Mulase: Category of vector bundles on algebraic curves and infinite dimensional Grassmannians, Intern. J. of Math. **1** (1990) 293–342.

[M2]   M. Mulase: A new super KP system and a characterization of the Jacobians of arbitrary algebraic super curves, to appear in J. Differ. Geom.

[MR]   M. Mulase and J. Rabin: Super Krichever functor, to appear in Intern. J. of Math.

[S]   I. Schur: Über vertauschbare lineare Differentialausdrücke, Sitzungsber. der Berliner Math. Gesel. **4** (1905) 2–8.

[Sc]   A. Schwarz: On solutions to the string equation, MSRI preprint 05429–91 (1991).

[SW]   G. B. Segal and G. Wilson: Loop groups and equations of KdV type, Publ. Math. I.H.E.S. **61** (1985) 5–65.

Contemporary Mathematics
Volume **136**, 1992

Splittable Jacobi Varieties

JOHN F. X. RIES

§0. Introduction. There are many interesting problems concerning the structure of Jacobi varieties $J(C)$ of special non-singular algebraic curves (or Riemann surfaces) C. The theta-divisor of $J(C)$ is irreducible, but is $J(C)$ splittable, that is, isogenous or isomorphic to a product of abelian varieties? If so, how is the polarization represented by data from the factors? Does $J(C)$ have other polarizations, other principal polarizations? These questions have received much study (see [**10,11,12,14,20,21,23,25,29**] and the references cited there).

In this paper we will survey some examples of splittable Jacobi varieties and examine three problems where this information could be used. Our examples will use the induced action of the automorphism group of C on the Jacobi variety of C to achieve a splitting. In this volume Earle gives a more general example in genus 3.

First, there are obvious uses in moduli problems. Suppose we are interested in curves C of genus g with a prescribed action of a group G as automorphisms of C. This may imply that $J(C)$ splits in a certain way. We can then parametrize all principally polarized abelian varieties (hereafter, ppav) with this action by G as polarization preserving automorphisms. This sets up a Schottky type problem in a moduli space which will be much smaller than the space of all ppavs of dimension g. We give a few examples of Jacobi varieties which are isomorphic to products.

For another example Riera and Rodríguez in [**23**] study the one-parameter family of ppavs of dimension 4 with $G = S_5$ action containing the Jacobi variety of Bring's curve. We will see that the Hurwitz curves of genus 7 ($G = SL(2, 2^3)$) and genus 14 ($G = PSL(2, 13)$) also belong to one-parameter families of ppavs whose automorphism group contains G.

Second, with a lot of information about $J(C)$, we can study certain properties of C. For example, suppose G is a group of automorphisms of C. The finite

---

1980 *Mathematics Subject Classification* (1985 *Revision*). Primary: 14H40, 14H35, 14K10.
This paper is in final form and no version of it will be submitted for publication elsewhere .

subgroups H of $J(C)$ invariant under the induced action of G on $J(C)$ classify unramified abelian covers D of C to which G lifts (not necessarily as a group). We get an extension

$$1 \longrightarrow \hat{H} \longrightarrow \mathcal{G} \longrightarrow G \longrightarrow 1$$

where $\mathcal{G}$ is a group of automorphisms of D and $\hat{H}$ is the dual group of H. An interesting problem is to determine if this extension splits. In [7,8] Cohen answered this question for the Klein-Hurwitz curve of genus 3 using presentations and matrix representations of the groups $\mathcal{G}$. We will reexamine this using an explicit splitting of the Jacobi variety.

Finally, we can study the Prym varieties and Prym map, for example, for cyclic unramified covers of hyperelliptic curves. We will examine the rank of the Prym map in this case.

Note that there are drawbacks to the methods presented here. For example, it is hard to see the curve C from the data of a polarization as an endomorphism from a ppav to its dual variety. Also it is difficult to relate theta-functions for the polarization of a ppav A to theta-functions of factors of A when A does not have the product polarization. It is easier to do this with an isogeny from a product of abelian varieties to A when the pull-back polarization on the product is the product polarization.

Notations:

  C non-singular algebraic curve of genus g over $\mathbb{C}$,

  ω_C is the canonical line bundle on C,

 $J(C)$ is the Jacobi variety of C,

  A_n is the group of points of order n in the abelian variety A,

 $\lambda_C : J(C) \to \hat{J}(C)$ is the principal polarization of $J(C)$ as the Jacobi variety of C,

 $K(\lambda) = \ker \lambda$, if $\lambda : A \to \hat{A}$ is a polarization of A,

  $\tilde{\epsilon} = \lambda_A^{-1}\hat{\epsilon}\lambda_B$, if (A, λ_A) and (B, λ_B) are ppavs and $\epsilon : A \to B$,

Aut(C) is the group of automorphisms of C; if $\alpha \in$ Aut(C), then we also let α
  denote the induced polarization preserving automorphism of $J(C)$,

 $[\mathbb{D}]$ is the linear equivalence class of the divisor $\mathbb{D}$ on C.

§1. First consider cyclic groups $G = \langle \alpha \mid \alpha^n = 1 \rangle \simeq \mathbb{Z}_n$. Suppose G acts on the curve C with $g = g(C) \geq 1$ and let $\pi : C \to C/G$ denote the quotient map. Then $\alpha^n - 1 = 0$ in End$(J(C))$.

Let $\phi_d(x)$ be the dth cyclotomic polynomial, so that $x^n - 1 = \prod_{d|n} \phi_d(x)$. These polynomials have integer coefficients and so $\phi_d(\alpha) \in$ End$(J(C))$. Let $A(d)$ be the connected component of the identity of $\ker \phi_d(\alpha)$. $A(d)$ is a (possibly trivial) subtorus of $J(C)$.

 REMARK. If $A(d) = \{0\}$, then $\phi_d(\alpha)$ is an isogeny of $J(C)$.

Since $T_0 A(d) \subset T_0 J(C) = H^0(C, \omega_C)^*$ is exactly the subspace where the matrix representing α has primitive dth roots of unity as eigenvalues, we see that $J(C)$ is isogenous to $\prod_{d|n} A(d)$. The main lemma is the following.

LEMMA. $A(n) \neq \{0\}$, that is, a matrix representing α always has a primitive nth root of unity as an eigenvalue. In fact, if π is branched over t points in C/G, then

$$dim\, A(n) = \frac{1}{2}(t + 2g(C/G) - 2)\phi(n),$$

where ϕ is Euler's phi-function.

This is very easy to prove using the decomposition of the action of G on $H_1(C, \mathbb{Z})$ in terms of local monodromies [5]. Since, for $d \mid n$, $J(C/\langle \alpha^d \rangle)$ is isogenous to $\prod_{e|d} A(e)$, an immediate consequence of this is that

$$A(d) = \{0\} \quad \text{iff} \quad g(C/\langle \alpha^d \rangle) = 0.$$

This then determines the minimal polynomial $f(x)$ of α in $\text{End}(J(C))$ as

$$f(x) = \prod_{\substack{d|n \\ g(C/\langle \alpha^d \rangle) \neq 0}} \phi_d(x).$$

Note that the minimal polynomial of α restricted to $A(d) \neq \{0\}$ is $\phi_d(x)$.

REMARK. Suppose π is determined by an epimorphism from Γ to G with torsion-free kernel, where Γ is a Fuchsian group with signature $[h; n_1, \cdots, n_t]$. Then $A(n)$ is the only non-trivial factor iff $h = 0$ and one of the following holds:
    (i) $n = p^k$, $n_1 = n_2 = p^k$, $n_3 = \cdots = n_t = p$ with p prime, $t \geq 3$ and $k \geq 1$, $g = \frac{1}{2}(t-2)(p-1)p^{k-1}$;
    (ii) $n = pq$, $n_1 = n_2 = p$, $n_3 = n_4 = q$ with p, q distinct primes, $g = (p-1)(q-1)$;
    (iii) $n = pq$, $n_1 = p$, $n_2 = pq$, $n_3 = q$ with p, q distinct primes, $g = \frac{1}{2}(p-1)(q-1)$.
Note that by the results in [26], the curves in case (ii) must all have an extra automorphism of order 2, generating with α a dihedral group of automorphisms of C. As we shall see, this implies that $J(C) = A(n)$ factors.

REMARK. Suppose $\beta \in \text{Aut}(C)$ normalizes $\langle \alpha \rangle$. Then β preserves all $A(d)$.

We will now examine generalizations of the examples in [10,25] where the Jacobi variety is isomorphic to a product. Suppose

$$G = \langle \alpha, \beta \mid \alpha^n = \beta^2 = 1 = \beta\alpha\beta\alpha \rangle \simeq D_{2n}.$$

Assume that n is an odd integer, $n \geq 3$. Then all elements of G of the form $\beta\alpha^m$ have order 2 and are conjugate in G to β. Suppose now that G acts on the curve C such that

$$g(C/\langle \alpha \rangle) = 0.$$

Then β has two fixed points and

$$g(C/\langle \beta \rangle) = \frac{1}{2}g(C).$$

Let $C_0 = C/\langle \beta \rangle$ and $\pi : C \to C_0$ be the quotient map. Then the transpose of π, $i = \tilde{\pi} : J(C_0) \to J(C)$, is an injection. We will identify $J(C_0)$ as a subtorus of $J(C)$ and drop the i unless needed.

Define $b : J(C_0) \times J(C_0) \to J(C)$ by

$$b(x, y) = x + (1 - \alpha)y.$$

Suppose $b(x, y) = 0$. Then

$$0 = x + (1 - \alpha)y.$$

Applying β to this equation we obtain

$$0 = x + (1 - \alpha^{-1})y.$$

Subtracting these two equations yields

$$\begin{aligned}
0 &= (\alpha^{-1} - \alpha)y \\
&= \alpha^{-1}(1 + \alpha)(1 - \alpha)y.
\end{aligned}$$

Since n is odd and $g(C/\langle \alpha \rangle) = 0$, we have that $1 + \alpha$ is a unit in $\mathbb{Z}[\alpha] \subset \text{End}(J(C))$. Also by the remarks above, $1 - \alpha$ is an isogeny of $J(C)$. Hence

$$H = \ker(1 - \alpha) \circ i$$

is a finite subgroup of $J(C_0)$ and

$$\ker b = \{0\} \times H.$$

Thus the induced map from $J(C_0) \times (J(C_0)/H)$ to $J(C)$ is an isomorphism.

We can say a little more. Let $\epsilon = \alpha + \alpha^{-1}$; then $2 - \epsilon = (1 - \alpha)(1 - \alpha^{-1})$. Since ϵ commutes with $1 + \beta$, it restricts to an automorphism of $J(C_0)$, which does not preserve the polarization of $J(C_0)$ unless $n = 3$. It is not hard to show that $\lambda_{C_0}(2 - \epsilon)$ is a polarization of $J(C_0)$, $\epsilon H \subset H$ and H is a maximal isotropic subgroup of $K(\lambda_{C_0}(2 - \epsilon)) \subset J(C_0)_n$. Thus $J(C_0)/H$ has an action by ϵ and a principal polarization, although it is not necessarily a Jacobi variety. With this data we can express the induced polarization and D_{2n} action on $J(C_0) \times (J(C_0)/H)$.

REMARK. In fact, $J(C_0)/H$ is isomorphic to $Prym(C_0, \pi)$ as ppavs.

Thus $J(C)$ belongs to a family of ppavs with D_{2n} action parametrized by (isomorphism classes of) quadruples $(A_0, \lambda_0, \iota : \mathbb{Z}[\eta] \to \text{End}(A_0), H)$ with the following properties. (A_0, λ_0) is a ppav. Suppose $f(T)$ is the minimal polynomial of α, $l = \frac{1}{2}\deg f$ and $Q(X)$ is the polynomial in $\mathbb{Z}[X]$ such that $T^l Q(T + T^{-1}) = f(T)$. Then $\mathbb{Z}[\eta] = \mathbb{Z}[X]/(Q(X))$ and ι is an injection such that if $\epsilon = \iota(\eta)$ then $\epsilon = \tilde{\epsilon}$. H is a maximal isotropic subgroup of $K(\lambda_0(2 - \epsilon))$.

REMARK. When $(A_0, \lambda_0) = (J(C_0), \lambda_{C_0})$ as above, the endomorphism ϵ comes from a correspondence $F : C \to C_0 \times C_0$ defined by $F(p) = (\pi(p), \pi(\alpha p))$. This in turn is determined by the existence of the function $C_0 = C/\langle \beta \rangle \to C/G = \mathbb{P}^1$ of degree n with a particular monodromy description and monodromy group D_{2n}. This raises several moduli questions.

> What is the dimension of a family of curves of genus g having a function of degree n with a given monodromy group $G \le S_n$?
>
> Does this imply any non-trivial endomorphisms of $J(C)$?
>
> What is the dimension of the family of Jacobi varieties $J(C)$ or ppavs A_0 with these endomorphisms?

Note that when $n = 3$, $\epsilon = -1$ is not a special endomorphism of $J(C_0)$, but the curves C_0 above are trigonal. We will return to these considerations in section 3. For another discussion see [16].

We could now apply the same reasoning to many more groups. We will illustrate this with the following examples.

Suppose k, n are positive integers such that $k^3 \equiv 1 \bmod n$ and $k \not\equiv 1 \bmod n$. We have $(k, n) = 1$. As in the case of dihedral groups, we will need in addition that

$$(k - 1, n) = 1.$$

As an automorphism of $\mathbb{Z}_n$, multiplication by k fixes only 0. It is easy to see that such k exists iff for any prime p dividing n, $p \equiv 1 \bmod 3$. We assume that k and n have the above properties. Then

$$k^2 + k + 1 \equiv 0 \bmod n.$$

Let

$$G = \langle \alpha, \beta \mid \alpha^n = \beta^3 = 1 = \beta \alpha \beta^2 \alpha^{-k} \rangle.$$

Then all elements of G of the form $\beta \alpha^m$ have order 3 and are conjugate to β. Suppose that G acts on the curve C such that $g(C/\langle \alpha \rangle) = 0$. Then β again has two fixed points and

$$g(C/\langle \beta \rangle) = \frac{1}{3} g(C).$$

Again, $1 - \alpha$ is an isogeny of $J(C)$ and $1 + \alpha + \cdots + \alpha^{n-1} = 0$ in $\mathrm{End}(J(C))$. In fact, since, for $d \mid n$, $d \ne n$, the group $\langle \beta, \alpha^d \rangle$ does not act on $\mathbb{P}^1$, this is the minimal polynomial of α.

Let $C_0 = C/\langle \beta \rangle$ and i denote the inclusion of $J(C_0)$ into $J(C)$. We again identify $J(C_0)$ as a subtorus of $J(C)$. Define $b : J(C_0) \times J(C_0) \times J(C_0) \to J(C)$ by

$$b(x, y, z) = x + (1 - \alpha)y + (1 - \alpha)^2 z.$$

Suppose

$$0 = x + (1 - \alpha)y + (1 - \alpha)^2 z.$$

Applying $\beta - 1$ and $\beta^2 - 1$ to this equation yields

$$0 = (\alpha - \alpha^k)(y + (2 - \alpha - \alpha^k)z)$$
$$0 = (\alpha - \alpha^{k^2})(y + (2 - \alpha - \alpha^{k^2})z).$$

Since $k - 1$ and $k^2 - 1$ are relatively prime to n, we can multiply these equations by appropriate units from $\mathbb{Z}[\alpha] \subset \text{End}(J(C))$ to obtain

$$0 = (1 - \alpha)(y + (2 - \alpha - \alpha^k)z)$$
$$0 = (1 - \alpha)(y + (2 - \alpha - \alpha^{k^2})z).$$

Subtracting these equations yields

$$0 = (1 - \alpha)(\alpha^k - \alpha^{k^2})z$$
$$= \text{unit} \cdot (1 - \alpha)^2 z.$$

Thus $b(x, y, z) = 0$ iff

$$x = 0, \quad (1 - \alpha)y = 0, \quad (1 - \alpha)^2 z = 0.$$

If we let, for $j = 1, 2, 3$,

$$H_j = \ker (1 - \alpha)^j \circ i,$$

then $H_1 \subset H_2 \subset H_3$ are finite subgroups of $J(C_0)$ and the induced map from $J(C_0) \times (J(C_0)/H_1) \times (J(C_0)/H_2)$ to $J(C)$ is an isomorphism.

Let $\epsilon = \alpha + \alpha^k + \alpha^{k^2}$. Since ϵ commutes with $1 + \beta + \beta^2$, it restricts to an endomorphism of $J(C_0)$ preserving each H_j. Hence ϵ acts on $J(C_0)/H_1$ and $J(C_0)/H_2$. Note that $\tilde\epsilon = \alpha^{-1} + \alpha^{-k} + \alpha^{-k^2}$ is the transpose of ϵ in $\text{End}(J(C))$ and

$$\tilde\epsilon - \epsilon = (1 - \alpha)(1 - \alpha^k)(1 - \alpha^{k^2}).$$

Also, the restriction of $\tilde\epsilon$ to $J(C_0)$ is the transpose of ϵ in $\text{End}(J(C_0))$. Hence $H_3 = \ker(\tilde\epsilon - \epsilon)$ in $J(C_0)$ and $J(C_0)/H_3 \simeq J(C_0)$. H_3 is a maximal isotropic subgroup for the polarization $\lambda_1 = (\tilde\epsilon - \epsilon)^*(\lambda_{C_0}) = \lambda_{C_0}(\epsilon - \tilde\epsilon)(\tilde\epsilon - \epsilon)$ of $J(C_0)$. Hence H_1 and H_2 are also isotropic subgroups and λ_1 descends to (generally non-principal) polarizations of $J(C_0)/H_1$ and $J(C_0)/H_2$.

For example, suppose $n = p$ is prime and the quotient map $C \to C/G$ has $(\beta, \beta^2\alpha^{-1}, \alpha)$ as a local monodromy description (see, for example [4,15,29]). Then $y^p = x(x-1)^k$ is an equation for C and $g(C) = \frac{1}{2}(p-1)$. Since $(1-\alpha)^{p-1} = \text{unit} \cdot p$ and $(\tilde\epsilon - \epsilon)^{\frac{1}{3}(p-1)} = \text{unit} \cdot p$, it is easy to see that, in this case,

$$H_1 = \{0\} \quad \text{and} \quad H_2 = H_3 \simeq \mathbb{Z}_p.$$

Thus $J(C)$ is isomorphic to $J(C_0) \times J(C_0) \times J(C_0)$. Note that when $n = 7$, C is the Klein-Hurwitz curve.

More generally, for $n = p$ prime and $C \to C/G$ branched over $t + 2$ points, 2 with local monodromy of order 3 and t with monodromy of order p, we have

$$H_1 \simeq \mathbb{Z}_p^{t-1}, \quad H_2 \simeq \mathbb{Z}_p^{2t-1} \quad \text{and} \quad H_3 \simeq \mathbb{Z}_p^{3t-2}.$$

This can be seen by deforming C to a reducible curve $C_1 \cup C_2 \cup C_2 \cup C_2$, where C_1 is the curve of the previous paragraph and C_2 is any curve of genus $\frac{1}{2}(t-1)(p-1)$ with an automorphism α_2 of order p such that $g(C_2/\langle \alpha_2 \rangle) = 0$.

We can describe some families of ppavs whose automorphism groups contain $S_n \times \mathbb{Z}_2$. (Note that all ppavs have -1 as a polarization preserving automorphism.) Suppose (A_0, λ_0) is a ppav of dimension d. Let $n \geq 3$ and $A = \prod_{i=1}^{n-1} A_0$. Let λ be the polarization

$$\lambda = \pi_1^*(\lambda_0) + \cdots + \pi_{n-1}^*(\lambda_0) + m^*(\lambda_0)$$

of A where π_i is the projection onto the ith factor and m is the addition map

$$m(x_1, \ldots, x_{n-1}) = x_1 + \cdots + x_{n-1}.$$

Then S_n acts as polarization preserving automorphisms of (A, λ) via the non-trivial component of the usual permutation representation. The transpose of m is the diagonal map

$$\tilde{m}(x) = (x, \ldots, x)$$

from A_0 to A. Hence $K(\lambda) = \tilde{m}((A_0)_n)$. If H_0 is any maximal isotropic subgroup of $K(n\lambda_0) = (A_0)_n$, then $H = \tilde{m}(H_0)$ is a maximal isotropic subgroup of $K(\lambda)$ preserved by S_n. Hence $(B, \lambda_B) = (A, \lambda)/H$ is a ppav of dimension $d(n-1)$ whose automorphism group contains S_n. The family of these abelian varieties is thus parametrized by (isomorphism classes of) triples (A_0, λ_0, H_0). Note that the abelian variety B is isomorphic to a product, even if the polarization λ_B is not. The map $f : A \to A$ defined by

$$f(x_1, \ldots, x_{n-1}) = (x_1 + x_{n-1}, \ldots, x_{n-2} + x_{n-1}, x_{n-1})$$

covers an isomorphism from $A_0 \times \cdots \times A_0 \times (A_0/H_0)$ to $A/H = B$.

For example, if $n = m^2$ and $H_0 = (A_0)_m$, this is a family of reducible ppavs with S_n action. If $n = 4$, $d = 1$ and $H_0 \simeq \mathbb{Z}_4$, this is the family of Jacobi varieties of curves of genus 3 whose automorphism groups contain S_4 studied in [19,25]. This family contains the Klein-Hurwitz curve and the Fermat quartic. If $n = 5$ and $d = 1$ (and so $H_0 \simeq \mathbb{Z}_5$), this is the one dimensional family of abelian varieties of dimension 4 with S_5 action containing the Jacobi variety of Bring's curve studied in [23]. If $n \geq 6$ and $d = 1$, then the automorphism group of (B, λ_B) is too large for it to be the Jacobi variety of a non-singular curve.

We conclude this section by considering the Jacobi varieties of the first few Hurwitz curves. The Klein-Hurwitz curve of genus 3 has already appeared. It's Jacobi variety can be shown to be isomorphic to a product of elliptic curves using the subgroups $Aff^+(1,7)$ or S_4 of its automorphism group. In the next section we will use the entire automorphism group $PSL(2,7)$ as in [11,25,29].

Macbeath showed in [20] that the Jacobi variety of Macbeath's curve C of genus 7 is isogenous to a product of elliptic curves using the subgroup $Aff(1,2^3)$ of its automorphism group $G = SL(2, 2^3)$. We can give an isogeny equivariant with respect to the entire automorphism group as follows. Suppose $\alpha, \beta, \gamma \in$

$SL(2, 2^3)$ are such that $\alpha\beta\gamma = 1 = \alpha^2 = \beta^3 = \gamma^7$. (Up to $\mathrm{Aut}(SL(2, 2^3))$ this triple is unique.) Then $E = C/\langle\beta\rangle$ is an elliptic curve. Identify $E = J(E)$ as a subtorus of $J(C)$. Let $A = \prod_{i=1}^{7} E$ and define $b : A \to J(C)$ by

$$b(x_1, \ldots, x_7) = x_1 + \gamma x_2 + \cdots + \gamma^6 x_7.$$

Using the character of the representation of G on $H^0(C, \omega_C)^*$ we find that

$$b^*(\lambda_C) = \lambda L,$$

where $\lambda = \pi_1^*(\lambda_E) + \cdots + \pi_7^*(\lambda_E)$ is the product polarization on A and $L \in M_7(\mathbb{Z}) \subseteq \mathrm{End}(A)$ is given by

$$L = \begin{pmatrix} 3 & -1 & -1 & 1 & 1 & -1 & -1 \\ -1 & 3 & -1 & -1 & 1 & 1 & -1 \\ -1 & -1 & 3 & -1 & -1 & 1 & 1 \\ 1 & -1 & -1 & 3 & -1 & -1 & 1 \\ 1 & 1 & -1 & -1 & 3 & -1 & -1 \\ -1 & 1 & 1 & -1 & -1 & 3 & -1 \\ -1 & -1 & 1 & 1 & -1 & -1 & 3 \end{pmatrix}.$$

We also have a representation $\rho : G \to SL(7, \mathbb{Z})$, determind by

$$\rho(\beta) = \begin{pmatrix} 1 & 0 & -1 & 0 & 1 & 0 & -1 \\ 0 & 0 & 0 & -1 & 0 & 1 & -1 \\ 0 & 0 & 0 & 0 & -1 & 1 & 0 \\ 0 & 0 & 0 & 0 & -1 & 0 & 0 \\ 0 & 0 & 0 & 1 & -1 & 0 & 0 \\ 0 & 0 & -1 & 1 & 0 & -1 & 0 \\ 0 & 1 & -1 & 0 & 1 & 0 & -1 \end{pmatrix}, \quad \rho(\gamma) = \begin{pmatrix} 0 & 0 & 0 & 0 & 0 & 0 & 1 \\ 1 & 0 & 0 & 0 & 0 & 0 & 0 \\ 0 & 1 & 0 & 0 & 0 & 0 & 0 \\ 0 & 0 & 1 & 0 & 0 & 0 & 0 \\ 0 & 0 & 0 & 1 & 0 & 0 & 0 \\ 0 & 0 & 0 & 0 & 1 & 0 & 0 \\ 0 & 0 & 0 & 0 & 0 & 1 & 0 \end{pmatrix},$$

such that for all $\delta \in G$, $b \circ \rho(\delta) = \delta \circ b$. Since these are polarization preserving automorphisms, we also have that, for all $\delta \in G$,

$$\rho(\delta)^t L \rho(\delta) = L.$$

Note that $\ker b$ must be a maximal isotropic subgroup of $K(\lambda L)$. In fact suppose $E_2 = \{0, \omega_1, \omega_2, \omega_3\}$. For $i = 1, 2, 3$ let

$$H_i = \{(x_1, \ldots, x_7) \in A : \text{for } 1 \leq j \leq 7, x_j = 0 \text{ or } \omega_i\},$$

$$H_i^0 = \{(x_1, \ldots, x_7) \in H_i : x_1 + \cdots + x_7 = 0\}.$$

H_i and H_i^0 are preserved by $\rho(G)$, H_i^0 is irreducible as a G-module and H_i^0 has no G-invariant complement in H_i. Then $K(\lambda L) = H_1^0 \oplus H_2^0$ as G-modules. If $\ker b = H_i^0$ for some i, then H_i/H_i^0 determines a $\mathbb{Z}_2$ in $J(C)$ fixed by G. Since G has no non-trivial extensions by $\mathbb{Z}_2$ and no non-trivial representations of dimension 6, no such $\mathbb{Z}_2$ exists. Also H_i^0 is isomorphic to $\mathbb{F}_8^2$ as G-modules. Thus $\ker b$ must be the graph in $\mathbb{F}_8^2 \oplus \mathbb{F}_8^2$ of scalar multiplication by an element of $\mathbb{F}_8$ which is neither 0 nor 1. (Note that then the permutations of $\{\omega_1, \omega_2, \omega_3\}$ are in one to one correspondence with $\mathbb{F}_8 \setminus \{0, 1\}$.) Since we could make this construction with any elliptic curve E and basis of E_2, we see that the Jacobi variety of Macbeath's curve lies in a one dimensional family of ppavs of dimension 7 whose automorphism groups contain $SL(2, 2^3)$.

REMARK. The same is true for the three curves of genus 14 whose automorphism groups are $PSL(2, 13)$.

§2. In this section we will examine the group of automorphisms induced on an unramified abelian cover of the Klein-Hurwitz curve. For an account of this topic, see [17].

Suppose $G \leq \text{Aut}(C)$ and $H \leq J(C)$ is a finite group preserved by the induced action of G. Suppose $p_0 \in C$, $z \in J(C)$ and define the usual embedding $\phi : C \to J(C)$ by

$$\phi(p) = [p - p_0] + z.$$

We identify C with $\phi(C)$. Then H determines an unramified abelian cover of C as follows. Letting $X = J(C)/H$, we obtain a sequence

$$1 \to H \to J(C) \to X \to 1.$$

Dualizing this sequence, letting $Y = \hat{X}$ and composing with λ_C^{-1}, we obtain

$$(1) \qquad\qquad\qquad 1 \to \hat{H} \to Y \xrightarrow{\psi} J(C) \to 1.$$

Let $D = \psi^{-1}(C)$. Translation by $\hat{H}$ in Y preserves D and $\psi : D \to C$ is the quotient map for this fixed point free action of $\hat{H}$.

Since G preserves H, the action of G on $J(C)$ lifts to an action of G on Y preserving $\hat{H}$. The action of G on $J(C)$ need not preserve C and so the action of G on Y need not preserve D. But we can change the action of G on $J(C)$ to an "affine" action which will preserve C. For $\eta \in G$ and $x \in J(C)$, let

$$U_\eta(x) = \eta x + u_\eta,$$

where $u_\eta = [\eta p_0 - p_0] + z - \eta z$. Then

$$(2) \qquad\qquad\qquad \phi \circ \eta = U_\eta \circ \phi.$$

It is clear that the collection $\{u_\eta : \eta \in G\}$ is a 1-cocycle determining an element Υ in $H^1(G, J(C))$. Hence the collection $\{U_\eta : \eta \in G\}$ forms a group of affine transformations of $J(C)$ isomorphic to G. From (2), we see that U_η maps C to itself. Note that changing ϕ by a translation changes the cocyle by a 1-coboundary.

We may extend the action of G from $J(C) = \text{Pic}_0(C)$ to all of $\text{Pic}(C)$.

LEMMA 1. Υ is trivial iff $\text{Pic}_1(C)^G \neq \emptyset$.

Now suppose that for each $\eta \in G$, v_η is any point in Y such that $\psi(v_\eta) = u_\eta$. Let V_η be the affine transformation of Y defined by $V_\eta(y) = \eta y + v_\eta$. Then

$U_\eta \circ \psi = \psi \circ V_\eta$ and so $\hat{H}$ and $\{V_\eta : \eta \in G\}$ generate a group $\mathcal{G}$ of affine transformations of Y preserving D. We have the sequence

$$(3) \qquad\qquad 1 \to \hat{H} \to \mathcal{G} \to G \to 1.$$

The sequence (1) is a short exact sequence of G (or $\mathbb{Z}[G]$) -modules. The long exact cohomology sequence associated to it contains

$$(4) \quad H^0(G, J(C)) \to H^1(G, \hat{H}) \to H^1(G, Y) \xrightarrow{\psi} H^1(G, J(C)) \xrightarrow{\delta} H^2(G, \hat{H}).$$

The next lemma follows from the the set-up above (see [**17**]).

LEMMA 2. $\delta(\Upsilon)$ is the class corresponding to the extension (3).

We now apply the above remarks to determine if the extension (3) is a semi-direct product for the Klein-Hurwitz curve C of genus 3, whose automorphism group is $G = PSL(2,7)$. Suppose $\alpha, \beta, \gamma \in G$ are such that $\alpha^2 = \beta^3 = \gamma^7 = \alpha\beta\gamma = 1$. Then $\theta = \gamma\beta\alpha$ has order 4, which completes a presentation of G. The quotient map $C \to C/G \simeq \mathbb{P}^1$ is branched over three points with (α, β, γ) as a monodromy description. If η is any element of G, then the genus of $C/\langle\eta\rangle$ is 1 if the order of η is 2, 3 or 4 and the genus is 0 if η has order 7.

LEMMA 3. In $End(J(C))$, let $\epsilon = \gamma + \gamma^2 + \gamma^4$. Then $\epsilon^2 + \epsilon + 2 = 0$, $\bar{\epsilon} = -1 - \epsilon$ and ϵ commutes with α and β.

Suppose $E = C/\langle\alpha\rangle$ and $\pi : C \to E$ is the quotient map. Then $g(E) = 1$. We identify $E = J(E)$ with image$(1 + \alpha)$ in $J(C)$. Since ϵ commutes with α, it restricts to a complex multiplication of E, which will be denoted by ϵ_0. Let ϵ_0 also denote the complex number $\frac{1}{2}(-1 + i\sqrt{7})$.

LEMMA 4. $E = \mathbb{C}/\mathbb{Z}[\epsilon_0]$ is the only elliptic curve with complex multiplication by ϵ_0 and $End(E) = \mathbb{Z}[\epsilon_0]$.

Let $A = E \times E \times E$ and $\lambda = \pi_1^*(\lambda_E) + \pi_2^*(\lambda_E) + \pi_3^*(\lambda_E)$, the product polarization on A. Define $b : J(C) \to A$ by

$$b(x) = (\pi(x), \pi(\beta x), \pi(\beta^2 x))$$

and define $\rho : G \to SL(3, \mathbb{Z}[\epsilon_0]) \subset End(A)$ by

$$\rho(\alpha) = \begin{pmatrix} 1 & 0 & 0 \\ \epsilon_0 & -1 & 0 \\ -1-\epsilon_0 & 0 & -1 \end{pmatrix} \quad \text{and} \quad \rho(\beta) = \begin{pmatrix} 0 & 1 & 0 \\ 0 & 0 & 1 \\ 1 & 0 & 0 \end{pmatrix}.$$

If

$$M = \begin{pmatrix} 2 & \epsilon_0 & -1-\epsilon_0 \\ -1-\epsilon_0 & 2 & \epsilon_0 \\ \epsilon_0 & -1-\epsilon_0 & 2 \end{pmatrix},$$

then λM is a principal polarization of A.

LEMMA 5. $b : (J(C), \lambda_C) \to (A, \lambda M)$ is an isomorphism of ppavs such that for all $\eta \in G$, $b \circ \eta = \rho(\eta) \circ b$.

We will abuse notation and identify $J(C)$ with A using b. Then $\mathrm{End}(J(C)) = M_3(\mathbb{Z}[\epsilon_0])$.

LEMMA 6. Suppose H is a finite subgroup of $J(C)$ preserved by G. Then there is a subgroup H_0 of E such that $H = H_0 \times H_0 \times H_0$ and $\epsilon_0 H_0 \leq H_0$.

Suppose H and H_0 are as in the above lemma. Then E/H_0 inherits the complex multiplication ϵ_0. Lemma 3 implies that $E/H_0 \simeq E$ and so there is an element $f(\epsilon_0) \in \mathbb{Z}[\epsilon_0]$ such that $H_0 = \ker f(\epsilon_0)$. Then $f(\epsilon) : J(C) \to J(C)$ has kernel H. Thus, in sequence (1), we may take $Y = J(C)$ and $\psi = f(\bar{\epsilon})$ has kernel $\hat{H} = \hat{H}_0 \times \hat{H}_0 \times \hat{H}_0$.

Now consider a 1-cocycle $\{w_\eta : \eta \in G\}$. Since $1 - \gamma$ is an isogeny of $J(C)$, we may change by a coboundary to obtain $w_\gamma = 0 = (0,0,0)^t$. Then a short calculation using the presentation of G shows that $w_\alpha = w_\beta = (0, e, -e)^t$, for some $e \in E$ such that $2(1 + 2\epsilon_0)e = 0$. Such a 1-cocycle is a 1-coboundary iff $(1 + 2\epsilon_0)e = 0$. Note that $(1 + 2\epsilon_0)^2 = -7$.

LEMMA 7. The map from E_2 to $H^1(G, J(C))$ which sends $\omega \in E_2$ to the class represented by the cocycle determined by $u_\alpha = u_\beta = (0, \omega, \omega)^t$ and $u_\gamma = 0$ is an isomorphism.

REMARK. Under this isomorphism, the map ψ in the sequence (4) becomes $f(\bar{\epsilon}_0) : E_2 \to E_2$.

Continuing with our choice of H as above, a similar calculation yields the following.

LEMMA 8. $H^1(G, \hat{H}) \simeq E_2 \cap \hat{H}_0$.

LEMMA 9. $H^0(G, J(C)) = J(C)^G = \{0\}$.

PROOF. For $K \leq G$, let $\iota(K) = \sum_{k \in K} k \in \mathrm{End}(J(C))$. Then $g(C/K) = 0$ iff $\iota(K) = 0$. Suppose $x \in J(C)^G$. Since $g(C/\langle \gamma \rangle) = 0$, we have $0 = \iota(K)x = 7x$. Also if $K \simeq S_3$, then $g(C/S_3) = 0$ and so $0 = \iota(K)x = 6x$. Together these imply that $x = 0$. ∎

Sequence (4) can now be rewritten as

$$(5) \qquad 0 \to E_2 \cap \hat{H}_0 \to E_2 \xrightarrow{\psi} E_2 \xrightarrow{\delta} H^2(G, \hat{H}).$$

Now suppose Υ is the class associated to the Klein-Hurwitz curve and ω and $\{u_\eta : \eta \in G\}$ correspond to Υ as in Lemma 7.

LEMMA 10. Υ is not trivial.

PROOF. Suppose $x = [p_1 + p_2 - p_3] \in \mathrm{Pic}_1(C)^G$. Note that $p_1 \neq p_3$ and $p_2 \neq p_3$ since no point of C is fixed by all of G. For any $\eta \in G$, $[p_1 + p_2 + \eta p_3] = [\eta p_1 + \eta p_2 + p_3]$. If $p_1 + p_2 + \eta p_3 = \eta p_1 + \eta p_2 + p_3$, then $p_3 = \eta p_3$, and this can happen for at most 7 elements of G. On the other hand, since C is not hyperelliptic, there is a unique differential on C with divisor $p_1 + p_2 + q_1 + q_2$. If $p_1 + p_2 + \eta p_3 \neq \eta p_1 + \eta p_2 + p_3$, then ηp_3 must be either q_1 or q_2. This can happen for at most 14 elements of G. Since these cases do not exhaust G, x cannot exist. ∎

REMARK. In fact, Burns shows in [6] that $\text{Pic}(C)^G \simeq \mathbb{Z}$, generated by a semi-canonical class in $\text{Pic}_2(C)$.

We can now prove the result of Cohen [8] mentioned above. Another proof can be given using the results in [17].

THEOREM. *Suppose H is a finite subgroup of $J(C)$ preserved by G. The extension (3) is split iff the order of H is odd.*

PROOF. Suppose H, and hence $\hat{H}_0$, have odd order. Then $E_2 \cap \hat{H}_0 = \{0\}$ and thus ψ is an isomorphism in (5). Hence $\delta(\Upsilon) = 0$.

Suppose $E_2 \subset \hat{H}_0$. Then $\psi = 0$ in (5) and hence δ is an isomorphism. Thus $\delta(\Upsilon) \neq 0$.

Suppose $E_2 = \{0, \omega_1, \omega_2, \omega_3\}$ such that $\epsilon_0 \omega_1 = \omega_1$ and $\epsilon_0 \omega_2 = 0$. The remaining possibilities for $E_2 \cap \hat{H}_0$ are $\{0, \omega_1\}$ and $\{0, \omega_2\}$. In both of these cases a short calculation shows that ω_3 is not in the image of ψ. Hence, if $\omega = \omega_3$, then $\delta(\Upsilon) \neq 0$.

Since all elements of order 2 in G are conjugate, $C/\langle \theta^2 \rangle \simeq E$. Since $g(C/\langle \theta \rangle) = 1$, $C/\langle \theta \rangle$ must be the quotient of E by a subgroup of order 2. We will show that $C/\langle \theta \rangle \simeq E/\langle \omega \rangle$. Applying $1 + \theta^2$ to $\phi(\theta p) = U_\theta \phi(p)$, we have

$$\begin{aligned}
(1 + \theta^2)\phi(\theta p) &= (1 + \theta^2)U_\theta \phi(p) \\
&= \theta(1 + \theta^2)\phi(p) + (1 + \theta^2)u_\theta \\
&= (1 + \theta^2)\phi(p) + (1 + \theta^2)u_\theta,
\end{aligned}$$
(6)

since θ acts as 1 on image$(1+\theta^2)$. Using the representation ρ, it is easy to see that image$(1+\theta^2) = \{0\} \times \{0\} \times E$ and $(1+\theta^2)u_\theta = (0, 0, (1+2\epsilon_0)\omega)^t = (0, 0, \omega)^t$. Thus equation (6) says that the automorphism induced on $C/\langle \theta^2 \rangle$ by θ is translation by ω.

REMARK. Since C is not hyperelliptic, the map $C^{(2)} \to J(C)$ which sends $p_1 + p_2$ to $\phi(p_1) + \phi(p_2)$ is an embedding. Also note that the map $C/\langle \theta \rangle = J(C/\langle \theta \rangle) \to J(C)$ necessarily has a kernel, since the maximal unramified abelian subcover of $C \to C/\langle \theta \rangle$ is nontrivial [1,25].

Since $\epsilon_0 : E \to E$ has kernel $\{0, \omega_1\}$ and $-1 - \epsilon_0$ has kernel $\{0, \omega_2\}$, we see that $E/\langle \omega_1 \rangle \simeq E/\langle \omega_2 \rangle \simeq E$. Thus if we show that $C/\langle \theta \rangle$ is not isomorphic to E, then we will be done. To see this, we will examine the family of curves of genus 3 whose automorphism groups contain S_4 studied by Kuribayashi and Sekita [19], that is, the family of plane quartics

$$X^4 + Y^4 + 1 + a(X^2Y^2 + X^2 + Y^2) = 0.$$

They show that this is the Klein-Hurwitz curve iff $a = 3\epsilon_0$ or $-3 - 3\epsilon_0$. We may take θ to be the automorphism of order 4 defined by $\theta(X, Y) = (-X^{-1}, YX^{-1})$. Then $C/\langle\theta^2\rangle$ has equation

$$V^2 = U^4 + \frac{2a}{a+2}U^2 + 1$$

and $C/\langle\theta\rangle$ has equation

$$V^2 = (U^2 - 1)(U^2 - \frac{a+1}{a+2}).$$

Taking $a = 3\epsilon_0$, we calculate the j-invariant of $C/\langle\theta^2\rangle$ to be -15^3 and the j-invariant of $C/\langle\theta\rangle$ to be $15^3 17^3$. Hence $C/\langle\theta^2\rangle$ is not isomorphic to $C/\langle\theta\rangle$. ∎

§3. In this section we will use the splitting of Prym varieties for cyclic unramified covers of hyperelliptic curves to examine the Prym map. Many of the results stated here are simple extensions of results in [2,3,9,13,14,22]. Curve still means non-singular curve unless otherwise stated.

Suppose $\pi : D \to C$ is the quotient map for the action of $G \leq \mathrm{Aut}(D)$ and $g = g(C) \geq 2$. Then we have induced maps $\pi : J(D) \to J(C)$ and $\tilde{\pi} : J(C) \to J(D)$. Note that $\tilde{\pi}^*(\lambda_D) = |G|\lambda_C$.

LEMMA 1 [1,25]. *Let G' be the commutator subgroup of G and G_0 be the (normal) subgroup of G generated by the elements of G having fixed points. Then $\ker\tilde{\pi}$ is canonically isomorphic to $(G/G_0G')\hat{\ }$.*

Now suppose that $G = \langle\alpha\rangle \simeq \mathbb{Z}_p$, where p is an odd prime. Recall from section 1 that $J(D)$ is isogenous to $A(1) \times A(p)$, with $A(1) = \tilde{\pi}(J(C))$. Let P denote $A(p)$ and $i : P \to J(D)$ the inclusion map. Then P inherits the polarization $\lambda_P = i^*(\lambda_D)$ from $J(D)$. The polarized abelian variety (P, λ_P) is called the Prym variety associated to the cover π and is denoted $Prym(C, \pi)$ or $Prym(\pi)$.

Let $F : J(C) \times P \to J(D)$ be the isogeny defined by $F = m \circ (\tilde{\pi} \times i)$. Suppose that π is branched over t points in C. Let $H = \ker\tilde{\pi}$. It is the subgroup of $J(C)$ determining π if $t = 0$ and $H = \{0\}$ if $t \neq 0$. Let $H^\perp$ be the orthogonal complement of H in $K(p\lambda_C)$. Then there is an isomorphism $\psi : H^\perp/H \to K(\lambda_P)$ giving the following.

LEMMA 2. *Suppose $\pi : D \to C$ is the quotient map for an action of G on D and $(P, \lambda_P) = Prym(\pi)$. Then*
   (i) $\dim P = \frac{1}{2}(2g - 2 + t)(p - 1)$,
   (ii) $K(\lambda_P) \simeq \mathbb{Z}_p^{2g-2\delta_{t,0}}$,
   (iii) $\ker F = \{(x, -\psi(\tilde{\pi}(x))) : x \in H^\perp\}$ is a maximal isotropic subgroup of $K(\pi_1^*(p\lambda_C) + \pi_2^*(\lambda_P))$.

REMARK. Since p is an odd prime, λ_P is never a multiple of a principal polarization on P.

The action of G on $J(D)$ preserves P. In fact, if $N(G)$ denotes the normalizer of G in $\mathrm{Aut}(D)$, then $N(G)$ restricts to polarization preserving automorphisms of $Prym(\pi)$. Thus β in $N(G)$ lead to splitting results for $Prym(\pi)$ in exactly the same way as they did for $J(C)$ in section 1.

We are interested in the map $Prym$ which associates the polarized abelian variety $Prym(\pi)$ to the quotient map π. Let $h = \dim P$. Let $\mathcal{A}_h(\lambda_P)$ denote the space of abelian varieties of dimension h with polarization of the same type as λ_P and $\mathcal{A}_h(\lambda_P, \mathbb{Z}_p)$ the space of triples (A, λ, G), where (A, λ) is in $\mathcal{A}_h(\lambda_P)$ and $G \simeq \mathbb{Z}_p$ acts by polarization preserving automorphisms in the same way as G acts on $Prym(\pi)$. That is, in a matrix representation of α on $T_0 J(D) = H^0(D, \omega_D)^*$, let n_k be the multiplicity of $e^{\frac{2\pi k i}{p}}$ as an eigenvalue. Then $n_0 = g$ and $n_1 + \cdots + n_{p-1} = h$. The sequence $(n_0, \ldots, n_{p-1})$ is determined up to permutation by the action of $\mathrm{Aut}(\mathbb{Z}_p)$ on the indices. By a result of Shimura [27],

$$\dim \mathcal{A}_h(\lambda_P, \mathbb{Z}_p) = \sum_{k=1}^{\frac{1}{2}(p-1)} n_k n_{p-k}.$$

Note that $\dim \mathcal{A}_h(\lambda_P) = \frac{1}{2}h(h+1)$.

Now let $\mathcal{R}_g^p$ denote the space of unramified G covers $\pi : D \to C$ where C is a curve of genus g. Then $n_0 = g$, $n_1 = \cdots = n_{p-1} = g - 1$, $h = (g-1)(p-1)$, $K(\lambda_P) \simeq \mathbb{Z}_p^{2g-2}$ and hence

$$\dim \mathcal{A}_h(\lambda_P, \mathbb{Z}_p) = \frac{1}{2}(g-1)^2(p-1).$$

Of course, the dimension of $\mathcal{R}_g^p$ is $3g - 3$. Then we have the map

$$Prym : \mathcal{R}_g^p \to \mathcal{A}_h(\lambda_P, \mathbb{Z}_p).$$

REMARK. $\dim \mathcal{R}_g^p \geq \dim \mathcal{A}_h(\lambda_P, \mathbb{Z}_p)$ iff $p = 3$ and $g = 2, 3, 4$ or $p = 5, 7$ and $g = 2$, with equality only if $p = 3, g = 4$ or $p = 7, g = 2$.

The Prym map can be extended to the space of so-called allowable covers of stable curves as in [2,9,13]. This space, which is denoted $\bar{\mathcal{R}}_g^p$, is a partial completion of $\mathcal{R}_g^p$ so that $Prym(\bar{\mathcal{R}}_g^p)$ is the closure of $Prym(\mathcal{R}_g^p)$ in $\mathcal{A}_h(\lambda_P, \mathbb{Z}_p)$. If $\pi : D \to C$ is the quotient map for a G action between stable curves of genera $(g-1)p+1$ and g, resp., then π is allowable iff

(i) if $q \in D$ is fixed by α, then q is a node, and, in order that the singularity and automorphism be smoothable, the rotation constants for the two branches must be inverses,

(ii) if $\Gamma(D)$ is the graph associated to D (for definitions, see [2,9,13]), then α acts as the identity on $H_1(\Gamma(D), \mathbb{Z})$.

These insure that $Prym(\pi)$ is an abelian variety.

EXAMPLE 1: WIRTINGER COVERS. Suppose X is any non-singular curve of genus $g - 1$ and q_1, q_2 are two distinct points of X. Let $C = X/q_1 \sim q_2$ and D be the p-gon formed by attaching together p copies of X so that q_2 on the i-th copy is identified with q_1 on the $i+1$-st copy. The automorphism α just rotates the p-gon. Then $\pi : D \to C = D/\langle \alpha \rangle$ is allowable and $Prym(\pi)$ is the polarized abelian variety (A, λ) of section 1, that is, $A = \prod_{i=1}^{p-1} J(X)$ and $\lambda = \pi_1^*(\lambda_X) + \cdots + \pi_{p-1}^*(\lambda_X) + m^*(\lambda_X)$. Recall that $\mathrm{Aut}(A, \lambda)$ contains $\mathbb{Z}_2 \times S_p$ and $K(\lambda) = \tilde{m}(J(X)_p)$. Note that $Prym(\pi)$ does not depend on q_1 and q_2.

EXAMPLE 2: ELLIPTIC TAILS. Similarly if D consists of p copies of X attached "evenly" around an elliptic curve E and α is the obvious rotation, then $Prym(\pi)$ is again (A, λ) as above.

LEMMA 3. *Suppose* (A_0, λ_0) *is a ppav with* A_0 *simple and* (A, λ) *is the polarized abelian variety formed from* (A_0, λ_0) *as above. Suppose* $\pi \in \bar{\mathcal{R}}_g^p$ *has* $Prym(\pi) = (A, \lambda)$. *Then there is a non-singular curve* X *of genus* $g - 1$ *such that* $(A_0, \lambda_0) = (J(X), \lambda_X)$ *and either* π *is a Wirtinger cover or* π *is a cover with an elliptic tail or* $p = 3$ *and* X *is trigonal.*

PROOF. The action of S_p on (A, λ) fixes every point of $K(\lambda)$. If we define an action of S_p on $J(C) \times A$ by having β in S_p act as 1 on $J(C)$, then S_p acts as polarization preserving automorphisms of $(J(C) \times A, p\lambda_C \times \lambda)$ preserving the maximal isotropic subgroup $\ker F$. Hence S_p descends to give polarization preserving automorphisms of the generalized Jacobi variety $J(D)$ of D. Let D_n and C_n denote the normalisations of D and C, resp. Then S_p also gives polarization preserving automorphisms of $J(D_n)$. Assume D is irreducible. Let K be the subgroup of S_p which comes from automorphisms of D_n. Then $G \leq A_p \leq K \leq G$ and so $\frac{1}{2} p! = p$. Hence $p = 3$. Also D has an automorphism β of order two which induces the automorphism -1 of $J(C)$ such that $X = D/\langle \beta \rangle$ is a non-singular curve of genus $g - 1$ with a g_3^1. Note that C_n has a g_2^1. Faber has examined the $p = 3$ case [13].

With the assumption that A_0 is simple, if D is reducible then an argument as in [2,9,13] yields the result. ∎

We can identify the cotangent spaces to these moduli spaces (or to suitable covers of them). Suppose $\pi \in \mathcal{R}_g^p$ and $(P, \lambda_P) = Prym(\pi) \in \mathcal{A}_h(\lambda_P, \mathbb{Z}_p)$. Let ρ be a line bundle on C of degree 0 representing a generator of the subgroup H of $J(C)$ determining π. Then

$$T_\pi^* \mathcal{R}_g^p \simeq H^0(D, \omega_D^2)^G \simeq H^0(C, \omega_C^2)$$

and

$$T_{Prym(\pi)}^* \mathcal{A}_h(\lambda_P, \mathbb{Z}_p) \simeq (Sym^2 \, T_0^* P)^G$$
$$\simeq \oplus_{k=1}^{\frac{1}{2}(p-1)} H^0(C, \omega_C \otimes \rho^k) \otimes H^0(C, \omega_C \otimes \rho^{-k}).$$

As in [3,9,13], the co-differential of $Prym$ is the cup-product

$$\mu : \oplus_{k=1}^{\frac{1}{2}(p-1)} H^0(C, \omega_C \otimes \rho^k) \otimes H^0(C, \omega_C \otimes \rho^{-k}) \to H^0(C, \omega_C^2).$$

The following lemma implies that $Prym$ is generically finite for $g \geq 5$ and $p \geq 3$.

LEMMA 4. *Suppose $\pi_0 : C \to X$ is a double cover of the curve X of genus 2 branched over $q_1, \ldots, q_{2r} \in X$ and determined by the line bundle δ of degree r on X such that $\delta^2 = \mathcal{O}_X(q_1 + \cdots + q_{2r})$. Let η be a line bundle of degree 0 on X such that $\eta^2 \neq \mathcal{O}_X$ and let $\rho = \pi_0^*(\eta)$. If $r \geq 3$ or $r = 2$ and $\delta^2 \neq \omega_X$, then up to replacing η by η^2, the cup-product map*

$$H^0(C, \omega_C \otimes \rho) \otimes H^0(C, \omega_C \otimes \rho^{-1}) \to H^0(C, \omega_C^2)$$

is surjective.

PROOF. With the hypotheses above, we have

$$H^0(C, \omega_C^2) \simeq H^0(X, \omega_X^2 \otimes \delta^2) \oplus H^0(X, \omega_X^2 \otimes \delta)$$

$$H^0(C, \omega_C \otimes \rho) \simeq H^0(X, \omega_X \otimes \eta) \oplus H^0(X, \omega_X \otimes \delta \otimes \eta)$$

$$H^0(C, \omega_C \otimes \rho^{-1}) \simeq H^0(X, \omega_X \otimes \eta^{-1}) \oplus H^0(X, \omega_X \otimes \delta \otimes \eta^{-1}).$$

Let ι denote the hyperelliptic involution of X. Then there are unique points p_1, p_2 such that $\omega_X \otimes \eta = \mathcal{O}_X(p_1 + p_2)$ and $\omega_X \otimes \eta^{-1} = \mathcal{O}_X(\iota p_1 + \iota p_2)$. If

$$(1) \qquad\qquad \{p_1, p_2\} \cap \{\iota p_1, \iota p_2\} \neq \emptyset,$$

then exactly one of the points, say p_1, is a Weierstrass point on X. In that case replacing η by η^2 replaces $p_1 + p_2$ by $2\, p_2$. Thus we may assume the intersection in (1) is empty.

Let ϕ_1 be the unique (up to scalar multiple) section of $\omega_X \otimes \eta$ and ϕ_2 the unique section of $\omega_X \otimes \eta^{-1}$. Then μ maps $\phi_1\, H^0(X, \omega_X \otimes \delta \otimes \eta^{-1}) \oplus \phi_2\, H^0(X, \omega_X \otimes \delta \otimes \eta)$ into $H^0(X, \omega_X^2 \otimes \delta)$. Since ϕ_1 and ϕ_2 vanish on disjoint divisors, the kernel of this map is isomorphic to $H^0(X, \delta)$. Counting dimensions, we see that this is onto. The base-point-free pencil trick then shows that $H^0(X, \omega_X \otimes \delta \otimes \eta) \otimes H^0(X, \omega_X \otimes \delta \otimes \eta^{-1})$ maps onto $H^0(X, \omega^2 \otimes \delta^2)$. ∎

We now return to the splitting results mentioned earlier. In particular, we are interested in covers of hyperelliptic curves. Let $\mathcal{H}_g^p$ denote the subspace of $\mathcal{R}_g^p$ of covers $\pi : D \to C$ such that C is hyperelliptic. The hyperelliptic involution of C lifts to an involution β of D and α, β generate a dihedral group $G_1 \simeq D_{2p}$ of automorphisms of D. Let $C_0 = D/\langle\beta\rangle$. Then $g_0 = g(C_0) = \frac{1}{2}(g - 1)(p - 1)$ and C_0 has a g_p^1, $f : C_0 \to \mathbb{P}^1$, whose monodromy group is D_{2p} and all local monodromies have order 2. Conversely the pair (C_0, f), where f has the above properties, has a curve D of genus $(g - 1)p + 1$ as "Galois closure". D is a 2-sheeted cover of C_0 and has a fixed-point-free automorphism α of order p such that $C = D/\langle\alpha\rangle$ is hyperelliptic. In the space $\mathcal{G}_{g_0}(g_p^1)$ of pairs (C_0, f) of curves of genus g_0 and maps $f : C_0 \to \mathbb{P}^1$ of degree p, let $\mathcal{G}_{g_0}(g_p^1, D_{2p})$ consist of the pairs (C_0, f) such that f has monodromy group D_{2p} or $\mathbb{Z}_p$. Let $\mathcal{G}_{g_0}^0(g_p^1, D_{2p})$ be the irreducible open subset of $\mathcal{G}_{g_0}(g_p^1, D_{2p})$ consisting of pairs (C_0, f) such that all local monodromies of f have order 2. Such an f is branched over $2g + 2$ points of $\mathbb{P}^1$. By considering covers of suitable $2g + 2$-pointed rational curves, we see that the other points of $\mathcal{G}_{g_0}(g_p^1, D_{2p})$ are limits of elements from $\mathcal{G}_{g_0}^0(g_p^1, D_{2p})$ where pairs of branch points "coalesce" to form ramification points of order p in f and rational components are blown down to obtain non-singular C_0. The associated covers π also deform to give an allowable cover $\pi : D \to C$. This includes all curves C_0 of genus g_0 having an automorphism of order p with $g + 1$ fixed points (hence $C_0/\langle \alpha_0 \rangle = \mathbb{P}^1$). D consists of two copies of C_0 with each fixed point of α_0 on the first copy of C_0 joined to its mate on the second copy and α acts by α_0 on one copy and by α_0^{-1} on the other. We have a bijection between $\mathcal{G}_{g_0}^0(g_p^1, D_{2p})$ and $\mathcal{H}_g^p$ and if $\bar{\mathcal{H}}_g^p$ denotes the closure of $\mathcal{H}_g^p$ in $\bar{\mathcal{R}}_g^p$, we have an inclusion of $\mathcal{G}_{g_0}(g_p^1, D_{2p})$ into $\bar{\mathcal{H}}_g^p$. Using the normal form for f given in [24], we see that $\bar{\mathcal{H}}_g^p$ also contains the hyperelliptic Wirtinger covers, that is, Wirtinger covers such that X is hyperelliptic and $p_1 = \iota p_2$, where ι is the hyperelliptic involution of X.

Recall from [24] the characterization of $Prym(\pi)$ for $\pi \in \mathcal{H}_g^p$. The endomorphism $\alpha + \alpha^{-1}$ of $J(D)$ commutes with $1 + \beta$ and so restricts to an automorphism ϵ of $J(C_0)$ such that $\tilde{\epsilon} = \epsilon$. Let $(P, \lambda_P) = Prym(\pi)$. Since $g(D/G_1) = 0$, we have that $(1 + \alpha + \cdots + \alpha^{p-1})(1 + \beta) = 0$ in $\text{End}(J(D))$ and so $J(C_0) \subset P$. If $b : J(C_0) \times J(C_0) \to J(D)$ is defined by $b(u, v) = u + \alpha v$, then b is an isomorphism onto P and

$$b^*(\lambda_P) = \begin{pmatrix} 2\lambda_{C_0} & \lambda_{C_0}\epsilon \\ \lambda_{C_0}\epsilon & 2\lambda_{C_0} \end{pmatrix}.$$

Note that $K(b^*(\lambda_P)) = \{(u, -u) : u \in J(C_0), (\epsilon - 2)u = 0\}$ and

$$\alpha \circ b = b \circ \begin{pmatrix} 0 & -1 \\ 1 & \epsilon \end{pmatrix}, \qquad \beta \circ b = b \circ \begin{pmatrix} 1 & \epsilon \\ 0 & -1 \end{pmatrix}.$$

Let $\gamma = e^{\frac{2\pi i}{p}}$ and $\mathfrak{o} = \mathbb{Z}[\gamma + \gamma^{-1}]$. Let $\mathcal{A}_{g_0}(\lambda_0, \mathfrak{o})$ be the space of triples (A_0, λ_0, R) where (A_0, λ_0) is a ppav of dimension g_0 and there is a homomorphism $j : \mathfrak{o} \to \text{End}(A_0)$ such that $R = j(\mathfrak{o})$, $\epsilon = j(\gamma + \gamma^{-1})$ acts as ϵ does on $J(C_0)$ and $\tilde{\epsilon} = \epsilon$. Let $\mathcal{A}_h(\lambda_P, D_{2p})$ be the space of triples (A, λ, G_1) where $(A, \lambda) \in \mathcal{A}_h(\lambda_P)$ and $G_1 \simeq D_{2p}$ acts as polarization preserving automorphisms in the same way as G_1 acts on $Prym(\pi)$ for $\pi \in \mathcal{H}_g^p$. The above splitting for $Prym(\pi)$ works for any element of $\mathcal{A}_h(\lambda_P, D_{2p})$ yielding an isomorphism $\Delta : \mathcal{A}_{g_0}(\lambda_0, \mathfrak{o}) \to \mathcal{A}_h(\lambda_P, D_{2p})$. By the results in [27],

$$\dim \mathcal{A}_{g_0}(\lambda_0, \mathfrak{o}) = \frac{1}{4}g(g-1)(p-1).$$

Thus the restriction of $Prym$ maps $\bar{\mathcal{H}}_g^p$ to $\mathcal{A}_h(\lambda_P, D_{2p})$. Because of the above isomorphism, we will view $Prym$ as mapping $\bar{\mathcal{H}}_g^p$ to $\mathcal{A}_{g_0}(\lambda_0, \mathfrak{o})$.

REMARK. An easy calculation shows that if $(A_0, \lambda_0, R) \in \mathcal{A}_{g_0}(\lambda_0, \mathfrak{o})$, $\text{End}(A_0) = R$ and $(A, \lambda, G_1) = \Delta(A_0, \lambda_0, R)$, then $\text{Aut}(A, \lambda) \simeq \mathbb{Z}_2 \times D_{2p}$. Hence the composition of Δ with the natural forgetful map $\mathcal{A}_h(\lambda_P, D_{2p}) \to \mathcal{A}_h(\lambda_P, \mathbb{Z}_p)$ is injective on the set where $\text{End}(A_0) \simeq \mathfrak{o}$.

LEMMA 5. *Suppose* $\pi \in \mathcal{R}_g^p$ *and* $Prym(\pi) = (A, \lambda, G)$, *where* $(A, \lambda, G_1) = \Delta(A_0, \lambda_0, R)$ *and* $G \leq G_1$. *Then* $\pi \in \mathcal{H}_g^p$.

PROOF. Suppose $\pi : D \to C$. Then any $\beta_1 \in G_1$ of order 2 acts as -1 on $K(\lambda)$. If we define β on $J(C) \times A$ to be $-1 \times \beta_1$, then β is a polarization preserving automorphism of $(J(C) \times A, p\lambda_C \times \lambda)$ preserving the maximal isotropic subgroup $\ker F$. Hence β descends to a polarization preserving automorphism of $J(D)$. The automorphism $-\beta$ cannot come from an automorphism of D, since it induces the identity on $J(C)$ and D is irreducible. Hence β comes from an automorphism of D, $\beta \in N(G)$ and so C is hyperelliptic. ∎

REMARK. $\dim \mathcal{A}_h(\lambda_P, \mathbb{Z}_p) = \dim \mathcal{A}_h(\lambda_P, D_{2p})$ iff $g = 2$. Also $\dim \mathcal{H}_g^p > \dim \mathcal{A}_{g_0}(\lambda_0, \mathfrak{o})$ iff $p = 3$ and $g = 2, 3, 4$ or $p = 5$ and $g = 2$. As noted before, equality holds only for $p = 7$ and $g = 2$.

We can again identify cotangent spaces to the various moduli spaces. Suppose $\pi \in \mathcal{H}_g^p$ and $(P, \lambda_P, G_1) = Prym(\pi) \in \mathcal{A}_h(\lambda_P, D_{2p})$, ι is the hyperelliptic involution of C and let ρ be a line bundle on C of degree 0 representing a generator of the subgroup H of $J(C)$ determining π. Then ι induces isomorphisms $H^0(C, \omega_C \otimes \rho^k) \to H^0(C, \omega_C \otimes \rho^{-k})$ and we have

$$T_\pi^* \mathcal{H}_g^p \simeq H^0(D, \omega_D^2)^{G_1} \simeq H^0(C, \omega_C^2)^{(\iota)}$$

and

$$T_{Prym(\pi)}^* \mathcal{A}_h(\lambda_P, D_{2p}) \simeq (Sym^2 T_0^* P)^{G_1}$$

$$\simeq \oplus_{k=1}^{\frac{1}{2}(p-1)} Sym^2 H^0(C, \omega_C \otimes \rho^k)$$

and the codifferential of $Prym$ restricted to $\mathcal{H}_g^p$ is $\mu \circ (1 \otimes \iota)$. Rather than try to determine the rank of this map, we will use the relationship of $Prym(\pi)$ to Jacobi varieties of special curves.

Finally let $\mathcal{M}_{g_0}$ be the space of non-singular curves of genus g_0 and $\mathcal{M}_{g_0}(\mathfrak{o})$ the space of pairs (C_0, R) where $C_0 \in \mathcal{M}_{g_0}$ and $R \leq End(J(C_0))$ is isomorphic to $\mathfrak{o}$ as above. Let $\Psi : \mathcal{M}_{g_0}(\mathfrak{o}) \to \mathcal{M}_{g_0}$ be the usual forgetful map. Then $\Psi(\mathcal{M}_{g_0}(\mathfrak{o}))$ is the space of non-singular curves of genus g_0 with a certain (singular, when $p \geq 5$) correspondence. We saw above that the g_p^1 f induced this correspondence on C_0 when $(C_0, f) \in \mathcal{G}_{g_0}(g_p^1, D_{2p})$. Since $Aut(C_0, f) = \{\delta \in Aut(C_0) : f \circ \delta = f\}$ is trivial for $(C_0, f) \in \mathcal{G}_{g_0}^0(g_p^1, D_{2p})$, a Hurwitz family exists [4,15] and we have a map $\Phi : \mathcal{G}_{g_0}(g_p^1, D_{2p}) \to \mathcal{M}_{g_0}(\mathfrak{o})$. This leads to the questions raised in section 1.

Thus we get a commutative diagram

$$
\begin{array}{ccc}
\bar{\mathcal{H}}_g^p & \xrightarrow{Prym} & \mathcal{A}_{g_0}(\lambda_0, \mathfrak{o}) \\
\uparrow & & \uparrow \\
\mathcal{G}_{g_0}(g_p^1, D_{2p}) & \xrightarrow{\Phi} \mathcal{M}_{g_0}(\mathfrak{o}) & \xrightarrow{\Psi} \mathcal{M}_{g_0}
\end{array}
$$

where the right vertical map sends (C_0, R) to $(J(C_0), R)$. Both vertical maps are injective. Thus $Prym$ is essentially equivalent to Φ.

LEMMA 6. *There are elements* $(C_0, f) \in \mathcal{G}_{g_0}(g_p^1, D_{2p})$ *such that* $J(C_0)$ *is simple and* $\text{End}(J(C_0)) \simeq \mathfrak{o}$. *Hence* Ψ *is generically injective.*

PROOF. We will fix p and use induction on g. For $g = 2$, consider the curve C_0 of genus $\frac{1}{2}(p - 1)$ with equation $y^2 = x^p - 1$. It is proved in [28] that $J(C_0)$ is simple and $\text{End}(J(C_0)) \simeq \mathbb{Z}[\gamma]$. Since most C_0 do not have an automorphism of order p, there must be C_0 with $\text{End}(J(C_0)) \simeq \mathfrak{o}$.

Assume first that for each π in (some cover of) $\mathcal{H}_g^p$, $J(C_0)$ is isogenous to $A(\pi) \times B(\pi)$ with A and B depending continuously on π. We can specialize π to a hyperelliptic Wirtinger cover and then $J(C_0)$ specializes to the product of $\frac{1}{2}(p - 1)$ copies of $J(X)$, where X is any hyperelliptic curve of genus $g - 1$. As remarked in [18] there are hyperelliptic curves of any genus with simple Jacobi varieties. Thus $A(\pi)$ and $B(\pi)$ specialize to products of $J(X)$. If $p = 3$ we are done. For $p \geq 5$, we can also specialize π as follows. Choose $\pi' \in \mathcal{H}_{g-1}^p$, $\pi' : D' \to C'$, $q \in C'$ and E any elliptic curve. Let $C = C' \amalg E/q \sim 0$ and attach p copies of E to $\pi'^{-1}(q)$ in D' to form D. Let $C_0' = D'/\langle\beta\rangle$. Then $J(C_0)$ specializes to $J(C_0') \times \prod_{i=1}^{\frac{1}{2}(p-1)} E$. Since this π and the hyperelliptic Wirtinger cover have a common specialization, $A(\pi) \times B(\pi)$ must specialize to a splitting of $J(C_0')$ with a division of the copies of E. This contradicts the induction assumption. Hence no splitting can work for all $J(C_0)$. In fact $J(C_0)$ must be simple on the complement of a countable union of proper subvarieties.

Now suppose that there is a ring $\mathfrak{o}_1$ containing $\mathfrak{o}$ such that for each π there is $R(\pi) \leq \text{End}(J(C_0))$ with $R(\pi) \simeq \mathfrak{o}_1$. Again consider the specialization in the paragraph above formed from $\pi' \in \mathcal{H}_{g-1}^p$. Assume that $\text{End}(J(C_0')) \simeq \mathfrak{o}$. Then the composition

$$\mathfrak{o} \to \mathfrak{o}_1 \to \mathfrak{o} \oplus M_{\frac{1}{2}(p-1)}(\text{End}(E)) \xrightarrow{proj_1} \mathfrak{o}$$

is the identity on $\mathfrak{o}$. But $\mathfrak{o}_1$ can have no zero-divisors since most $J(C_0)$ are simple. Hence $\mathfrak{o}_1 \simeq \mathfrak{o}$. ∎

REMARK. Suppose $(C_0, R) \in \mathcal{M}_{g_0}(\mathfrak{o})$ and $\text{End}(J(C_0)) \simeq \mathfrak{o}$. Then all f such that $(C_0, f) \in \mathcal{G}_{g_0}(g_p^1, D_{2p})$ induce the same ring of correspondences on $J(C_0)$.

REMARK. Since p is prime, if C_0 has two (inequivalent) g_p^1's, then $g_0 = \frac{1}{2}(g - 1)(p - 1) \leq (p - 1)^2$, that is, $g \leq 2p - 1$. Hence, for $g \geq 2p$, $\Psi \circ \Phi$ is injective.

Now we will consider the rank of $Prym$ restricted to the hyperelliptic locus. Let $\mathcal{G}$ denote $\mathcal{G}_{g_0}(g_p^1, D_{2p})$. Suppose first that $p = 3$. Then $\mathfrak{o} = \mathbb{Z}$ and $(\Psi \circ \Phi)(\mathcal{G})$ is the locus of trigonal curves of genus $g_0 = g - 1$. Hence

$$\text{rank } Prym = \dim \Phi(\mathcal{G}) = \begin{cases} 1, & \text{if } g = 2 \\ 3, & \text{if } g = 3 \\ 6, & \text{if } g = 4 \\ 2g - 1, & \text{if } g \geq 5. \end{cases}$$

Comparing this to $\dim \mathcal{A}_{g-1}(\lambda_0)$ and using the remark above we obtain the following.

THEOREM. *Prym restricted to $\bar{\mathcal{H}}_g^3$ is surjective for $g = 2, 3, 4$, generically two to one for $g = 5$ and generically injective for $g \geq 6$.*

PROOF. We need only consider the case $g = 5$. Since $\dim \mathcal{G} = \dim \Phi(\mathcal{G})$, for the generic point $(J(C_0), \lambda_{C_0})$ in $Prym(\mathcal{H}_5^3)$, $Prym^{-1}(J(C_0), \lambda_{C_0}) \subset \mathcal{H}_5^3$. Then Φ is generically two to one since the generic curve C_0 of genus 4 has exactly two g_3^1's. ∎

Now assume $p \geq 5$. Suppose π is a hyperelliptic Wirtinger cover. Let X be a hyperelliptic curve of genus $g - 1$ and choose (a non-Weierstrass point, if $g \geq 3$) $q_1 \in X$. D is a p-gon formed from p copies of X by identifying ιq_1 on the i-th copy of X with q_1 on the $i+1$-st. The automorphism β just flips the p-gon, that is, q on the i-th copy of X is sent to ιq on the $p - i$-th copy. $D/\langle \beta \rangle$ consists of $\frac{1}{2}(p-1)$ copies of X and one copy of $\mathbb{P}^1$, which gets blown down to obtain a stable curve in $\bar{\mathcal{M}}_{g_0}$, the space of stable curves of genus g_0. The locus of such curves is contained in the closure of $(\Psi \circ \Phi)(\mathcal{G})$ in $\bar{\mathcal{M}}_{g_0}$ and has dimension 1 if $g = 2, p = 5$ and dimension $2g - 2$ otherwise. Since $(\Psi \circ \Phi)(\mathcal{G})$ certainly contains non-singular curves, we have

$$\dim (\Psi \circ \Phi)(\mathcal{G}) \geq \begin{cases} 2, & \text{if } g = 2, p = 5 \\ 2g - 1, & \text{if } g = 2, p \geq 7 \text{ or } g \geq 3, p \geq 5. \end{cases}$$

THEOREM. *For $p \geq 5$, Prym restricted to $\bar{\mathcal{H}}_g^p$ is generically finite, except in the one case when $g = 2, p = 5$, where Prym generically has one dimensional fibers.*

PROOF. We only need to consider the exceptional case. Note that in this one case, the choice of $q_1 \in X$, an elliptic curve, is "lost" when $\mathbb{P}^1$ is blown down. All the curves in $(\Psi \circ \Phi)(\mathcal{G})$ have a singular correspondence. Since the generic curve of genus 2 has no singular correspondences [18], $\dim \Phi(\mathcal{G}) \leq 2$. ∎

REMARK. Recall the remark above that, if $g \geq 2p$, then Φ must be injective. This bound can be considerably improved. Suppose (C_0, f_1) and (C_0, f_2) are in $\mathcal{G}_{g_0}(g_p^1, D_{2p})$. Let $\pi_{01} : D_1 \to C_0$ and $\pi_{02} : D_2 \to C_0$ be the associated 2-sheeted covers of C_0. Suppose f_1 and f_2 induce the same ring of correspondences of C_0. Then for some k between 1 and $\frac{1}{2}(p-1)$, the maps $D_1 \to C_0 \times C_0$ and $D_2 \to C_0 \times C_0$ defined by sending $q \in D_1$ to $(\pi_{01}(q), \pi_{01}(\alpha q))$ and $q \in D_2$ to $(\pi_{02}(q), \pi_{02}(\alpha^k q))$, resp., induce the same endomorphism of $J(C_0)$. For $p \in C_0$, let $\mathbb{D}_i(p)$, $i = 1, 2$, be the divisors of degree 2 on C_0 defined by

$$p \times \mathbb{D}_i(p) = D_i \cap (p \times C_0).$$

Then this common endomorphism is determined by sending the divisor class of degree 0, $[p - q]$, in $J(C_0)$ to $[\mathbb{D}_1(p) - \mathbb{D}_1(q)] = [\mathbb{D}_2(p) - \mathbb{D}_2(q)]$. This implies that, for all $p, q \in C_0$, $\mathbb{D}_1(p) + \mathbb{D}_2(q)$ is linearly equivalent to $\mathbb{D}_1(q) + \mathbb{D}_2(p)$. If D_1 does not equal D_2, then C_0 has a g_4^1 (possibly with base-points). Since $(4, p) = 1$, we have $g_0 = \frac{1}{2}(g-1)(p-1) \leq (4-1)(p-1)$, that is, $g \leq 7$. Thus, if $g \geq 8$, then Φ is injective.

The case $g = 2, p = 5$ is exceptional for another reason. (For a fuller account of this topic, see [16].) Suppose $f : C_0 \to \mathbb{P}^1$ is a g_n^1 on a curve of genus $g_0 \geq 2$ with monodromy group G. Suppose f is branched over t points in $\mathbb{P}^1$. Speaking loosely, the space of all possible deformations of this map is paramatrized by a Hurwitz space $\mathcal{H}$, which is a finite unbranched cover of $(\mathbb{P}^1)^{(t)} - \Delta_t$, the space of (unordered) sets of t distinct points from $\mathbb{P}^1$. The question is: How many different curves are paramatrized by this? If $\delta \in \mathrm{Aut}(\mathbb{P}^1)$, then $\delta \circ f$ is really the same as f. So the space of inequivalent (C_0, f)'s has dimension $t - 3$. Under the right conditions, there is a morphism $\Phi : \mathcal{H} \to \mathcal{M}_{g_0}$ and the question becomes: What is the dimension of $\Phi(\mathcal{H})$?

EXAMPLE. If $f : C_0 \to \mathbb{P}^1$ is the quotient map for the action of $G \leq \mathrm{Aut}(C_0)$, then $\dim \Phi(\mathcal{H}) = t - 3$.

EXAMPLE. If $f : C_0 \to \mathbb{P}^1$ is a simple cover, that is, for all $p \in \mathbb{P}^1$, $f^{-1}(p)$ consists of n or $n - 1$ points, then $G = S_n$ and $\dim \Phi(\mathcal{H}) = \min\{t - 3, 3g_0 - 3\}$.

The map f is called primitive if f cannot be factored as $C_0 \xrightarrow{f_1} X \xrightarrow{f_2} \mathbb{P}^1$ for some curve X. This is equivalent to the condition that if $\pi_0 : D \to C_0$ is the Galois closure of f and $G_0 \leq G$ is such that $C_0 = D/G_0$, then there are no groups K strictly between G_0 and G. Zariski conjectures in [30] that if $t \geq 3g_0$ and f is primitive, then $\dim \Phi(\mathcal{H}) = 3g_0 - 3$. As we have seen, $\mathcal{G}_2(g_5^1, D_{10})$ is 3-dimensional since $t = 6$, but $\dim \Phi(\mathcal{G}_2(g_5^1, D_{10})) = 2$.

References

1. R. D. M. Accola, *Riemann surfaces, theta-functions and abelian automorphism groups*, Lecture Notes in Math., vol. 483, Springer-Verlag, Berlin-Heidelberg-New York, 1975.
2. A. Beauville, *Prym varieties and the Schottky problem*, Invent. Math. **41** (1977), 149-196.
3. ———, *Variétiés de Prym et Jacobiennes intermédiares*, Ann. Sci. École Norm. Sup. **10** (1977), 309-391.
4. R. Biggers, M. Fried, *Relations between moduli spaces of covers of $\mathbb{P}^1$ and representations of the Hurwitz monodromy group*, J. für d. reine Math. **235** (1982), 87-121.
5. S. A. Broughton, *The homology and higher representations of the automorphism group of a Riemann surface*, Trans. Amer. Soc. **300** (1987), 153-158.
6. D. Burns, *On the geometry of elliptic modular surfaces and representations of finite groups*, Lecture Notes in Math., vol. 1008, 1983, pp. 1-29.
7. J. Cohen, *On Hurwitz extensions of $PSL(2,7)$*, Math. Proc. Camb. Phil. Soc. **86** (1979), 395-400.
8. ———, *On covering Klein's curve and generating projective groups*, The Geometric Vein: the Coxeter Festschrift (C. Davis, ed.), Springer-Verlag, Berlin-Heidelberg-New York, 1982, pp. 511-518.
9. R. Donagi, R. C. Smith, *The structure of the Prym map*, Acta Math. **146** (1981), 25-102.
10. C. Earle, *Some Jacobian varieties which split*, Lecture Notes in Math., vol. 747, 1979, pp. 101-107.
11. ———, *H. E. Rauch, function theorist*, Differential Geometry and Complex Analysis (I. Chavel, H. M. Farkas, eds.), Springer-Verlag, Berlin-Heidelberg-New York, 1985, pp. 15-31.
12. ———, *Some Riemann surfaces whose Jacobians have strange product structures*, in this volume.

13. C. Faber, *Prym varieties of triple cyclic covers*, Math. Z. **199** (1988), 61-79.

14. J. Fay, *Theta-functions on Riemann Surfaces*, Lecture Notes in Math., vol. 352, Springer-Verlag, Berlin-Heidelberg-New York, 1973.

15. M. Fried, *Fields of definition of function fields and Hurwitz families - groups as Galois groups*, Comm. Alg. **5** (1977), 17-82.

16. _____, *Combinatorial computation of moduli dimension of Nielsen classes of covers*, Contemp. Math. **89** (1989), 61-79.

17. M. Fried, H. Völklein, *Unramified abelian extensions of Galois covers*, Theta-functions Bowdoin 1987 (L. Ehrenpreis, R. C. Gunning, eds.), Proc. Sym. Pure Math., vol. 49, 1989, pp. 675-693.

18. S. Koizumi, *The ring of algebraic correspondences on a generic curve of genus g*, Nagoya Math. J. **60** (1976), 173-180.

19. A. Kuribayashi, E. Sekita, *On a family of Riemann surfaces I.*, Bull. Facul. Sci. Eng. Chuo U. **22** (1979), 107-129.

20. A. M. Macbeath, *On a curve of genus 7*, Proc. Lond. Math. Soc. **15** (1965), 527-542.

21. H. H. Martens, *Riemann matrices with many polarizations*, Complex Analysis and its Applications, vol. III, International Atomic Energy Agency, Vienna, 1976, pp. 35-48.

22. D. Mumford, *Prym varieties I*, Contributions to Analysis, Academic Press, New York, 1974, pp. 325-350.

23. G. Riera, R. Rodríguez, *The period matrix of Bring's curve*, Pacific Jour. Math. (to appear).

24. J. F. X. Ries, *The Prym variety for a cyclic unramified cover of a hyperelliptic Riemann surface*, J. für d. reine Math. **340** (1983), 59-69.

25. _____, *The splitting of some Jacobi varieties using their automorphism groups*, preprint.

26. _____, *Subvarieties of moduli space determined by finite groups acting on surfaces*, Trans. Amer. Math. Soc. (to appear).

27. G. Shimura, *On analytic families of polarized abelian varieties and automorphic functions*, Annals Math. **78** (1963), 149-192.

28. G. Shimura, Y. Taniyama, *Complex Multiplication of Abelian Varieties*, Math. Soc. Japan, Tokyo, 1961.

29. C. L. Tretkoff, M. D. Tretkoff, *Combinatorial group theory, Riemann surfaces and differential equations*, Contemp. Math. **33** (1984), 467-519.

30. O. Zariski, *On the moduli of algebraic functions possessing a given monodromie group*, Collected Papers vol III, Topology of Curves and Surfaces and Special Topics in the theory of Algebraic Varieties, MIT Press, 1978, pp. 155-175.

MATHEMATICS DEPARTMENT, SUNY AT BINGHAMTON, BINGHAMTON, NY 13902

Contemporary Mathematics
Volume **136**, 1992

Theta Divisors for Vector Bundles

MONTSERRAT TEIXIDOR-I-BIGAS AND LORING W. TU

ABSTRACT. Let $\mathcal{M}_{n,d}$ be the moduli space of semistable bundles of rank n and degree d over a curve of genus ≥ 2. We give an upper bound for the dimension of the space of theta functions of any level on $\mathcal{M}_{n,d}$. We also show that every theta diviosr on $\mathcal{M}_{n,d}$ is ample. These two results imply as a corollary an upper bound for the self-intersection number of the theta divisor when the moduli space is smooth.

1 Introduction

The theta functions that play such an important role in number theory may be viewed as sections of a line bundle $[\Theta]$ on the Jacobian $J(C)$ of a smooth projective curve C. This line bundle $[\Theta]$ is associated to a divisor Θ in $J(C)$, usually called a *theta divisor*, which is simply the zero set of a theta function on $J(C)$. Let g be the genus of C. Among the properties of a theta divisor, four are particularly noteworthy for their many applications:

(1) The space of theta functions of level k is k^g-dimensional:
$$\dim H^0(J(C), [\Theta]^k) = k^g.$$

(2) The line bundle $[\Theta]$ on the Jacobian $J(C)$ is ample.

(3) (Poincaré's formula) If Θ^g denotes the self-intersection of the theta divisor with itself g times, then $\Theta^g = g!$.

(4) A theta divisor is irreducible.

Since the Jacobian $J(C)$ may be interpreted as the moduli space of line bundles of a fixed degree on C, a natural generalization would be to consider the moduli space $\mathcal{M}_{n,d}$ of semistable vector bundles of rank n and degree d on C. In [DN] Drezet and Narasimhan construct divisors Θ_F on $\mathcal{M}_{n,d}$, associated to suitably chosen vector bundles F on C. These divisors Θ_F generalize in a natural way the theta divisors on the Jacobian of C. The irreducibility of Θ_F

1991 Mathematics Subject Classification. Primary 14H60.
The first author was partially supported by an NSF grant.
This paper is in final form and no version of it will be submitted for publication elsewhere.

has been proven by Feinberg [F]. The goal of this paper is to prove the analogues of the other three theorems above for vector bundles.

Throughout this paper C will be a smooth irreducible projective curve of genus $g \geq 2$ over the complex numbers. We denote by $\mathcal{M}_{n,d}(C)$ or simply $\mathcal{M}_{n,d}$ the moduli space of equivalence classes of *semistable* vector bundles of rank n and degree d on C. Similarly, $\mathcal{M}_{n,d}^{\text{stab}}$ denotes the moduli space of isomorphism classes of *stable* vector bundles of rank n and degree d on C. For a vector bundle F on C, define

$$\Theta_F^{\text{stab}} = \{E \in \mathcal{M}_{n,d}^{\text{stab}} \mid h^0(E \otimes F) \geq 1\}$$

and

$$\Theta_F = \text{closure of } \Theta_F^{\text{stab}} \text{ in } \mathcal{M}_{n,d}.$$

Let

$$n' = \frac{1}{\text{GCD}(n,d)} n \quad \text{and} \quad d' = \frac{1}{\text{GCD}(n,d)}[n(g-1) - d].$$

Drezet and Narasimhan showed in [DN] the existence of vector bundles F of rank n' and degree d' on C such that Θ_F is a divisor in $\mathcal{M}_{n,d}$. Because $\mathcal{M}_{n,d}$ is locally factorial, the Weil divisor Θ_F is a Cartier divisor and therefore corresponds to a line bundle $[\Theta_F]$ on $\mathcal{M}_{n,d}$. Our results are as follows.

Theorem 1 (Theta functions of level k) *Except for the case $g = n = 2$,*

$$\dim H^0(\mathcal{M}_{n,d}, [\Theta_F]^k) \leq (n'k)^{\dim \mathcal{M}_{n,d}} = (n'k)^{n^2(g-1)+1}$$

for any positive integer k.

In analogy with the line bundle case we call a section of $[\Theta_F]^k$ a *theta function of level k* on $\mathcal{M}_{n,d}$.

Theorem 2 (Ampleness of a theta divisor) *The line bundle $[\Theta_F]$ on the moduli space $\mathcal{M}_{n,d}$ is ample.*

Theorem 3 (Self-intersection of a theta divisor) *Assume n and d coprime, and set*

$$m = \dim \mathcal{M}_{n,d} = n^2(g-1) + 1.$$

Then

$$\Theta_F^m \leq m! n^m.$$

We know that the upper bound in Theorem 1 is attained in the following two cases:

(i) $n = 1$: the moduli space $\mathcal{M}_{1,d}$ is the Jacobian $J(C)$ and the upper bound reduces to k^g.

(ii) $k = 1$ and d is a multiple of n: the upper bound reduces to 1. The equality

$$\dim H^0(\mathcal{M}_{n,n(g-1)}, [\Theta_\mathcal{O}]) = 1$$

is proved in [BNR].

In Theorem 3 the upper bound is attained for $n = 1$, since in this case $m = \dim \mathcal{M}_{1,d} = g$ and $\Theta^g = g!$ is simply Poincaré's formula.

Instead of $\mathcal{M}_{n,d}$ we may consider only the semistable bundles whose determinant is isomorphic to a given line bundle $L \in J_d(C)$:

$$\mathcal{M}_{n,L} := \{E \in \mathcal{M}_{n,d} \mid \det E = L\}.$$

As before, we can define a theta divisor $\Theta_{F,L}$ in $\mathcal{M}_{n,L}$. Lately there has been considerable activity on the problem of computing $\dim H^0(\mathcal{M}_{n,L}, [\Theta_{F,L}]^k)$ (see for example [V], [Bo], [BNR], [DN], [B]). E. Verlinde gives a conjectural formula for this dimension in [V], in the context of conformal field theory. Bott and Szenes [Bo] have recast Verlinde's formula in the framework of representation theory. Most of the work on theta divisors for vector bundles that has been done concerns the fixed-determinant moduli space $\mathcal{M}_{n,L}$. For the space whose dimension we compute in Theorem 1 there is so far not even a conjectural formula in the literature. Anecdotal evidence suggests that the actual dimension in Theorem 1 will be a very involved polynomial in k. Our upper bound has the virtue of easy computability.

The proof of Theorem 1, uses the technique of spectral curves from the paper [BNR] of Beauville, Narasimhan, and Ramanan. While their paper deals with theta functions of level 1 and degree $n(g - 1)$, we allow all levels and all degrees, at the expense of getting an upper bound instead of an exact formula. We apply the newly-minted twisted Brill-Noether theory of Ghione and Hirschowitz to relate the theta divisor in $\mathcal{M}_{n,d}$ to a "twisted" theta divisor in the Jacobian $J(\tilde{C})$ of the spectral curve. This then allows us to bound the dimension of the theta functions of level k on $\mathcal{M}_{n,d}$ by the dimension of the space of sections of an associated line bundle on $J(\tilde{C})$. This last number is easily computed using the Riemann-Roch theorem for an abelian variety.

To prove Theorem 2 we follow a suggestion of Aaron Bertram to examine the ampleness of the line bundle on $\mathcal{M}_{n,d}$ by pulling it back to the finite cover $\tau : \mathcal{M}_L \times J_0(C) \to \mathcal{M}_{n,d}$. We then apply Ghione's twisted Brill-Noether result ([G]) and the seesaw theorem ([M], p. 54) to identify the pullback bundle $\tau^*[\Theta_F]$ on $\mathcal{M}_L \times J_0(C)$. It turns out to be a tensor product of ample line bundles on the two factors. The ampleness of $[\Theta_F]$ readily follows.

Theorem 3 on the self-intersection number of the theta divisor on a smooth $\mathcal{M}_{n,d}$ is a consequence of the Hirzebruch-Riemann-Roch formula, which relates the Euler characteristic $\chi([\Theta_F]^k)$ to intersection numbers on a smooth $\mathcal{M}_{n,d}$. With the aid of Theorems 1 and 2, a comparison of the two sides of the Hirzebruch-Riemann-Roch formula as k tends to infinity yields the desired inequality.

The case of an elliptic curve is quite different from curves of higher genus because over an elliptic curve there are no stable bundles except when the rank and the degree are relatively prime. Some of the definitions need to be modified, but again there is a natural notion of a theta divisor. For an elliptic curve, answers to all the questions considered in this paper may be found in [T].

Acknowledgment. We are indebted to Aaron Bertram for his help with Theorem 2, to Ron Donagi for explaining to us spectral curves, and to Raoul Bott,

Burt Feinberg, and Andras Szenes for very helpful conversations.

Notation and Conventions.

$[D]$ = the line bundle associated to the divisor D

$h^q(E) = \dim H^q(C, E)$ for a vector bundle E on a curve C

$J_d = J_d(C) = \mathcal{M}_{1,d} = \{$ isomorphism classes of line bundles of degree d on $C \}$

$\mathcal{M}_d = \mathcal{M}_{n,d}$ = moduli space of equivalence classes of semistable vector bundles of rank n and degree d on the curve C

$\mathcal{M}_d^{\mathrm{stab}} = \mathcal{M}_{n,d}^{\mathrm{stab}}$ = moduli space of isomorphism classes of stable vector bundles of rank n and degree d on the curve C

$\mathcal{M}_L = \mathcal{M}_{n,L}$ = subvariety of $\mathcal{M}_{n,d}$ consisting of bundles whose determinant bundle is isomorphic to L

$r_E = \mathrm{rk}\ E$ = rank of the vector bundle E

$\mu_E = (\deg E)/(\mathrm{rk}\ E)$ = slope of the vector bundle E

$\mathcal{O}$ = the trivial line bundle on the curve C

Θ = a theta divisor in the Jacobian $J(C)$

$\chi(E) = \sum(-1)^q \dim H^q(C, E)$ = Euler characteristic of the bundle E

2   Background materials

We recall in this section some concepts and results that will be needed in the proofs to follow.

2.1   Stable and semistable bundles

Fix a smooth irreducible projective curve C of genus $g \geq 2$.

Definition. The *slope* μ_E of a vector bundle E of rank n and degree d on the curve C is defined to be

$$\mu_E = \frac{\deg E}{\mathrm{rk}\ E} = \frac{d}{n}.$$

The bundle E is said to be *stable* if for every proper subbundle $F \subset E$,

$$\mu_F < \mu_E.$$

It is *semistable* if the strict inequality above is replaced by $\leq$.

It follows immediately from the definition that if n and d are coprime, then the notions of stability and semistability for E coincide.

Mumford proved that the isomorphism classes of stable bundles of rank n and degree d on C form a moduli space $\mathcal{M}_{n,d}^{\text{stab}}$, which is a smooth quasiprojective variety. The space $\mathcal{M}_{n,d}^{\text{stab}}$ is not compact except when n and d are coprime. Seshadri showed that $\mathcal{M}_{n,d}^{\text{stab}}$ can be compactified by adding to it *equivalence classes* of semistable bundles.

The semistable bundles of a fixed slope μ on the curve C form an abelian category $\mathcal{C}_\mu$, and the simple objects in the category $\mathcal{C}_\mu$ are precisely the stable bundles ([S], pp. 17-18). By the Jordan-Hölder theorem for an abelian category, every semistable bundle of slope μ has a filtration

$$E = E_0 \supset E_1 \supset \ldots \supset 0$$

with all $E_i \in \mathcal{C}_\mu$, such that the successive quotients E_i/E_{i+1} are stable bundles of slope μ; moreover, although the filtration for a given $E \in \mathcal{C}_\mu$ need not be unique, the associated graded object $\text{Gr}(E) := \oplus E_i/E_{i+1}$ is unique up to isomorphism.

Definition. Two semistable bundles E and E' of rank n and degree d on C are said to be *equivalent* if their associated graded bundles $\text{Gr}(E)$ and $\text{Gr}(E')$ are isomorphic.

If E is stable, then $\text{Gr}(E) = E$. Hence, two stable bundles are equivalent if and only if they are isomorphic.

Let $\mathcal{M}_{n,d}(C)$ be the moduli space of equivalence classes of semistable bundles of rank n and degree d over C. We will often omit to write C if it is understood that the curve is fixed.

Facts.

1. The moduli space $\mathcal{M}_{n,d}$ of semistable bundles is an irreducible normal projective variety of dimension $n^2(g-1) + 1$.

2. The moduli space $\mathcal{M}_{n,d}^{\text{stab}}$ of stable bundles is an open subset in $\mathcal{M}_{n,d}$ and is smooth.

3. $\mathcal{M}_{n,d}^{\text{stab}} = \mathcal{M}_{n,d}$ if and only if n and d are coprime.

4. The moduli space $\mathcal{M}_{n,d}$ of semistable bundles is smooth precisely in the following cases:

   (i) n and d are coprime.

   (ii) $g = 2, n = 2$, and d is even.

5. If $\mathcal{M}_{n,d}$ is singular, then its singular locus is precisely the complement of $\mathcal{M}_{n,d}^{\text{stab}}$. Moreover, the codimension of the singular locus in $\mathcal{M}_{n,d}$ is at least 2.

2.2   Spectral curves

Fix a smooth projective curve C of genus $g \geq 2$, and fix a positive integer n. In [BNR] Beauville, Narasimhan, and Ramanan construct an n-sheeted branched covering $\pi : \tilde{C} \to C$ such that a general semistable bundle E of rank n on C is the direct image $\pi_* \xi$ of a line bundle ξ upstairs on $\tilde{C}$. There is a definite relationship between the degrees of E and ξ, which we shall derive below. This curve $\tilde{C}$ is called a *spectral curve* of C associated to vector bundles of rank n. For a given n, there are many spectral curves $\tilde{C}$, but they all have the same genus. In fact, the genus $\tilde{g}$ of the spectral curve $\tilde{C}$ turns out to be the same as the dimension of the moduli space $\mathcal{M}_{n,d}$:

$$\tilde{g} = \mathrm{genus}(\tilde{C}) = \dim \mathcal{M}_{n,d} = n^2(g-1) + 1.$$

By the Leray spectral sequence, because the fibers of $\pi : \tilde{C} \to C$ are 0-dimensional,

$$H^i(\tilde{C}, \xi) = H^i(C, \pi_* \xi).$$

Hence, $\chi(\xi) = \chi(\pi_* \xi) = \chi(E)$ and the degree δ of the line bundle ξ may be related to the degree d of E by the Riemann-Roch formula:

$$\begin{aligned}
\chi(\xi) &= \chi(E), \\
\deg \xi - (\tilde{g} - 1) &= \deg E - n(g-1), \\
\delta := \deg \xi &= d + (\tilde{g} - 1) - n(g-1) \\
&= d + (n^2 - n)(g-1).
\end{aligned}$$

Because the direct image of a line bundle need not be semistable, the induced map

$$\pi_* : J_\delta(\tilde{C}) \dashrightarrow \mathcal{M}_{n,d}$$

is only a *rational* map; that is, it is defined only on a Zariski open subset of $J_\delta(\tilde{C})$.

The main result of Beauville, Narasimhan, and Ramanan may be stated as follows:

Theorem 4 ([BNR]) *For a fixed positive integer n, there exists a spectral curve $\pi : \tilde{C} \to C$, such that the induced rational map from line bundles of degree δ to vector bundles of degree d,*

$$\pi_* : J_\delta(\tilde{C}) \dashrightarrow \mathcal{M}_{n,d},$$

is dominant.

Here "dominant" means that the image of π_* contains a dense open subset of $\mathcal{M}_{n,d}$.

We define the *semistable locus* J^{ss} of $J_\delta(\tilde{C})$ to be

$$J^{\mathrm{ss}} = \{L \in J_\delta(\tilde{C}) \mid \pi_* L \text{ is semistable on } C\}.$$

Then J^{ss} is precisely the domain of π_* and there is a *regular* dominant map

$$\pi_* : J^{ss} \to \mathcal{M}_{n,d}.$$

We call $J_\delta(\tilde{C}) - J^{ss}$ the *nonsemistable locus* of $J_\delta(\tilde{C})$ and denote it by J^{nss}. Similarly, the *stable locus* J^{stab} is

$$J^{stab} = \{L \in J_\delta(\tilde{C}) \mid \pi_*L \text{ is stable on } C\},$$

and the *nonstable locus* J^{ns} is $J_\delta(\tilde{C}) - J^{stab}$.

Proposition 5 ([BNR], Prop. 5.1(c) and Remark 5.3)

   (i) *The codimension of the nonsemistable locus J^{nss} in $J_\delta(\tilde{C})$ is at least 2.*

   (ii) *Except when $g = n = 2$, the codimension of the nonstable locus J^{ns} in $J_\delta(\tilde{C})$ is at least 2.*

2.3   Theta divisors in the moduli space of semistable bundles

The problem of finding divisors in the moduli space $\mathcal{M}_{n,d}$ of semistable vector bundles can be better understood in the larger context of finding cycles in $\mathcal{M}_{n,d}$. We may try to characterize cycles in $\mathcal{M}_{n,d}$ by the number of independent sections that a vector bundle possesses. However, $\mathcal{M}_{n,d}$ presents a peculiar problem in that equivalent bundles need not have isomorphic space of sections. For this reason we prefer to start with $\mathcal{M}_{n,d}^{stab}$. Define

$$W_{n,d}^{r,stab} = \{E \in \mathcal{M}_{n,d}^{stab} \mid h^0(E) \geq r+1\},$$

and

$$W_{n,d}^r = \text{closure of } W_{n,d}^{r,stab} \text{ in } \mathcal{M}_{n,d}.$$

In analogy with the line bundle case we call these $W_{n,d}^r$'s *Brill-Noether loci*. For $n = 1$ the study of these subvarieties constitutes the heart of classical Brill-Noether theory, and the results are by now fairly complete [ACGH]. For vector bundles of higher ranks, however, even such basic questions as the existence, dimension, connectedness, irreducibility, singularities, and Chow classes of the Brill-Noether loci remain largely uncharted territories. We do know that $W_{n,d}^{r,stab}$ is the image of a determinantal variety and as such, its expected codimension in $\mathcal{M}_{n,d}^{stab}$ is:

$$
\begin{aligned}
\text{expected codim } W_{n,d}^r &= h^0(E)h^1(E) \\
&= (r+1)(r+1-\chi(E)) \\
&= (r+1)(r+1-d+n(g-1)),
\end{aligned}
$$

for any vector bundle E of rank n and degree d with $h^0(E) = r+1$. From this formula it follows that the expected codimension of $W_{n,d}^r$ is one if and only if

$r = 0$ and $d = n(g - 1)$. Indeed, it is shown in [Su] that $W^0_{n,n(g-1)}$ is a divisor in $\mathcal{M}_{n,n(g-1)}$.

To obtain divisors in $\mathcal{M}_{n,d}$ for degrees other than $d = n(g - 1)$, one can "twist" the Brill-Noether loci $W^r_{n,d}$ by a vector bundle of a suitably chosen rank n' and degree d'. For any vector bundle F over the curve C, define

$$W^{r,\mathrm{stab}}_{n,d}(F) = \{E \in \mathcal{M}^{\mathrm{stab}}_{n,d} \mid h^0(E \otimes F) \geq r + 1\},$$

and

$$W^r_{n,d}(F) = \text{closure of } W^{r,\mathrm{stab}}_{n,d}(F) \text{ in } \mathcal{M}_{n,d}.$$

We call either of these a *twisted Brill-Noether locus*. As above,

$$
\begin{aligned}
\text{expected codim } W^{r,\mathrm{stab}}_{n,d}(F) &= (r+1)(r+1-\chi(E \otimes F)) \\
&= (r+1)(r+1 - nd' - n'd + nn'(g-1)),
\end{aligned}
$$

where $\chi(E \otimes F)$ is computed from the Riemann-Roch formula

$$\chi(E \otimes F) = \deg(E \otimes F) - \mathrm{rk}\ (E \otimes F)(g - 1).$$

If n' and d' are chosen in such a way that $\chi(E \otimes F) = 0$, then the expected codimension of $W^{r,\mathrm{stab}}_{n,d}(F)$ and hence also of $W^r_{n,d}$ will be 1. For example, a possible pair (n', d') is

$$n' = \frac{1}{\mathrm{GCD}(n,d)}n, \quad d' = \frac{1}{\mathrm{GCD}(n,d)}[n(g-1) - d]. \tag{1}$$

Naturally, any multiple of this pair will also do. Fix a vector bundle F on C such that $\chi(E \otimes F) = 0$ for any $E \in \mathcal{M}_{n,d}$ and set

$$\Theta^{\mathrm{stab}}_F := W^{0,\mathrm{stab}}_{n,d}(F) := \{E \in \mathcal{M}^{\mathrm{stab}}_{n,d} \mid h^0(E \otimes F) \geq 1\},$$

and

$$\Theta_F := \text{closure of } \Theta^{\mathrm{stab}}_F \text{ in } \mathcal{M}_{n,d}.$$

By ([DN], p. 90) if F is a general stable bundle of rank n' and degree d' as in (1), then Θ_F is a divisor in $\mathcal{M}_{n,d}$. For such a bundle F we will call Θ_F a *theta divisor* of the moduli space $\mathcal{M}_{n,d}$.

2.4   Twisted Brill-Noether theory for line bundles

When the twisted Brill-Noether loci $W^r_{n,d}(F)$ defined in §2.3 are specialized to line bundles, we will suppress the subscript $n = 1$ and simply write

$$W^r_d(F) = \{L \in J_d(C) \mid h^0(L \otimes F) \geq r + 1\}.$$

For $d = g - 1$ and F a vector bundle of degree $d' = 0$, we set

$$\Theta_F = W^0_{g-1}(F) = \{L \in J_{g-1}(C) \mid h^0(L \otimes F) \geq 1\}.$$

By the choice of d and d', for $L \in J_{g-1}(C)$ the Riemann-Roch theorem gives

$$\chi(L \otimes F) = \deg(L \otimes F) - (\text{rk } F)(g - 1) = 0.$$

So the expected codimension of Θ_F is 1.

When $n = 1$ and $d = g - 1$ it follows from ([DN], p. 55 and p. 90) that Θ_F is a divisor for a general stable bundle F of rank $n' = 1$ and degree $d' = 0$. Since our F is of arbitrary rank, we still need to show that Θ_F is a divisor. This will follow from several results of twisted Brill-Noether theory for line bundles proven in Ghione [G] and Hirschowitz [H]. We quote here the special cases we need.

Ghione's Theorem ([G], [L]). *Let C be a smooth projective curve of genus g and let F be a vector bundle of degree 0 on C.*

(i) *The subvariety $\Theta_F \subset J_{g-1}(C)$ is nonempty and therefore has dimension $\geq g - 1$.*

(ii) *If Θ_F is a divisor, then the class of Θ_F in the Chow ring of $J_{g-1}(C)$ is $(\text{rk } F)\Theta$, where Θ is the class of the Riemann theta divisor.*

In the theorem above, Ghione assumes that F is general in a suitable sense, but Lazarsfeld shows that this assumption is not necessary ([L], Th. 2.6).

Hirschowitz's Theorem ([H], §4.6). *On a smooth curve the tensor product of two general stable bundles E and F is nonspecial; this means $h^1(E \otimes F) = 0$.*

Corollary. *On a smooth curve C of genus g, if F is a general stable vector bundle of degree 0, then Θ_F is a divisor in $J_{g-1}(C)$.*

Proof of Corollary. By Ghione,

$$\dim \Theta_F \geq g - 1.$$

As computed earlier, for any $L \in J_{g-1}(C), \chi(L \otimes F) = 0$. Since F is assumed to be general, if L is also general, then by Hirschowitz's theorem,

$$h^0(L \otimes F) = \chi(L \otimes F) + h^1(L \otimes F) = 0.$$

Thus, Θ_F is not the entire Jacobian and so must be a divisor. ∎

3   Theta functions of level k

In this section we prove Theorem 1, which gives an upper bound for the dimension of $H^0(\mathcal{M}_{n,d}, [\Theta_F]^k)$. We begin by determining the inverse image of the theta divisor under π_*.

Lemma 6 *For any $g \geq 1$, the pullback of the bundle $[\Theta_F]$ by the map $\pi_* : J^{ss} \to \mathcal{M}_{n,d}$ is $[\Theta_{\pi \cdot F}]|_{J^{ss}}$.*

We denote the pullback bundle by $\pi^*[\Theta_F]$ instead of $\pi_*^*[\Theta_F]$.

Proof. By the correspondence between Weil divisors and line bundles on the smooth variety J^{ss} to identify the line bundle $\pi^*[\Theta_F]$, we may just as well identify the divisor $\pi_*^{-1}(\Theta_F)$. This is easy to do on the stable locus:

$$
\begin{aligned}
\pi_*^{-1}(\Theta_F^{\mathrm{stab}}) &= \{L \in J^{\mathrm{stab}} \mid \pi_* L \in \Theta_F^{\mathrm{stab}}\} \\
&= \{L \in J^{\mathrm{stab}} \mid h^0(C, (\pi_* L) \otimes F) \geq 1\} \\
&= \{L \in J^{\mathrm{stab}} \mid h^0(C, \pi_*(L \otimes \pi^* F)) \geq 1\} \\
&= \{L \in J^{\mathrm{stab}} \mid h^0(\tilde{C}, L \otimes \pi^* F) \geq 1\} \\
&= \Theta_{\pi^* F} \cap J^{\mathrm{stab}}.
\end{aligned}
$$

By Proposition 5, J^{ns} has codimension ≥ 2 in J^{ss}, since J^{ss} and $J_\delta(\tilde{C})$ have the same dimension. It follows that $\pi_*^{-1}(\Theta_F)$ and $\Theta_{\pi^* F} \cap J^{\mathrm{ss}}$ agree outside a subvariety of codimension ≥ 2 and therefore must be equal:

$$
\pi_*^{-1}(\Theta_F) = \Theta_{\pi^* F} \cap J^{\mathrm{ss}}.
$$

■

Assuming that $(g, n) \neq (2, 2)$, we prove Theorem 1 now. The map

$$
\pi_* : J^{\mathrm{ss}} \to \mathcal{M}_{n,d}
$$

induces a pullback map on the space of sections

$$
\pi^* : H^0(\mathcal{M}_{n,d}, [\Theta_F]^k) \to H^0(J^{\mathrm{ss}}, \pi^*[\Theta_F]^k)
$$

by sending s to $s \circ \pi_*$. Because π_* is dominant, this π^* on sections is injective, and so

$$
h^0(\mathcal{M}_{n,d}, [\Theta_F]^k) \leq h^0(J^{\mathrm{ss}}, \pi^*[\Theta_F]^k) \tag{2}
$$

By Lemma 6,

$$
\pi^*[\Theta_F]^k = [\Theta_{\pi^* F}]^k|_{J^{\mathrm{ss}}}. \tag{3}
$$

Because codim $J_\delta(\tilde{C}) - J^{\mathrm{ss}} \geq 2$ (Prop. 5), by the Levi extension theorem

$$
H^0(J^{\mathrm{ss}}, [\Theta_{\pi^* F}]^k|_{J^{\mathrm{ss}}}) \simeq H^0(J_\delta(\tilde{C}), [\Theta_{\pi^* F}]^k). \tag{4}
$$

To compute the dimension of this last space, we first apply Ghione's theorem

$$
[\Theta_{\pi^* F}] \simeq [n'\Theta] \simeq [\Theta]^{n'}
$$

and then the Riemann-Roch theorem for abelian varieties:

$$
\begin{aligned}
\chi(J_\delta(\tilde{C}), [\Theta_{\pi^* F}]^k) &= \chi(J_\delta(\tilde{C}), \Theta^{n'k}) \\
&= (n'k\Theta)^{\tilde{g}}/\tilde{g}! \\
&= (n'k)^{\tilde{g}}, \tag{5}
\end{aligned}
$$

where $\Theta^{\tilde{g}} = \tilde{g}!$ by Poincaré's formula. Because $[\Theta]^{n'k}$ is ample and the canonical bundle K of $J_\delta(\tilde{C})$ is trivial, the Kodaira vanishing theorem gives

$$H^p(J_\delta(\tilde{C}), [\Theta]^{n'k}) = H^p(J_\delta(\tilde{C}), K \otimes [\Theta]^{n'k}) = 0 \tag{6}$$

for all $p \geq 1$, where K is the canoncal bundle on $J_\delta(\tilde{C})$. Combining (2), (3), (4), (5), and (6) we finally obtain

$$h^0(\mathcal{M}_{n,d}, [\Theta_F]^k) \leq (n'k)^{\tilde{g}}.$$

■

4 The ampleness of a theta divisor

As before, let C be a smooth irreducible projective curve of genus g, and let $\mathcal{M}_{n,d}(C)$ be the moduli space of semistable curves of rank n and degree d on C. We sometimes write $\mathcal{M}_d$ or $\mathcal{M}_{n,d}$ instead of $\mathcal{M}_{n,d}(C)$. Similarly, if $\mathcal{M}_{n,L}(C)$ is the subvariety of $\mathcal{M}_{n,d}(C)$ consisting of bundles with whose determinant is isomorphic to L, we write $\mathcal{M}_L$ or $\mathcal{M}_{n,L}$ instead of $\mathcal{M}_{n,L}(C)$. We prove now that the line bundle $[\Theta_F]$ corresponding to a theta divisor Θ_F in $\mathcal{M}_{n,d}$ is ample.

Lemma 7 *If E and E' are equivalent semistable bundles and M is any line bundle on C, then $E \otimes M$ and $E' \otimes M$ are also equivalent semistable bundles.*

Proof. Let

$$E = E_0 \supset E_1 \supset E_2 \supset \ldots \supset 0$$

and

$$E' = E_0' \supset E_1' \supset E_2' \supset \ldots \supset 0$$

be Jordan-Hölder filtrations of E and E'. Then $\{E_i \otimes M\}$ and $\{E_i' \otimes M\}$ are Jordan-Hölder filtrations of $E \otimes M$ and $E' \otimes M$ respectively. Because

$$\frac{E_i \otimes M}{E_{i+1} \otimes M} \simeq \frac{E_i}{E_{i+1}} \otimes M,$$

(which we can prove by tensoring the exact sequence

$$0 \to E_{i+1} \to E_i \to \frac{E_i}{E_{i+1}} \to 0$$

by M), and because

$$\oplus \frac{E_i}{E_{i+1}} \simeq \oplus \frac{E_i'}{E_{i+1}'}$$

by the equivalence of E and E', we conclude that

$$\oplus \frac{E_i \otimes M}{E_{i+1} \otimes M} \simeq \oplus \left(\frac{E_i}{E_{i+1}} \otimes M \right) \simeq \oplus \left(\frac{E_i'}{E_{i+1}'} \otimes M \right) \simeq \frac{E_i' \otimes M}{E_{i+1}' \otimes M}.$$

Therefore, $E \otimes M$ and $E' \otimes M$ are equivalent.   ∎

Fix a line bundle L of degree d on the curve C. By Lemma 7 the tensor product map

$$\tau : \mathcal{M}_L \times J_0(C) \rightarrow \mathcal{M}_{n,d}$$
$$(E, M) \mapsto E \otimes M$$

is well-defined on the equivalence classes of semistable vector bundles.

Proposition 8 *The map* $\tau : \mathcal{M}_L \times J_0(C) \rightarrow \mathcal{M}_{n,d}$ *is surjective and is an* n^{2g}- *sheeted unbranched covering.*

Proof. We first prove the surjectivity of τ. Let $E' \in \mathcal{M}_{n,d}$. To find $(E, M) \in \mathcal{M}_L \times J_0(C)$ such that

$$E \otimes M \simeq E', \tag{7}$$

we take the determinant of both sides of (7):

$$L \otimes M^n \simeq \det E'$$

and solve for M:

$$M^n \simeq (\det E') \otimes L^{-1}. \tag{8}$$

Since $(\det E') \otimes L^{-1}$ has degree 0, it is an element of $J_0(C)$; hence, M is simply an n-torsion point of the abelian variety $J_0(C)$ of dimension g. There are n^{2g} such n-torsion points. For each such M, the vector bundle E is obtained from (7):

$$E \simeq E' \otimes M^{-1}.$$

Let's check that E indeed has determinant L:

$$\det E = (\det E') \otimes M^{-n} = L \text{ by (8).}$$

If $M_i, i = 1, \ldots, n^{2g}$, are the n-torsion points of $J_0(C)$ and

$$E_i = E' \otimes M_i^{-1},$$

then $(E_i, M_i), i = 1, \ldots, n^{2g}$, are the inverse images of E' under τ. They are all distinct because the M_i's are all distinct. Therefore, $\tau : \mathcal{M}_L \times J_0(C) \rightarrow \mathcal{M}_{n,d}$ is not only surjective, but is an n^{2g}-sheeted unbranched covering.   ∎

We recall here a standard theorem concerning ample line bundles.

Theorem 9 (EGA, III, 2.6.2) *Let* $f : \tilde{X} \rightarrow X$ *be a finite surjective morphism of quasiprojective varieties, and* η *a line bundle on* X. *Then* $f^*\eta$ *is ample on* $\tilde{X}$ *if and only if* η *is ample on* X.

On account of this theorem and Proposition 8, in order to show that $[\Theta_F]$ is ample on $\mathcal{M}_{n,d}$, it suffices to show that $\pi^*[\Theta_F]$ is ample on $\mathcal{M}_L \times J_0(C)$.

THETA DIVISORS FOR VECTOR BUNDLES 339

Proposition 10 *Let Θ_F be a theta divisor on $\mathcal{M}_{n,d}$ and let $p_{\mathcal{M}}$ and p_J be the projections of $\mathcal{M}_L \times J_0(C)$ to its two factors. The pullback of the line bundle $[\Theta_F]$ from $\mathcal{M}_{n,d}$ by*

$$\tau : \mathcal{M}_L \times J_0(C) \to \mathcal{M}_{n,d}$$

can be decomposed as

$$\tau^*[\Theta_F] \simeq p_{\mathcal{M}}^*[\Theta_{F,L}] \otimes p_J^*[nn'\Theta].$$

Proof. We first check the restriction of $\tau^*[\Theta_F]$ to a general fiber of $p_{\mathcal{M}}$: $\mathcal{M}_L \times J_0(C) \to \mathcal{M}_{n,d}$. Let $E \in \mathcal{M}_L^{\text{stab}}$ be a general element. Since $\tau^{-1}(\Theta_F)$ is a divisor in $\mathcal{M}_L \times J_0(C)$, its restriction to a general fiber $\{E\} \times J_0(C)$ of $p_{\mathcal{M}}$ is a divisor in $\{E\} \times J_0(C)$. Then

$$\begin{aligned}
\tau^{-1}(\Theta_F) \cap (\{E\} \times J_0(C)) &= \{(E,M) \mid M \in J_0(C), E \otimes M \in \Theta_F^{\text{stab}}\} \\
&= \{M \in J_0(C) \mid h^0(E \otimes M \otimes F) \geq 1\} \\
&= \Theta_{E \otimes F}.
\end{aligned}$$

By Ghione's theorem, $[\Theta_{E \otimes F}] \simeq [nn'\Theta]$, so as line bundles on $J_0(C)$,

$$\tau^*[\Theta_F]|_{\{E\} \times J_0(C)} \simeq [nn'\Theta].$$

Next let $M \in J_0(C)$ be a general element, and consider the restriction of $\tau^{-1}(\Theta_F)$ to the fiber $\mathcal{M}_L \times \{M\}$ of p_J:

$$\begin{aligned}
\tau^{-1}(\Theta_F) \cap (\mathcal{M}_L^{\text{stab}} \times \{M\}) &= \{(E,M) \mid E \in \mathcal{M}_L^{\text{stab}}, E \otimes M \in \Theta_F^{\text{stab}}\} \\
&= \{E \in \mathcal{M}_L^{\text{stab}} \mid h^0(E \otimes F \otimes M) \geq 1\} \\
&= \Theta_{F \otimes M, L} \cap \mathcal{M}_L^{\text{stab}}.
\end{aligned}$$

For a general $M \in J_0(C)$, the intersection $\tau^{-1}(\Theta_F) \cap (\mathcal{M}_L \times \{M\})$ is a divisor in $\mathcal{M}_L$. Since this divisor and the divisor $\Theta_{F \otimes M, L}$ agree on $\mathcal{M}_L^{\text{stab}}$, whose complement in $\mathcal{M}_L$ has codimension ≥ 2, as line bundles on $\mathcal{M}_L$

$$\tau^*[\Theta_F]|_{\mathcal{M}_L \times \{M\}} \simeq [\Theta_{F \otimes M, L}].$$

By Drezet and Narasimhan [DN, Th. B, p.55], the line bundle $[\Theta_{F \otimes M, L}]$ on $\mathcal{M}_L$ is independent of $F \otimes M$, so we may write

$$[\Theta_{F \otimes M, L}] \simeq \Theta_{F, L}.$$

Consider the line bundle

$$N = \tau^*[\Theta_F] \otimes p_{\mathcal{M}}^*[\Theta_{F,L}]^{-1} \otimes p_J^*[nn'\Theta]^{-1}$$

on $\mathcal{M}_L \times J_0(C)$. The restriction of N to a general vertical fiber and a general horizontal fiber of $\mathcal{M}_L \times J_0(C)$ are both trivial. By the seesaw theorem ([M], p. 54), N is the trivial line bundle. ∎

We can now finally prove Theorem 2 that the line bundle $[\Theta_F]$ on $\mathcal{M}_{n,d}$ is ample. As noted earlier, $[\Theta_F]$ on $\mathcal{M}_{n,d}$ is ample if and only if its pullback $\tau^*[\Theta_F]$

to $\mathcal{M}_L \times J_0(C)$ is ample, because $\tau : \mathcal{M}_L \times J_0(C) \to \mathcal{M}_{n,d}$ is a finite map. By Drezet and Narasimhan ([DN], Th. B) the group $\text{Pic}(\mathcal{M}_L)$ is isomorphic to $\mathbf{Z}$, generated by $[\Theta_{F,L}]$. Since $[\Theta_{F,L}]$ has sections ([BNR], Th. 3), it must be the ample generator. Since $[\Theta_{F,L}]$ is ample on $\mathcal{M}_L$ and $[nn'\Theta]$ is ample on $J_0(C)$, the tensor product $\tau^*\Theta_F = p_{\mathcal{M}}^*[\Theta_{F,L}] \otimes p_J^*[nn'\Theta]$ is ample on $\mathcal{M}_L \times J_0(C)$ by the Segre embedding.

5   The self-intersection number of a theta divisor

Theorem 1 gives us an upper bound on $h^0([\Theta_F]^k)$. On the other hand, in case $\mathcal{M}_{n,d}$ is smooth, the Hirzebruch-Riemann-Roch theorem also gives a formula for the Euler characteristic $\chi([\Theta_F]^k)$. By comparing these two expressions we will find an upper bound on the self-intersection number Θ_F^m, where $m = \dim \mathcal{M}_{n,d}$.

In this section assume that $\mathcal{M}_{n,d}$ is smooth. Recall that the Chern character of a complex line bundle L over a space X is by definition

$$\text{ch}(L) := \sum_{i=0}^{\infty} \text{ch}_i(L) := e^{c_1(L)} = \sum_{i=0}^{\infty} \frac{1}{i!} c_1(L)^i \ \in \ H^*(X; \mathbf{Z}).$$

Thus,

$$\text{ch}_i(L) = \frac{1}{i!} c_1(L)^i \ \in \ H^{2i}(X; \mathbf{Z}).$$

Fix a nonnegative integer k. By the Hirzebruch-Riemann-Roch formula, if X is an algebraic manifold of dimension m, then

$$
\begin{aligned}
\chi(L^k) &= [\text{ch}(L^k) \, \text{td}(X)]_m \\
&= \sum_{i=0}^{m} \text{ch}_i(L^k) \, \text{td}_{m-i}(X) \\
&= \sum_{i=0}^{m} \frac{1}{i!} c_1(L)^i k^i \, \text{td}_{m-i}(X) \\
&= \frac{c_1(L)^m}{m!} k^m + \text{ terms involving lower powers of } k.
\end{aligned}
$$

Now take X to be the smooth moduli space $\mathcal{M}_{n,d}$ and L to be the line bundle corresponding to an ample theta divisor Θ_F. Because $[\Theta_F]$ is ample by Theorem 2, for k sufficiently large

$$h^q([\Theta_F]^k) = 0 \quad \text{for all } q \geq 1$$

by Serre's theorem. Hence, for k sufficiently large,

$$
\begin{aligned}
h^0([\Theta_F]^k) &= \chi([\Theta_F]^k) \\
&= \frac{\Theta_F^m}{m!} k^m + \text{ terms involving lower powers of } k \\
&\leq (n'k)^m \qquad \text{(by Theorem 1).}
\end{aligned}
$$

This is possible only if the coefficient $\frac{\Theta_F^m}{m!}$ of k^m in the formula for $h^0([\Theta_F]^k)$ above is at most $(n')^m$. Therefore,

$$\Theta^m \leq m!(n')^m.$$

This completes the proof of Theorem 3.

References

[ACGH]   E. Arbarello, M. Cornalba, P. A. Griffiths, and J. Harris, *Geometry of Algebraic Curves*, Grundlehren 267, Springer-Verlag, New York, 1985.

[BNR]   A. Beauville, M. S. Narasimhan, S. Ramanan, Spectral curves and the generalized theta divisor, J. reine angew. Math. **398** (1989), 169-179.

[B]   A. Bertram, A partial verification of the Verlinde formulae for vector bundles of rank 2, preprint.

[Bo]   R. Bott, Stable bundles revisited, to appear in *Surveys in Differential Geometry*, Supplement to J. of Diff. Geometry.

[DN]   J.-M. Drezet and M. S. Narasimhan, Groupe de Picard des variétés de modules de fibrés semi-stables sur les courbes algébriques, Invent. Math. **97** (1989), 53-94.

[F]   B. Feinberg, The irreducibility of the generalized theta divisor, to appear.

[G]   F. Ghione, Un problème du type Brill-Noether pour les fibrés vectoriels, in *Algebraic Geometry – Open Problems, Ravello 1982*, Lect. Notes in Math. 997 (1983), 197-209.

[GD]   A. Grothendieck and J. Dieudonné, *Eléments de Géometrie Algébrique* (EGA), III, Publ. Math. I.H.E.S. 11 (1961).

[H]   A. Hirschowitz, Problèmes de Brill-Noether en rang supérieur, preprint.

[L]   R. Lazarsfeld, Some applications of the theory of positive vector bundles, in *Complete Intersections, Acireale 1983*, Lect. Notes in Math. 1092 (1984), 29-61.

[M]   D. Mumford, *Abelian Varieties*, second edition, Oxford University Press, Oxford, 1974.

[S]   C. S. Seshadri, *Fibrés Vectoriels sur les Courbes Algébriques*, Astérisque 96, Société Mathématique de France, Paris, 1982.

[Su]   N. Sundaram, Special divisors and vector bundles, Tohoku Math. Jour. **39** (1987), 175-213.

[T]   L. Tu, Semistable bundles over an elliptic curve, to appear in Advances in Math.

[V]   E. Verlinde, Fusion rules and modular transformations in 2d conformal field theory, Nucl. Phys. **B 300** (1988), 360-376.

DEPARTMENT OF MATHEMATICS, TUFTS UNIVERSITY, MEDFORD, MA 02155

Recent Titles in This Series

(Continued from the front of this publication)

(See the AMS catalog for earlier titles)